T0296265

CAMBRIDGE LIBRARY COLLECTION

Books of enduring scholarly value

Technology

The focus of this series is engineering, broadly construed. It covers technological innovation from a range of periods and cultures, but centres on the technological achievements of the industrial era in the West, particularly in the nineteenth century, as understood by their contemporaries. Infrastructure is one major focus, covering the building of railways and canals, bridges and tunnels, land drainage, the laying of submarine cables, and the construction of docks and lighthouses. Other key topics include developments in industrial and manufacturing fields such as mining technology, the production of iron and steel, the use of steam power, and chemical processes such as photography and textile dyes.

Electromagnetic Theory

Oliver Heaviside (1850–1925) was a scientific maverick and a gifted self-taught electrical engineer, physicist and mathematician. He patented the co-axial cable, pioneered the use of complex numbers for circuit analysis, and reworked Maxwell's field equations into a more concise format. In 1891 the Royal Society made him a Fellow for his mathematical descriptions of electromagnetic phenomena. Along with Arthur Kennelly, he also predicted the existence of the ionosphere. Often dismissed by his contemporaries, his work achieved wider recognition when he received the inaugural Faraday Medal in 1922. Published 1893 this is the first of three volumes that bring together Heaviside's contributions to electromagnetic theory. It introduces the subject at length, and features his first description of vector analysis and the reworking of Maxwell's field equations into the form we know today.

Cambridge University Press has long been a pioneer in the reissuing of out-of-print titles from its own backlist, producing digital reprints of books that are still sought after by scholars and students but could not be reprinted economically using traditional technology. The Cambridge Library Collection extends this activity to a wider range of books which are still of importance to researchers and professionals, either for the source material they contain, or as landmarks in the history of their academic discipline.

Drawing from the world-renowned collections in the Cambridge University Library, and guided by the advice of experts in each subject area, Cambridge University Press is using state-of-the-art scanning machines in its own Printing House to capture the content of each book selected for inclusion. The files are processed to give a consistently clear, crisp image, and the books finished to the high quality standard for which the Press is recognised around the world. The latest print-on-demand technology ensures that the books will remain available indefinitely, and that orders for single or multiple copies can quickly be supplied.

The Cambridge Library Collection will bring back to life books of enduring scholarly value (including out-of-copyright works originally issued by other publishers) across a wide range of disciplines in the humanities and social sciences and in science and technology.

Electromagnetic Theory

VOLUME 1

OLIVER HEAVISIDE

CAMBRIDGE
UNIVERSITY PRESS

CAMBRIDGE UNIVERSITY PRESS

Cambridge, New York, Melbourne, Madrid, Cape Town,
Singapore, São Paolo, Delhi, Tokyo, Mexico City

Published in the United States of America by Cambridge University Press, New York

www.cambridge.org
Information on this title: www.cambridge.org/9781108032155

© in this compilation Cambridge University Press 2011

This edition first published 1893
This digitally printed version 2011

ISBN 978-1-108-03215-5 Paperback

ELECTROMAGNETIC THEORY.

BY

OLIVER HEAVISIDE.

VOLUME I.

LONDON :
"THE ELECTRICIAN" PRINTING AND PUBLISHING COMPANY,
LIMITED,
SALISBURY COURT, FLEET STREET, E.C.

1893.

Printed and Published by
" THE ELECTRICIAN " PRINTING AND PUBLISHING CO., LIMITED,
1, 2, and 3, Salisbury Court, Fleet Street,
London, E.C.

PREFACE.

THIS work was originally meant to be a continuation of the
series "Electromagnetic Induction and its Propagation,"
published in *The Electrician* in 1885-6-7, but left unfinished.
Owing, however, to the necessity of much introductory
repetition, this plan was at once found to be impracticable,
and was, by request, greatly modified. The result is something approaching a connected treatise on electrical theory,
though without the strict formality usually associated with
a treatise. As critics cannot always find time to read more
than the preface, the following remarks may serve to direct
their attention to some of the leading points in this volume.

The first chapter will, I believe, be found easy to read,
and may perhaps be useful to many men who are accustomed
to show that they are practical by exhibiting their ignorance
of the real meaning of scientific and mathematical methods
of enquiry.

The second chapter, pp. 20 to 131, consists of an outline
scheme of the fundamentals of electromagnetic theory from
the Faraday-Maxwell point of view, with some small modifications and extensions upon Maxwell's equations. It is done
in terms of my rational units, which furnish the only way of
carrying out the idea of lines and tubes of force in a consistent and intelligible manner. It is also done mainly in
terms of vectors, for the sufficient reason that vectors are
the main subject of investigation. It is also done in the
duplex form I introduced in 1885, whereby the electric and

A

magnetic sides of electromagnetism are symmetrically exhibited and connected, whilst the "forces" and "fluxes"
are the objects of immediate attention, instead of the
potential functions which are such powerful aids to obscuring
and complicating the subject, and hiding from view useful
and sometimes important relations.

The third chapter, pp. 132 to 305, is devoted to vector
algebra and analysis, in the form used by me in my former
papers. As I have at the beginning and end of this chapter
stated my views concerning the unsuitability of quaternions
for physical requirements, and my preference for a vector
algebra which is based upon the vector and is dominated by
vectorial ideas instead of quaternionic, it is needless to say
more on the point here. But I must add that it has been
gratifying to discover among mathematical physicists a considerable and rapidly growing appreciation of vector algebra
on these lines; and moreover, that students who had found
quaternions quite hopeless could understand my vectors very
well. Regarded as a treatise on vectorial algebra, this chapter has manifest shortcomings. It is only the first rudiments
of the subject. Nevertheless, as the reader may see from the
applications made, it is fully sufficient for ordinary use in
the mathematical sciences where the Cartesian mathematics
is usually employed, and we need not trouble about more
advanced developments before the elements are taken up.
Now, there are no treatises on vector algebra in existence yet,
suitable for mathematical physics, and in harmony with the
Cartesian mathematics (a matter to which I attach the
greatest importance). I believe, therefore, that this chapter
may be useful as a stopgap.

The fourth chapter, pp. 306 to 466, is devoted to the
theory of plane electromagnetic waves, and, being mainly
descriptive, may perhaps be read with profit by many who
are unable to tackle the mathematical theory comprehensively. It may be also useful to have results of mathematical

reasoning expanded into ordinary language for the benefit of mathematicians themselves, who are sometimes too apt to work out results without a sufficient statement of their meaning and effect. But it is only introductory to plane waves. Some examples in illustration thereof have been crowded out, and will probably be given in the next volume. I have, however, included in the present volume the application of the theory (in duplex form) to straight wires, and also an account of the effects of self-induction and leakage, which are of some significance in present practice as well as in possible future developments. There have been some very queer views promulgated officially in this country concerning the speed of the current, the impotence of self-induction, and other material points concerned. No matter how eminent they may be in their departments, officials need not be scientific men. It is not expected of them. But should they profess to be, and lay down the law outside their knowledge, and obstruct the spreading of views they cannot understand, their official weight imparts a fictitious importance to their views, and acts most deleteriously in propagating error, especially when their official position is held up as a screen to protect them from criticism. But in other countries there is, I find, considerable agreement with my views.

Having thus gone briefly through the book, it is desirable to say a few words regarding the outline sketch of electromagnetics in the second chapter. Two diverse opinions have been expressed about it. On the one hand, it has been said to be too complicated. This probably came from a simple-minded man. On the other hand, it has been said to be too simple. This objection, coming from a wise man, is of weight, and demands some notice.

Whether a theory can be rightly described as too simple depends materially upon what it professes to be. The phenomena involving electromagnetism may be roughly divided into two classes, primary and secondary. Besides the main

primary phenomena, there is a large number of secondary
ones, partly or even mainly electromagnetic, but also trenching
upon other physical sciences. Now the question arises whether
it is either practicable or useful to attempt to construct a
theory of such comprehensiveness as to include the secondary
phenomena, and to call it the theory of electromagnetism. I
think not, at least at present. It might perhaps be done if
the secondary phenomena were thoroughly known; but their
theory is so much more debatable than that of the primary
phenomena that it would be an injustice to the latter to too
closely amalgamate them. Then again, the expression of the
theory would be so unwieldy as to be practically useless; the
major phenomena would be apparently swamped by the minor.
It would, therefore, seem best not to attempt too much, but
to have a sort of abstract electromagnetic scheme for the
primary phenomena only, and have subsidiary extensions
thereof for the secondary. The theory of electromagnetism
is then a primary theory, a skeleton framework corresponding
to a possible state of things simpler than the real in innu-
merable details, but suitable for the primary effects, and
furnishing a guide to special extensions. From this point of
view, the theory cannot be expressed too simply, provided it
be a consistent scheme, and be sufficiently comprehensive to
serve for a framework. I believe the form of theory in the
second chapter will answer the purpose. It is especially
useful in the duplex way of exhibiting the relations, which is
clarifying in complicated cases as well as in simple ones. It
is essentially Maxwell's theory, but there are some differences.
Some are changes of form only; for instance, the rationalisation effected by changing the units, and the substitution of
the second circuital law for Maxwell's equation of electromotive force involving the potentials, etc. But there is one
change in particular which raises a fresh question. What is
Maxwell's theory? or, What should we agree to understand
by Maxwell's theory?

The first approximation to the answer is to say, There is
Maxwell's book as he wrote it ; there is his text, and there
are his equations : together they make his theory. But when
we come to examine it closely, we find that this answer is
unsatisfactory. To begin with, it is sufficient to refer to
papers by physicists, written say during the twelve years
following the first publication of Maxwell's treatise, to see
that there may be much difference of opinion as to what his
theory is. It may be, and has been, differently interpreted by
different men, which is a sign that it is not set forth in a perfectly clear and unmistakeable form. There are many obscurities and some inconsistencies. Speaking for myself, it was
only by changing its form of presentation that I was able to
see it clearly, and so as to avoid the inconsistencies. Now
there is no finality in a growing science. It is, therefore,
impossible to adhere strictly to Maxwell's theory as he gave it
to the world, if only on account of its inconvenient form.
But it is clearly not admissible to make arbitrary changes in
it and still call it his. He might have repudiated them
utterly. But if we have good reason to believe that the
theory as stated in his treatise does require modification to
make it self-consistent, and to believe that he would have
admitted the necessity of the change when pointed out to him,
then I think the resulting modified theory may well be called
Maxwell's.

Now this state of things is exemplified by his celebrated
circuital law defining the electric current in terms of magnetic
force. For although he did not employ the other, or second
circuital law, yet it may be readily derived from his equation
of electromotive force ; and when this is done, and the law
made a fundamental one, we readily see that the change it
suffers in passing from the case of a stationary to that of a
moving medium should be necessarily accompanied by a
similar change in the first, or Maxwell's circuital law. An
independent formal proof is unnecessary ; the similarity of

form and of the conditions of motion show that Maxwell's auxiliary term in the electromotive force, viz., $V\mathbf{qB}$ (the motional electric force), where $\mathbf{q}$ is the velocity of the medium and $\mathbf{B}$ the induction, requires the use of a similar auxiliary term in the first circuital law, viz., $V\mathbf{Dq}$, the motional magnetic force, $\mathbf{D}$ being the displacement. And there is yet another change sometimes needed. For whilst $\mathbf{B}$ is circuital, so that a convective magnetic current does not appear in the second circuital equation, $\mathbf{D}$ is not always circuital, and convective electric current must therefore appear in the first circuital equation. For the reason just mentioned, it is the theory as thus modified that I consider to represent the true Maxwellian theory, with the other small changes required to make a fit. But further than this I should not like to go, because, having made a fit, it is not necessary, and because it would be taking too great a liberty to make additions without the strongest reason to consider them essential.

The following example, which has been suggested to me by remarks in Prof. Lodge's recent paper on "Aberration Problems," referring to a previous investigation of Prof. J. J. Thomson, will illustrate the matter in question. It is known that if V be the speed of light through ether, the speed through a stationary transparent body, say water, is V/μ, if μ is the refractive index. Now what is the speed when the water is itself moving in the same direction as the light waves? This is a very old problem. Fresnel considered that the external ether was stationary, and that the ether was μ^2 times as dense in the water as outside, and that, when moving, the water only carried forward with it the extra ether it contained (or equivalently). This makes the speed of light referred to the external ether be $V/\mu + v(1 - \mu^{-2})$, if v is the speed of the water. The experiments of Fizeau and Michelson have shown that this result is at least approximately true, and there is other evidence to support Fresnel's hypothesis, at least in a generalised form. But, in the case

of water, the additional speed of light due to the motion of
the water might be $\frac{1}{2}v$ instead of $(1 - \mu^{-2})\,v$, without much
disagreement. Now suppose we examine the matter electromagnetically, and enquire what the increased speed through
a moving dielectric should be. If we follow Maxwell's
equations literally, we shall find that the extra speed is $\frac{1}{2}v$,
provided v/V is small. This actually seems to corroborate
the experimental results. But the argument is entirely a
deceptive one. Maxwell's theory is a theory of propagation
through a simple medium. Fundamentally it is the ether,
but when we pass to a solid or liquid dielectric it is still to be
regarded as a simple medium in the same sense, because the
only change occurring in the equations is in the value of one
or both ethereal constants, the permittivity and inductivity—
practically only the first. Consequently, if we find, as above,
that when the medium is itself moved, its velocity is *not*
superimposed upon that of the velocity of waves through the
medium at rest, the true inference is that there is something
wrong with the theory. For all motion is relative, and it is
an axiomatic truth that there should be superimposition of
velocities, so that $V/\mu + v$ should be the velocity in the above
case according to any rational theory of propagation through
a simple medium, the extra velocity being the full v, instead
of $\frac{1}{2}v$. And, as a matter of fact, if we employ the modified
or corrected circuital law above referred to, we do obtain full
superimposition of velocities.

This example shows the importance of having a simply
expressed and sound primary theory. For if the auxiliary
hypotheses required to explain outstanding or secondary phenomena be conjoined to an imperfect primary theory we shall
surely be led to wrong results. Whereas if the primary theory
be good, there is at least a chance of its extension by auxiliary
hypotheses being also good. The true conclusion from Fizeau
and Michelson's results is that a transparent medium like
water cannot be regarded as (in the electromagnetic theory)

a simple medium like the ether, at least for waves of light, and that a secondary theory is necessary. Fresnel's sagacious speculation is justified, except indeed as regards its form of expression. The ether, for example, may be identical inside and outside the body, and the matter slip through it without sensibly affecting it. At any rate the evidence that this is the case preponderates, the latest being Prof. Lodge's experiments with whirling discs, though on the other hand must not be forgotten the contrary conclusion arrived at by Michelson as to the absence of relative motion between the earth and surrounding ether. But if the ether be stationary, Fresnel's speculation is roughly equivalent to supposing that the molecules of transparent matter act like little condensers in increasing the permittivity, and that the matter, when in motion, only carries forward the increased permittivity. But however this matter may be finally interpreted, we must have a clear primary theory that can be trusted within its limits. Whether Maxwell's theory will last, as a sufficient and satisfactory primary theory upon which the numerous secondary developments required may be grafted, is a matter for the future to determine. Let it not be forgotten that Maxwell's theory is only the first step towards a full theory of the ether; and, moreover, that no theory of the ether can be complete that does not fully account for the omnipresent force of gravitation.

There is one other matter that demands notice in conclusion. It is not long since it was taken for granted that the common electrical units were correct. That curious and obtrusive constant 4π was considered by some to be a sort of blessed dispensation, without which all electrical theory would fall to pieces. I believe that this view is now nearly extinct, and that it is well recognised that the 4π was an unfortunate and mischievous mistake, the source of many evils. In plain English, the common system of electrical units involves an irrationality of the same kind as would be brought into the

metric system of weights and measures, were we to define
the unit area to be the area, not of a square with unit side,
but of a circle of unit diameter. The constant π would then
obtrude itself into the area of a rectangle, and everywhere
it should not be, and be a source of great confusion and
inconvenience. So it is in the common electrical units,
which are truly irrational. Now, to make a mistake is easy
and natural to man. But that is not enough. The next
thing is to correct it. When a mistake has once been started,
it is not necessary to go on repeating it for ever and ever
with cumulative inconvenience.

The B. A. Committee on Electrical Standards had to do
two kinds of work. There was the practical work of making
standards from the experimentally found properties of matter
(and ether). This has been done at great length, and with
much labour and success. But there was also the theoretical
work of fixing the relations of the units in a convenient,
rational, and harmonious manner. This work has not yet
been done. To say that they ought to do it is almost a
platitude. Who else should do it? To say that there is
not at present sufficient popular demand for the change does
not seem very satisfactory. Is it not for leaders to lead?
And who should lead but the men of light and leading who
have practical influence in the matter?

Whilst, on the one hand, the immense benefit to be gained
by rationalising the units requires some consideration to fully
appreciate, it is, on the other hand, very easy to overestimate
the difficulty of making the change. Some temporary inconvenience is necessary, of course. For a time there would be
two sorts of ohms, &c., the old style and the new (or rational).
But it is not a novelty to have two sorts of ohms. There
have been several already. Remember that the number of
standards in present existence is as nothing to the number
going to be made, and with ever increasing rapidity, by reason
of the enormously rapid extension of electrical industries.

Old style instruments would very soon be in a minority, and then disappear, like the pins. I do not know that there is a more important practical question than this one of rational-ising the units, on account of its far-reaching effect, and think that whilst the change could be made now with ease (with a will, of course), it will be far more troublesome if put off until the general British units are reformed; even though that period be not so distant as it is customary to believe. Electricians should set a good example.

The reform which I advocate is somewhat similar to the important improvement made by chemists in *their* units about a quarter of a century ago. One day our respected master informed us that it had been found out that water was not HO, as he had taught us before, but something else. It was henceforward to be H_2O. This was strange at first, and inconvenient, for so many other formulæ had to be altered, and new books written. But no one questions the wisdom of the change. Now observe, here, that the chemists, when they found that their atomic weights were wrong, and their formulæ irrational, did not cry " Too late," ignore the matter, and ask Parliament to legalise the old erroneous weights ! They went and set the matter right. *Verb. sap.*

December 16, 1893.

CONTENTS.

———◆———

CHAPTER III.

THE ELEMENTS OF VECTORIAL ALGEBRA AND ANALYSIS.
(Pages 132 to 305.)

CHAPTER IV.

THEORY OF PLANE ELECTROMAGNETIC WAVES.
(Pages 306 to 466.)

A*

CORRECTIONS.

Page 82, equation (26). In case of eolotropy, add to the right
  member the term **S**a, the scalar product of the torque **S**,
  per unit volume, and the spin a.

Page 186, § 125. I regret to have misrepresented Dr. C. V.
  Burton's notion of a moving strain-figure. He does it
  entirely by conservative elastic forces, and my objection
  does not apply.

Page 188, equation (143). On the right side, for **D** read D,
  three times.

Page 243. The sectional number is 153A.

CHAPTER I.

INTRODUCTION.

§ **1.** *Preliminary Remarks.*—The main object of the series of articles of which this is the first, is to continue the work entitled " Electromagnetic Induction and its Propagation," commenced in *The Electrician* on January 3, 1885, and continued to the 46th Section in September, 1887, when the great pressure on space and the want of readers appeared to necessitate its abrupt discontinuance. (A straggler, the 47th Section, appeared December 31, 1887.) Perhaps there were other reasons than those mentioned for the discontinuance. We do not dwell in the Palace of Truth. But, as was mentioned to me not long since, " There is a time coming when all things shall be found out." I am not so sanguine myself, believing that the well in which Truth is said to reside is really a bottomless pit.

The particular branch of the subject which I was publishing in the summer of 1887 was the propagation of electromagnetic waves along wires through the dielectric surrounding them. This is itself a large and many-sided subject. Besides a general treatment, its many-sidedness demands that special cases of interest should receive separate full development. In general, the mathematics required is more or less of the character sometimes termed transcendental. This is a grandiloquent word, suggestive of something beyond human capacity to find out ; a word to frighten timid people into believing that it is all speculation, and therefore unsound. I do not know where transcendentality begins. You can find it in arithmetic. But never mind the word. What is of more importance is the fact that the interpretation of transcendental formulæ is sometimes

B

very laborious. Now the real object of true naturalists, in
Sir W. Thomson's meaning of the word, when they employ
mathematics to assist them, is not to make mathematical exercises (though that may be necessary), but to find out the connections of known phenomena, and by deductive reasoning, to
obtain a knowledge of hitherto unknown phenomena. Anything, therefore, that aids this, possesses a value of its own wholly
apart from immediate or, indeed, any application of the kind
commonly termed practical. There is, however, practicality in
theory as well as in practice. The very useful word " practician " has lately come into use. It supplies a want, for it is
evident the moment it is mentioned that a practician need not
be a practical man ; and that, on the other hand, it may happen
occasionally that a man who is not a practician may still be
quite practical.

§ 2. Now, I was so fortunate as to discover, during the
examination of a practical telephonic problem, that in a certain
case of propagation along a conducting circuit through a conducting dielectric, the transcendentality of the mathematics
automatically vanished, by the distorting effects on an electromagnetic wave, of the resistance of the conductor, and of the
conductance of the dielectric, being of opposite natures, so that
they neutralised one another, and rendered the circuit non-
distortional or distortionless. The mathematics was reduced,
in the main, to simple algebra, and the manner of transmission
of disturbances could be examined in complete detail in an
elementary manner. Nor was this all. The distortionless
circuit could be itself employed to enable us to understand the
inner meaning of the transcendental cases of propagation, when
the distortion caused by the resistance of the circuit makes the
mathematics more difficult of interpretation. For instance, by
a study of the distortionless circuit we are enabled to see not
only that, but also *why*, self-induction is of such great importance in the transmission of rapidly-varying disturbances in
preserving their individuality and preventing them from being
attenuated to nearly nothing before getting from one end of a
long circuit to the other ; and *why* copper wires are so successful in, and iron wires so prejudicial to effective, long-distance
telephony. These matters were considered in Sections 40 to

45 (June, July, August, 1887) of the work I have referred to, and Sections 46, 47 contain further developments.

§ 3. But that this matter of the distortionless circuit has, directly, important practical applications, is, from the purely scientific point of view, a mere accidental circumstance. Perhaps a more valuable property of the distortionless circuit is, that it is the Royal Road to electromagnetic waves in general, especially when the transmitting medium is a conductor as well as a dielectric. I have somewhat developed this matter in the *Phil. Mag.*, 1888-9. Fault has been found with these articles that they are hard to read. They were harder, perhaps, to write. The necessity of condensation in a journal where space is so limited and so valuable, dealing with all branches of physical science, is imperative. What *is* an investigator to do, when he can neither find acceptance of matter in a comparatively elementary form by journals of a partly scientific, partly technical type, with many readers, nor, in a more learned form, by a purely scientific journal with comparatively few readers, and little space to spare? To get published at all, he must condense greatly, and leave out all explanatory matter that he possibly can. Otherwise, he may be told his papers are more fit for publication in book form, and are therefore declined.

There is a third course, of course, viz., to keep his investigations to himself. But that does not answer, in a general way, though it may do so sometimes. It is like putting away seed in a mummy case, instead of planting it, and letting it take its chance of growing to a useful plant. There is nothing like publication and free criticism for utility. I can see only one good excuse for abstaining from publication when no obstacle presents itself. You may grow your plant yourself, nurse it carefully in a hot-house, and send it into the world full-grown. But it cannot often occur that it is worth the trouble taken. As for the secretiveness of a Cavendish, that is utterly inexcusable; it is a sin. It is possible to imagine the case of a man being silent, either from a want of confidence in himself, or from disappointment at the reception given to, and want of appreciation of, the work he gives to the world; few men have an unbounded power of persistence; but to make valuable discoveries, and to hoard them up as Cavendish did, without any

B 2

valid reason, seems one of the most criminal acts such a man
could be guilty of. This seems strong language, but as Prof. Tait
tells us that it is almost criminal not to know several foreign
languages, which is a very venial offence in the opinion of others,
it seems necessary to employ strong language when the crimi-
nality is more evident. (*See*, on this point, the article in *The
Electrician*, November 14, 1890. It is both severe and logical.)

§ 4. I had occasion, just lately, to use the word "naturalist."
The matter involved here is worthy of parenthetical consideration. Sir W. Thomson does not like "physicist," nor, I think,
"scientist" either. It must, however, be noted that the
naturalist, as at present generally understood, is a student of
living nature only. He has certainly no exclusive right to so
excellent a name. On the other hand, the physicist is a
student of inanimate nature, in the main, so that he has no
exclusive right to the name, either. Both are naturalists. But
their work is so different, and their type of mind also so
different, that it seems very desirable that their names should
be differentiated, and that "naturalist," comprehending both,
should be subdivided. Could not one set of men be induced to call
themselves organists? We have organic chemistry, and organisms,
and organic science ; then why not organists ? Perhaps, however, organists might not care to be temporarily confounded
with those members of society who earn their living by setting a
cylinder in rotatory motion. If so, there is another good name,
viz., vitalist, for the organist, which would not have any ludic-
rous association. Then about the other set of men. Are they
not essentially students of the properties of matter, and therefore materialists? That "materialist" is the right name is
obvious at a glance. Here, however, a certain suppositious
evil association of the word might militate against its adoption.
But this would be, I think, an unsound objection, for I do not
think there is, or ever was, such a thing as a materialist, in
the supposed evil sense. Let that notion go, and the valuable
word "materialist" be put to its proper use, and be dignified
by association with an honourable body of men.

Buffon, Cuvier, Darwin, were typical vitalists.

Newton, Faraday, Maxwell, were typical materialists.

All were naturalists. For my part I always admired the old-fashioned term "natural philosopher." It was so dignified, and raised up visions of the portraits of Count Rumford, Young, Herschel, Sir H. Davy, &c., usually highly respectable-looking elderly gentlemen, with very large bald heads, and much wrapped up about the throats, sitting in their studies pondering calmly over the secrets of nature revealed to them by their experiments. There are no natural philosophers now-a-days. How is it possible to be a natural philosopher when a Salvation Army band is performing outside; joyously, it may be, but not most melodiously? But I would not disparage their work; it may be far more important than his.

§ 5. Returning to electromagnetic waves. Maxwell's inimitable theory of dielectric displacement was for long generally regarded as a speculation. There was, for many years, an almost complete dearth of interest in the unverified parts of Maxwell's theory. Prof. Fitzgerald, of Dublin, was the most prominent of the very few materialists (if I may use the word) who appeared to have a solid faith in the electromagnetic theory of the ether; thinking about it and endeavouring to arrive at an idea of the nature of diverging electromagnetic waves, and how to produce them, and to calculate the loss of energy by radiation. An important step was then made by Poynting, establishing the formula for the flow of energy. Still, however, the theory wanted experimental proof. Three years ago electromagnetic waves were nowhere. Shortly after, they were everywhere. This was due to a very remarkable and unexpected event, no less than the experimental discovery by Hertz, of Karlsruhe (now of Bonn), of the veritable actuality of electromagnetic waves in the ether. And it never rains but it pours; for whilst Hertz with his resonating circuit was working in Germany (where one would least expect such a discovery to be made, if one judged only by the old German electro-dynamic theories), Lodge was doing in some respects similar work in England, in connection with the theory of lightning conductors. These researches, followed by the numerous others of Fitzgerald and Trouton, J. J. Thomson, &c., have dealt a death-blow to the electro-dynamic speculations of the Weber-Clausius type (to mention only the first and one of the last), and have given to Maxwell's theory just what was wanted in its higher parts, more experimental basis. The interest excited has been immense, and the theorist can now write about electromagnetic waves without incurring the reproach that he is working out a mere paper theory. The speedy recognition of Dr. Hertz by the Royal Society is a very unusual testimony to the value of his researches.

At the same time I may remark that to one who had carefully examined the nature of Maxwell's theory, and looked into its consequences, and seen how *rationally* most of the phenomena of electromagnetism were explained by it, and how it furnished the only approximately satisfactory (paper) theory of light known; to such a one Hertz's demonstration came as a matter of course—only it came rather unexpectedly.

§ 6. It is not by any means to be concluded that Maxwell spells Finality. There is no finality. It cannot even be accurately said that the Hertzian waves prove Maxwell's dielectric theory completely. The observations were very rough indeed, when compared with the refined tests in other parts of electrical science. The important thing proved is that electromagnetic waves in the ether at least approximately in accordance with Maxwell's theory are a reality, and that the Faraday-Maxwellian method is the correct one. The other kind of electrodynamic speculation is played out completely. There will be plenty of room for more theoretical speculation, but it must now be of the Maxwellian type, to be really useful.

§ 7. In what is to follow, the consideration of electromagnetic waves will (perhaps) occupy a considerable space. How much depends entirely upon the reception given to the articles. Mathematics is at a discount, it seems. Nevertheless, as the subject is intrinsically a mathematical one, I shall not scruple to employ the appropriate methods when required. The reader whose scientific horizon is bounded entirely by commercial considerations may as well avoid these articles. Speaking without prejudice, matter more to his taste may perhaps be found under the heading TRADE NOTICES.* *Sunt quos curriculo.*

I shall, however, endeavour to avoid investigations of a complex character; also, when the methods and terms used are not

---

* Referring to *The Electrician*, in which this work first appeared.

generally known I shall explain them. Considering the lapse of time since the discontinuance of E.M.I. and its P. it would be absurd to jump into the middle of the subject all at once. It must, therefore, be gradually led up to. I shall, therefore, in the next place make a few remarks upon mathematical investigations in general, a subject upon which there are many popular delusions current, even amongst people who, one would think, should know better.

§ 8. There are men of a certain type of mind who are never wearied with gibing at mathematics, at mathematicians, and at mathematical methods of inquiry. It goes almost without saying that these men have themselves little mathematical bent. I believe this to be a general fact ; but, as a fact, it does not explain very well their attitude towards mathematicians. The reason seems to lie deeper. How does it come about, for instance, that whilst they are themselves so transparently ignorant of the real nature, meaning, and effects of mathematical investigation, they yet lay down the law in the most confident and self-satisfied manner, telling the mathematician what the nature of his work is (or rather is not), and of its erroneousness and inutility, and so forth ? It is quite as if they knew all about it.

It reminds one of the professional paradoxers, the men who want to make you believe that the ratio of the circumference to the diameter of a circle is 3, or 3·125, or some other nice easy number (any but the right one) ; or that the earth is flat, or that the sun is a lump of ice ; or that the distance of the moon is exactly 6 miles 500 yards, or that the speed of the current varies as the square of the length of the line. They, too, write as if they knew all about it ! Plainly, then, the anti-mathematician must belong to the same class as the paradoxer, whose characteristic is to be wise in his ignorance, whereas the really wise man is ignorant in his wisdom. But this matter may be left for students of mind to settle. What is of greater importance is that the anti-mathematicians sometimes do a deal of mischief. For there are many of a neutral frame of mind, little acquainted themselves with mathematical methods, who are sufficiently impressible to be easily taken in by the gibers and to be prejudiced thereby ; and, should they possess some mathematical bent, they may be hindered

by their prejudice from giving it fair development. We
cannot all be Newtons or Laplaces, but that there is an
immense amount of moderate mathematical talent lying latent
in the average man I regard as a fact ; and even the moderate
development implied in a working knowledge of simple algebraical equations can, with common-sense to assist, be not
only the means of valuable mental discipline, but even be of
commercial importance (which goes a long way with some
people), should one's occupation be a branch of engineering for
example.

§ 9. " Mathematics is gibberish." Little need be said about
this statement. It is only worthy of the utterly illiterate.
" What is the use of it ? It is all waste of time. Better be
doing something useful. Why, you might be inventing a new
dynamo in the time you waste over all that stuff." Now,
similar remarks to these I have often heard from fairly intelligent and educated people. *They* don't see the use of it, that is
plain. That is nothing ; what is to the point is that they conclude that it is of no use. For it may be easily observed that
the parrot-cry " What's the use of it ?" does not emanate in a
humble spirit of inquiry, but on the contrary, quite the reverse.
You can see the nose turn up.

But what is the use of it, then ? Well, it is quite certain that
if a person has no mathematical talent whatever he had really
better be doing something " useful," that is to say, something
else than mathematics, (inventing a dynamo, for instance,) and
not be wasting his time in (so to speak) trying to force a crop
of wheat on the sands of the sea-shore. This is quite a personal
question. Every mind should receive fair development (in
good directions) for what it is capable of doing fairly well.
People who do not cultivate their minds have no conception of
what they lose. They become mere eating and drinking and
money-grabbing machines. And yet they seem happy ! There
is some merciful dispensation at work, no doubt.

" Mathematics is a mere machine. You can't get anything
out of it that you don't put in first. You put it in, and then
just grind it out again. You can't discover anything by
mathematics, or invent anything. You can't get more than a
pint out of a pint pot." And so forth.

It is scarcely credible to the initiated that such statements could be made by any person who could be said to have an intellect. But I have heard similar remarks from really talented men, who might have fair mathematical aptitude themselves, though quite undeveloped. The fact is, the statements contain at once a profound truth, and a mischievous fallacy. That the fallacy is not self-evident affords an excuse for its not being perceived even by those who may (perhaps imperfectly) recognise the element of truth. But as regards the truth mentioned, I doubt whether the caviller has generally any distinct idea of it either, or he would not express it so contemptuously along with the fallacy.

§ 10. By any process of reasoning whatever (not fancy) you cannot get any results that are not implicitly contained in the material with which you work, the fundamental data and their connections, which form the basis of your inquiry. You may make mistakes, and so arrive at erroneous results from the most correct data. Or the data may be faulty, and lead to erroneous conclusions by the most correct reasoning. And in general, if the data be imperfect, or be only true within certain limits hardly definable, the results can have but a limited application. Now all this obtains exactly in mathematical reasoning. It is in no way exempt from the perils of reasoning in general. But why the mathematical reasoning should be singled out for condemnation as mere machine work, dependent upon what the machine is made to do, with a given supply of material, is not very evident. The cause lies deep in the nature of the anti-mathematician ; he has not recognised that all reasoning must be, in a sense, mechanical, else it is not sound reasoning at all, but vitiated by fancy.

Mathematical reasoning is, fundamentally, not different from reasoning in general. And as by the exercise of the reason discoveries can be made, why not by mathematical reasoning? Whatever were Newton and the long array of mathematical materialists who followed him doing all the time? Making discoveries, of course, largely assisted by their mathematics. I say nothing of the pure mathematicians. Their discoveries are extensions of the field of mathematics itself—a perfectly limitless field. I refer only to students of Nature on its material side, who have employed mathematics expressly for the purpose of making discoveries. Some of the unmathematical believe that the mathematician is merely engaged in counting or in doing long sums; this probably arises from reminiscences of their schooldays, when they were flogged over fractions. Now this is only a part of his work, a sometimes necessary and very disagreeable part, which he would willingly hand over to a properly trained computator. This part of the work only concerns the size of the effects, but it is the effects themselves to which I refer when I speak of discoveries.

§ 11. Mathematics is reasoning about quantities. Even if qualities are in question, it is their quantities that are subjected to the mathematics. If there be something which cannot be reduced to a quantity, or more generally to a definite function, no matter how complex and involved, of any number of other quantities which can be measured (either actually, or in imagination), then that something cannot be accurately reasoned about, because it is in part unknown. Not unknown in the sense in which a quantity is said to be unknown in algebra, when it is virtually known because virtually expressible in terms of known quantities, but literally unknown by the absence of sufficient quantitative connection with the known. Thus only the known can be accurately reasoned about. But this includes, it will be observed, everything that can be deduced from the known, without appeal to the unknown. The unknown is not necessarily unknowable; fresh knowns may make the former unknowns become also known. The distinction is a very important one. The limits of human knowledge are ever shifting. But there must be an ultimate limit, because we are a part of Nature, and cannot go beyond it. Beyond this limit, the Unknown becomes the Unknowable, which it is of little service to discuss, though it will always be a favourite subject of speculation. But whatever is in this Universe can be (or might be) found out, and therefore does not belong to the unknowable. Thus the constitution of the middle of the sun, or of the ether, or the ultimate nature of magnetisation, or of universal gravitation, or of life, are not unknowable; and this statement is true, even though they should never be discovered. There are no inscrutables in Nature. By Faith only can we go beyond—as far and where we please.

Human nature, or say a man, is a highly complex quantity. We are compelled to take him in parts, and consider this or that quality, and imagine it measured and brought into proper connection with other qualities and external influences. Yet a man, if we only knew him intimately enough, could be formularised, and have his whole life-history developed. Even the universe itself, if every law in action were thoroughly known, could have its history, past, present, and to come, formularised down to the minutest particulars, provided no discontinuity or special act of creation occur. But even the special act of creation could be formularised, and its effects deduced, if we knew in what it consisted. And special acts of creation might be going on continuously, involving continuous changes in the laws of nature, and could be formularised, if the acts of creation were known, or the—so to speak—law of the discontinuities. The case is somewhat analogous to that of impressed forces acting upon a dynamical system. The behaviour of the system is perfectly definite and formularisable so long as no impressed forces act, and ceases to be definite if unknown impressed forces act. But if the forces be also known, then the course of events is again definitely formularisable. The assumption of a special act of creation, either now or at any time, is merely a confession of ignorance. We have no evidence of any such discontinuities. We cannot prove that there have never been any; nor can we prove that the sun will not rise to-morrow, or that the clock will not wind itself up again when the weight has run down.

§ 12. Nearly all the millions, or rather billions, of human beings who have peopled this earth have been content to go through life taking things as they found them, and without any desire to understand what is going on around them. It is exceedingly remarkable that the scientific spirit (asking how it is done), which is so active and widespread at the present day, should be of such recent origin. With a few exceptions, it hardly existed amongst the Ancients (who would be more appropriately termed the Youngsters). It is a very encouraging fact for evolutionists, leading them to believe that the evolution of man is not played out ; but that man is capable, intellectually, of great development, and that the general standard will be far higher in the future than at present.

Now, in the development of our knowledge of the workings of Nature out of the tremendously complex assemblage of phenomena presented to the scientific inquirer, mathematics plays in some respects a very limited, in others a very important part. As regards the limitations, it is merely necessary to refer to the sciences connected with living matter, and to the ologies generally, to see that the facts and their connections are too indistinctly known to render mathematical analysis practicable, to say nothing of the complexity. Facts are of not much use, considered as facts. They bewilder by their number and their apparent incoherency. Let them be digested into theory, however, and brought into mutual harmony, and it is another matter. Theory is the essence of facts. Without theory scientific knowledge would be only worthy of the madhouse.

In some branches of knowledge, the facts have been so far refined into theory that mathematical reasoning becomes applicable on a most extensive scale. One of these branches is Electromagnetism, that most extensive science which presents such a remarkable two-sidedness, showing the electric and the magnetic aspects either separately or together, in stationary conditions, and a third condition when the electric and magnetic forces act suitably in dynamical combination, with equal development of the electric and magnetic energies, the state of electromagnetic waves.

It goes without saying that there are numerous phenomena connected with electricity and magnetism which are very imperfectly understood, and which have not been formularised, except perhaps in an empirical manner. Such is particularly the case where the sciences of Electricity and Chemistry meet. Chemistry is, so far, eminently unmathematical (and therefore a suitable study for men of large capacity, who may be nearly destitute of mathematical talent—but this by the way), and it appears to communicate a part of its complexity and vagueness to electrical science whenever electrical phenomena which we can study are accompanied by chemical changes. But generally speaking, excepting electrolytic phenomena and other complications (*e.g.*, the transport of matter in rarefied media when electrical discharges occur), the phenomena of electromagnetism are, in the main, remarkably well known, and amenable to mathematical treatment.

§ 13. Ohm (a distinguished mathematician, be it noted) brought into order a host of puzzling facts connecting electromotive force and electric current in conductors, which all previous electricians had only succeeded in loosely binding together qualitatively under some rather vague statements. Even as late as 20 years ago, "quantity" and "tension" were much used by men who did not fully appreciate Ohm's law. (Is it not rather remarkable that some of Germany's best men of genius should have been, perhaps, unfairly treated? Ohm; Mayer; Reis; even von Helmholtz has mentioned the difficulty he had in getting recognised. But perhaps it is the same all the world over.) Ohm found that the results could be summed up in such a simple law that he who runs may read it, and a schoolboy now can predict what a Faraday then could only guess at roughly. By Ohm's discovery a large part of the domain of electricity became annexed to theory. Another large part became virtually annexed by Coulomb's discovery of the law of inverse squares, and completely annexed by Green's investigations. Poisson attacked the difficult problem of induced magnetisation, and his results, though differently expressed, are still the theory, as a most important first approximation. Ampère brought a multitude of phenomena into theory by his investigations of the mechanical forces between conductors supporting currents and magnets. Then there were the remarkable researches of Faraday, the prince of experimentalists, on electrostatics and electrodynamics and the induction of currents. These were rather long in being brought from the crude experimental state to a compact system, expressing the real essence. Unfortunately, in my opinion, Faraday was not a mathematician. It can scarcely be doubted that had he been one, he would have been greatly assisted in his researches, have saved himself much useless speculation, and would have anticipated much later work. He would, for instance, knowing Ampère's theory, by his own results have readily been led to Neumann's theory, and the connected work of Helmholtz and Thomson. But it is perhaps too much to expect a man to be both the prince of experimentalists and a competent mathematician.

Passing over the other developments which were made in the theory of electricity and magnetism, without striking new departures, we come to about 1860. There was then a collection of detached theories, but loosely connected, and embedded in a heap of unnecessary hypotheses, scientifically valueless, and entirely opposed to the spirit of Faraday's ways of thinking, and, in fact, to the spirit of the time. All the useless hypotheses had to be discarded, for one thing ; a complete and harmonious theory had to be made up out of the useful remainder, for another ; and, in particular, the physics of the subject required to be rationalised, the supposed mutual attractions or repulsions of electricity, or of magnetism, or of elements of electric currents upon one another, abolished, and electromagnetic effects accounted for by continuous actions through a medium, propagated in time. All this, and much more, was done. The crowning achievement was reserved for the heaven-sent Maxwell, a man whose fame, great as it is now, has, comparatively speaking, yet to come.

§ 14. It will have been observed that I have said next to nothing upon the study of pure mathematics ; this is a matter with which we are not concerned. But that I have somewhat dilated (and I do not think needlessly) upon the advantages attending the use of mathematical methods by the materialist to assist him in his study of the laws governing the material universe, by the proper co-ordination of known and the discovery of unknown (but not unknowable) phenomena.

It was discovered by mathematical reasoning that when an electric current is started in a wire, it begins entirely upon its skin, in fact upon the outside of its skin ; and that, in consequence, sufficiently rapidly impressed fluctuations of the current keep to the skin of the wire, and do not sensibly penetrate to its interior.

Now very few (if any) unmathematical electricians can understand this fact ; many of them neither understand it nor believe it. Even many who do believe it do so, I believe, simply because they are told so, and not because they can in the least feel positive about its truth of their own knowledge. As an eminent practician remarked, after prolonged scepticism, " When Sir W. Thomson says so, who can doubt it ? " What a world of worldly wisdom lay in that remark !

Now I *do* admire this characteristically stubborn English way of being determined not to be imposed upon by any absurd theory that goes against all one's most cherished convictions, and which cannot be properly understood without mathematies. For without the mathematics, and with only the sure knowledge of Ohm's law and the old-fashioned notions concerning the function of a conducting wire to guide one, no one would think of such a theory. It is quite preposterous from this point of view. Nevertheless, it is true ; and the view was not put forward as a hypothesis, but as a plain matter of fact.

The case in question is one in which we can be very sure of all the fundamental data of any importance, and the laws concerned. We can, for instance, by straightforward experiment, especially with properly constructed induction balances admitting of exact interpretation of results, readily satisfy ourselves that a high degree of accuracy must obtain not merely for Ohm's law, but also for Faraday's law of E.M.F. in circuits, and even in iron for Poisson's law of induced magnetisation, within certain limits. We have, therefore, all the conditions wanting for the successful application of mathematical reasoning of a precise character, and justification for the confidence that mathematicians can feel in the results theoretically deduced in a legitimate manner, however difficult it may be to give an easily intelligible account of their meaning to the unmathematical.

This, however, I will say for the sceptic who has the courage of his convictions, and writes openly against what is, to him, pure nonsense. He is doing, in his way, good service in the cause of truth and the advancement of scientific knowledge, by stimulating interest in the subject and causing people to inquire and read and think about these things, and form their own judgment if possible, and modify their old views if they should be found wanting. Nothing is more useful than open and free criticism, and the truly earnest and disinterested student of science always welcomes it.

§ 15. The following may assist the unmathematical reader to an understanding of the subject. It is not demonstrative, of course, but is merely descriptive. If, however, it be translated into mathematical language and properly worked out, it will be found to be demonstrative, and to lead to a complete theory of the functions of wires in general.

Start with a very long solenoid of fine wire in circuit with a source of electrical energy. Let the material inside the solenoid be merely air, that is to say, ether and air. If we examine the nature of the fluctuations of current in the coil in relation to the fluctuations of impressed force on it, we find that the current in the coil behaves as if it were a material fluid possessing inertia and moving against resistance. The fanatics of Ohm's law do not usually take into account the inertia. It is as if the current in the coil could not move without simultaneously setting into rotation a rigid material core filling the solenoid, and free to rotate on its axis.

If we take the air out of the solenoid and substitute any other non-conducting material for a core the same thing happens: only the inertia varies with the material, according to its magnetic inductivity.

But if we use a conducting core we get new phenomena, for we find that there is no longer a definite resistance and a definite inertia. There is now frictional resistance in the core, and this increases the effective resistance of the coil. At the same time the inertia is reduced.

On examining the theory of the matter (on the basis of Ohm's law and Faraday's law applied to the conducting core) we find that we can now account for things (in our analogy) by supposing that the rigid solid core first used is replaced by a viscous fluid core, like treacle. On starting a current in the coil it cannot now turn the core round bodily at once, but only its external portion. In time, however, if the source of current be steadily operating, the motion will penetrate throughout the viscous core, which will finally move as the former rigid core did. If, however, the current in the coil fluctuate in strength very rapidly, the corresponding fluctuations of motion in the core will be practically confined to its skin. The effective inertia is reduced because the core does not move as a rigid body; the effective resistance is increased by the viscosity generating heat.

Now, returning to the solenoid, we have perfect symmetry with respect to its axis, since the core is supposed to be exceedingly long, and uniformly lapped with wire. The situation of the source of energy, as regards the core itself, is plainly on its boundary, where the coil is placed ; and it is therefore a matter of common-sense that in the communication of energy to the core either when the current is steady or when it varies, the transfer of energy takes place transversely, that is, from the boundary to the axis, in planes perpendicular to the axis, and therefore perpendicular to the current in the core itself. This is confirmed by the electromagnetic equations.

But the electromagnetic equations go further than this, and assert that the transference of energy in any isotropic electrical conductor always takes place across the lines of conduction current, and not merely in the case of a core uniformly lapped with wire, where it is nearly self-evident that it must be so. This is a very important result, being the post-finger pointing to a clear understanding of electromagnetism. Passing to the case of a very long straight round wire supporting an electric current, we are bound to conclude that the transference of energy takes place transversely, not longitudinally ; that is, across the wire instead of along it. The source of energy must, therefore, first supply the dielectric surrounding the wire before the substance of the wire itself can be influenced ; that is, the dielectric must be the real primary agent in the electromagnetic phenomena connected with the electric current in the wire.

Beyond this transverse transference of energy, there does not, however, at first sight, appear to be much analogy between the case of the solenoid with a core and the straight wire in a dielectric. The source of energy in one case is virtually brought right up to the surface of the core in a uniform manner. But in the other case the source of energy—the battery, for instance— may be miles away at one end of the wire, and there is no immediately obvious uniform application of the source to the skin of the wire. But observe that in the former case the magnetic force is axial, and the electric current circular, whilst in the latter case the electric current (in the straight wire) is axial and the magnetic force circular. Now an examination of the electromagnetic equations shows that the conditions of propagation of axial magnetic force and circular current are the

c

same as those for axial current and circular magnetic force. We therefore further conclude that (with the exchanges made) the phenomena concerned in the core of the solenoid and in the long straight wire are of the same character.

Furthermore, if we go into detail, and consider the influence of the surroundings of the wire (which go to determine the value of the inductance of the circuit) we shall find that not only is the character of the phenomena the same, but that they may be made similar in detail (so as to be represented by similar curves, for example).

The source of energy, therefore, is virtually transferred instantly from its real place to the whole skin of the wire, over which it is uniformly spread, just as in the case of the conducting core within a solenoid.

§ 16. So far we can go by Ohm's law and Faraday's law of E.M.F., and, if need be, Poisson's law of induced magnetisation (or its modern equivalent practically). But it is quite impossible to stop here. Even if we had no knowledge of electrostatics and of the properties of condensers, we should, by the above course of inquiry, be irresistibly led to a theory of transmission of electrical disturbances through a medium surrounding the wire, instead of through the wire. Maxwell's theory of dielectric displacement furnishes what is wanted to explain results which are in some respects rather unintelligible when deduced in the above manner without reference to electrostatic phenomena.

We learn from it that the battery or other source of energy acts upon the dielectric primarily, producing electric displacement and magnetic induction; that disturbances are propagated through the dielectric at the speed of light; that the manner of propagation is similar to that of displacements and motions in an incompressible elastic solid; that electrical conductors act, as regards the internal propagation, not as conductors but rather as obstructors, though they act as conductors in another sense, by guiding the electromagnetic waves along definite paths in space, instead of allowing them to be immediately spread away to nothing by spherical enlargement at the speed of light; that when we deal with steady states, or only slowly varying states, involving immensely great wave length in the dielectric, the resulting magnetic phenomena are just such as would arise were the speed of propagation infinitely great instead of being finite; that if we make our oscillations faster, we shall begin to get signs of propagation in the manner of waves along wires, with, however, great distortion and attenuation by the resistance of the wires; that if we make them much faster we shall obtain a comparatively undistorted transmission of waves (as in long-distance telephony over copper wires of low resistance); and that if we make our oscillations very fast indeed, we shall have practically mere skin conduction of the waves along the wires at the speed of light (as in some of Lodge's lightning-conductor experiments, and more perfectly with Hertzian waves).

Now all these things have been worked out theoretically, and, as is now well known, most of them have been proved experimentally; and yet I hear someone say that Hertz's experiments don't prove anything in particular!

Lastly, from millions of vibrations per second, proceed to billions, and we come to light (and heat) radiation, which are, in Maxwell's theory, identified with electromagnetic disturbances. The great gap between Hertzian waves and waves of light has not yet been bridged, but I do not doubt that it will be done by the discovery of improved methods of generating and observing very short waves.

CHAPTER II.

OUTLINE OF THE ELECTROMAGNETIC CONNECTIONS.

Electric and Magnetic Force; Displacement and Induction; Elastivity and Permittivity, Inductivity and Reluctivity.

§ 20. Our primary knowledge of electricity, in its quantitative aspect, is founded upon the observation of the mechanical forces experienced by an electrically charged body, by a magnetised body, and by a body supporting electric current. In the study of these mechanical forces we are led to the more abstract ideas of electric force and magnetic force, apart from electrification, or magnetisation, or electric current, to work upon and produce visible effects. The conception of fields of force naturally follows, with the mapping out of space by means of lines or tubes of force definitely distributed. A further and very important step is the recognition that the two vectors, electric force and magnetic force, represent, or are capable of measuring, the actual physical state of the medium concerned, from the electromagnetic point of view, when taken in conjunction with other quantities experimentally recognisable as properties of matter, showing that different substances are affected to different extents by the same intensity of electric or magnetic force. Electric force is then to be conceived as producing or being invariably associated with a flux, the electric displacement; and similarly magnetic force as producing a second flux, the magnetic induction.

If $\mathbf{E}$ be the electric force at any point and $\mathbf{D}$ the displacement, we have

$$\mathbf{D} = c\mathbf{E}; \quad \cdots \cdots \quad (1)$$

and similarly, if **H** be the magnetic force and **B** the induction, then

$$\mathbf{B} = \mu\mathbf{H}. \qquad \qquad (2)$$

Here the ratios c and μ represent physical properties of the medium. The one (μ), which indicates capacity for supporting magnetic induction, is its inductivity ; whilst the other, indicating the capacity for permitting electric displacement, is its permittivity (or permittancy). Otherwise, we may write

$$\mathbf{E} = c^{-1}\mathbf{D}, \qquad \qquad (3)$$

$$\mathbf{H} = \mu^{-1}\mathbf{B} ; \qquad \qquad (4)$$

and now the ratio c^{-1} is the elastivity and μ^{-1} is the reluctivity (or reluctancy). Sometimes one way is preferable, sometimes the other.

Electric and Magnetic Energy.

§ 21. All space must be conceived to be filled with a medium which can support displacement and induction. In the former aspect only it is a dielectric. It is, however, equally necessary to consider the magnetic side of the matter, and we may, without coining a new word, generally understand by a dielectric a medium which supports both the fluxes mentioned.

Away from matter (in the ordinary sense) the medium concerned is the ether, and μ and c are absolute constants. The presence of matter, to a first approximation, merely alters the value of these constants. The permittivity is always increased, so far as is known. On the other hand, the inductivity may be either increased or reduced, there being a very small increase or decrease in most substances, but a very large increase in a few, the so-called magnetic metals. The range within which the proportionality of flux to force obtains is then a limited one, which, however, contains some important practical applications.

When the fluxes vary, their rates of increase $\dot{\mathbf{B}}$ and $\dot{\mathbf{D}}$ are the velocities corresponding to the forces **E** and **H**, provided no other effects are produced. The activity of **E** is $\mathbf{E}\dot{\mathbf{D}}$, and that of **H** is $\mathbf{H}\dot{\mathbf{B}}$. The work spent in producing the fluxes (not counting what may be done simultaneously in other ways) is, therefore—

$$U = \int_o^D \mathbf{E}\, d\mathbf{D}, \qquad \qquad T = \int_o^B \mathbf{H}\, d\mathbf{B}, \qquad \ldots \quad (5)$$

where U is the electric and T the magnetic energy per unit volume.

When μ and c are constants, these give

$$U = \tfrac{1}{2}ED = \tfrac{1}{2}cE^2, \quad . \quad . \quad . \quad . \quad (6)$$

$$T = \tfrac{1}{2}HB = \tfrac{1}{2}\mu H^2, \quad . \quad . \quad . \quad . \quad (7)$$

to express the energy stored in the medium, electric and magnetic respectively.

When μ and c are not constants, the previous expressions (5) will give definite values to the energy provided there be a definite relation between a force and a flux. If, however, there be no definite relation (which means that other circumstances than the value of the force control the value of the flux), the energy stored will not be strictly expressible in terms simply of the force and the flux, and there will be usually a waste of energy in a cyclical process, as in the case of iron, so closely studied by Ewing. This does not come within the scope of a precise mathematical theory, which must of necessity be a sort of skeleton framework, with which complex details have to be separately adjusted in the most feasible manner that presents itself.

Eolotropic Relations.

§ 22. But a precise theory nevertheless admits of considerable extension from the above with μ and c regarded as scalar constants. All bodies are strained more or less, and are thereby usually made eolotropic, even if they be not naturally eolotropic. The force and the flux are not then usually concurrent, or identical in direction. But at any point in an eolotropic substance there are always (if force and flux be proportional) three mutually perpendicular axes of concurrence—the principal axes—when we have (referring to displacement)

$$D_1 = c_1 E_1, \qquad D_2 = c_2 E_2, \qquad D_3 = c_3 E_3,$$

if the c's are the principal permittivities, the E's the corresponding effective components of the electric force, and the D's those of the displacement. If **E** be parallel to a principal axis, so is **D**. In general, by compounding the force components, we obtain **E** the actual force, and by compounding the flux components obtain **D** the displacement to correspond, which can only concur with **E** in the above-mentioned special

cases if the principal permittivities be all different. But should
a pair be equal, then **E** and **D** concur in the plane containing
the equal permittivities, for the permittivity is the same for
any axis in this plane. The energy stored is still half the scalar product of the
force and the flux, or $\frac{1}{2}$**ED**, understanding that the scalar
product of two vectors, which is the product of their tensors
(or magnitudes) when they concur, is the same multiplied by
the cosine of their included angle in the general case.

Vector-analysis is, I think, most profitably studied in the
concrete application to physical questions, for which, indeed, it
is specially adapted. Nevertheless, it will be convenient, a
little later, to give a short account of the very elements of the
subject, in order not to have to too frequently interrupt our
electromagnetic arguments by mathematical explanations. In
the meantime, consider the dielectric medium further.

**Distinction between Absolute and Relative Permittivity or
Inductivity.**

§ 23. The two quantities c and μ are to be regarded as
known data, given over all space, usually absolute constants ;
but when the simpler properties of the ether are complicated
by the presence of matter, then varying in value from place
to place in isotropic but heterogenous substances ; or, in case
of eolotropy, the three principal values must be given, as well
as the direction of their axes, for every point considered.
Keeping to the case of isotropy, the *ratios* c/c_0 and μ/μ_0 of the
permittivity and inductivity of a body to that of the standard
ether are the specific inductive capacities, electric and magnetic
respectively ; and are mere numerics, of course. They do not
express physical properties themselves except in the limited
sense of telling us how many times as great something is in
one case than in another. This is an important point. It is
like the difference between density and specific gravity. It is
possible to so choose the electric and magnetic units that $\mu = 1$,
$c = 1$ in ether ; then μ and c in all bodies are mere numerics.
But although this system (used by Hertz) has some evident
recommendations, I do not think its adoption is desirable, at
least at present. I do not see how it is possible for any
medium to have less than two physical properties effective in

the propagation of waves.   If this be admitted, I think it may
also be admitted to be desirable to explicitly admit their existence and symbolise them (not as mere numerics, but as physical
magnitudes in a wider sense), although their precise interpretation may long remain unknown.

If, for example, H be imagined to be the velocity of a substance, then $\frac{1}{2}\mu H^2$ is its kinetic energy, and μ its density.
And if E be a torque, then c^{-1} (the elastivity) is the corresponding coefficient of elasticity, the rigidity, or *quasi*-rigidity,
as the case may be ; whilst c is the coefficient of compliance, or
the compliancy; and $\frac{1}{2}cE^2$ is the stored energy of the strain.

Dissipation of Energy.   The Conduction-current ; Conductivity and Resistivity.   The Electric Current.

§ 24.   Besides influencing the values of the ether constants as
above described, we have also to admit that in certain kinds
of matter, when under the influence of electric force, energy is
dissipated *continuously*, besides being stored.   These are called
electrical conductors.   When the conduction is of the simplest
(metallic) type, the waste of energy takes place at a rate proportional to the square of the electric force.   Thus, if Q_1 be
the Joulean waste,

$$Q_1 = kE^2 = EC, \quad \cdots \quad (8)$$

if
$$C = kE. \quad \cdots \quad (9)$$

This new flux C is the conduction current, and k is the conductivity (electric).   Its reciprocal is the resistivity.

The termination -ivity is used in connection with specific
properties.   It does not always sound well at first, but that
wears off.   Sometimes the termination -ancy does as well.

The conductivity k is constant (at one temperature), or is a
linear operator, as in the previous cases with respect to μ and
c.   The dissipation of energy does not imply its destruction,
but simply its rejection or waste, so far as the special electromagnetic affairs we are concerned with.   The conductor is
heated, and the heat is radiated or conducted away.   This is
also (most probably) an electromagnetic process, but of a
different order.   Only in so far as the effect of the heat alters
the conductivity, &c., or, by differences of temperature, causes

thermo-electric force, are we concerned with energy wasted according to Joule's law.

The activity of the electric force, when there is waste, as well as storage, is

$$E(C + \dot{D}) = Q_1 + \dot{U}. \quad \ldots \quad (10)$$

The sum $C + \dot{D}$ is the electric current, when the medium is at rest. When it is in motion, a further term has sometimes to be added, viz., the convection current.

Fictitious Magnetic Conduction-current and Real Magnetic Current.

§ 25. If a substance were found which could not support magnetic force without a continuous dissipation of energy, such a substance would (by analogy) be a magnetic conductor. Let, for instance,

$$K = g H, \quad \ldots \quad (11)$$

then K is the density of the magnetic conduction current, and the rate of waste of energy is

$$Q_2 = HK = g H^2. \quad \ldots \quad (12)$$

The activity of the magnetic force is now

$$H(K + \dot{B}) = Q_2 + \dot{T}. \quad \ldots \quad (13)$$

Compare this equation with (10). The magnetic current is $K + \dot{B}$.

As there is (I believe) no evidence that the property symbolised by g has any existence, it is needless to invent a special name for it or its reciprocal, but to simply call g the magnetic conductivity. The idea of a magnetic current is a very useful one, nevertheless. The magnetic current $\dot{B}$ is of course real; it is the part K that is speculative. It plays an important part in the theory of the transmission of waves in conductors.

Forces and Fluxes.

§ 26. So far we have considered the two forces, electric and magnetic, producing four fluxes, two involving storage and two waste of energy, and we have defined the terminology when the state of things at a point is concerned. We reckon forces per

unit length, fluxes per unit area, and energies or wastes per
unit volume.   It is thus a unit cube that is referred to, whose
edge, side, and volume are utilised.   But a unit cube does not
mean a cube whose edge is 1 centim. or any other concrete
length ; it may indeed be of any size if the quantities concerned
are uniformly distributed throughout it, but as they usually
vary from place to place, the unit cube of reference should be
imagined to be infinitely small.   The next step is to display the
equivalent relations, and develop the equivalent terminology,
when any finite volume is concerned, in those cases that admit
of the same simple representation in the form of linear equations.

Line-integral of a Force.   Voltage and Gaussage.

§ 27. The line-integral of the electric force from one point to
another along a stated path is the electromotive force along
that path ; this was abbreviated by Fleeming Jenkin to E.M.F.
He was a practical man, as well as a practician.   When expressed in terms of a certain unit called the volt, electromotive
force may be, and often is, called the voltage.   This is much
better than " the volts."   I think, however, that it may often
be conveniently termed the voltage irrespective of any particular unit.   We might put it in this way.   Volta was a
distinguished man who made important researches connected
with electromotive force, which is, therefore, called voltage,
whilst a certain unit of voltage is called a volt.   At any rate,
we may try it and see how it works.

The line-integral of the magnetic force from one point to
another along a stated path is sometimes called the magnetomotive force.   The only recommendation of this cumbrous
term is that it is correctly correlated with the equally cumbrous electromotive force.   Magnetomotive force may be called
the gaussage [pr. gowsage], after Gauss, who distinguished himself in magnetic researches ; and a certain unit of gaussage
may be called a gauss [pr. gowce].   I believe this last has
already been done, though it has not been formally sanctioned.
Gaussage may also be experimented with.

The voltage or the gaussage along a line is the sum of the
effective electric or magnetic forces along the line ; the effective force being merely the tangential component of the real

force.   Thus, electric force is the voltage per unit length, and
magnetic force the gaussage per unit length along lines of force.

Surface-integral of a Flux.   Density and Intensity.

§ 28. Next as regards the fluxes, when considered with
reference to any area.   The flux through an area is the sum
of the effective fluxes through its elementary units of area ;
the effective flux being the normal component of the flux or
the component perpendicular to the area.   We do not, I think,
need a number of new words to distinguish fluxes through a
surface from fluxes per unit surface.   Thus, we may speak of
the induction through a surface (or through a circuit bounding
it) ; or of the current through a surface (as across the section of
a wire); or of the displacement through a surface (as in a condenser), without any indefiniteness, meaning in all cases the
surface integral of the flux in question.

In contradistinction to this, it may be sometimes convenient
to speak of the density of the current, or of the induction, or
of the displacement, that is, the amount per unit area.
Similarly, we may sometimes speak of the intensity of the
electric or magnetic force, using "density" for a flux and
"intensity" for a force.

It may be observed by a thoughtful reader that there is a
good deal of the conventional in thus associating one set of
vectors with a line, and another set with a surface, and other
quantities with a volume.   It is, however, of considerable practical utility to carry out these distinctions, at least in a mathematical treatment.   But it should never be forgotten that
electric force, equally with displacement, is distributed throughout *volumes*, and not merely along lines or over areas.

Conductance and Resistance.

§ 29. Conductivity gives rise to conductance, and resistivity
to resistance.   For explicitness, let a conducting mass of any
shape be perfectly insulated, except at two places, A and B, to
be conductively connected with a source of voltage.   Let the
voltage established between A and B through the conductor be
V, and let it be the same by any path.   This will be the case
when the current is steady.   Also let C be this steady current,
in at A and out at B.   We shall have

$$V = RC, \qquad C = KV, \quad . \quad . \quad . \quad . \quad (1)$$

where R and K are constants, the resistance and the conductance, taking the place of resistivity and conductivity, when voltage takes the place of electric force, and the current that of current-density. The activity of the impressed voltage is

$$VC = RC^2 = KV^2, \quad . \quad . \quad . \quad . \quad (2)$$

and represents the Joulean waste per second in the whole conductor, or the volume-integral of **EC** or of kE^2 before considered.

Permittance and Elastance.

§ 30. Permittivity gives rise to permittance, and elasticity to elastance. To illustrate, for the conductor, substitute a nonconducting dielectric, leaving the terminals and external arrangements as before. We have now a charged condenser. Displacement, *i.e.*, the time-integral of the current, takes the place of current in the last case, and we now have

$$D = SV, \qquad V = S^{-1}D, \quad . \quad . \quad . \quad (3)$$

if D is the displacement, S the permittance, and its reciprocal the elastance of the condenser. This elastance has been called the stiffness of the condenser by Lord Rayleigh. It is the elastic resistance to displacement. The displacement is the measure of the charge of the condenser.

The total energy in the condenser is

$$\tfrac{1}{2}VD = \tfrac{1}{2} SV^2 = \tfrac{1}{2} S^{-1}D^2, \quad . \quad . \quad . \quad (4)$$

i.e., half the product of the force (total) and the flux (total), between and at the terminals ; it is also the volume-integral of the energy-density, or $\tfrac{1}{2}\Sigma cE^2$.

As the dielectric is supposed to be a non-conductor, the current is $\dot{D}$ or $S\dot{V}$, and only exists when the charge is varying. But it may also be conducting. If so, let the conductance be K, making the conduction current be $C = KV$. The true current (that is, *the* current) is now the sum of the conduction and displacement currents. Say,

$$T = C + \dot{D} = \left(K + S \frac{d}{dt} \right) V. \quad . \quad . \quad . \quad (5)$$

This is the characteristic equation of a condenser. It comes to the same thing if the condenser be non-conducting, but be shunted by a conductance, K. In a conducting dielectric the permittivity and the conductivity are therefore in parallel arc, as it were. It was probably by a consideration of conduction in a leaky condenser that Maxwell was led to his inimitable theory of the dielectric, by which he boldly cut the Gordian knot of electromagnetic theory.

The activity of the terminal voltage we find by multiplying (5) by V, giving

$$V\Gamma = VC + V\dot{D},$$

$$= RC^2 + \frac{d}{dt}\left(\tfrac{1}{2} SV^2\right), \quad \cdots \quad (6)$$

representing the waste in Joulean heating and the rate of increase of the electric energy. Each of these quantities is the sum of the same quantities per unit volume throughout the substance concerned.

Permeance, Inductance and Reluctance.

§ 31. Permeability gives rise to permeance, inductivity to inductance, and reluctivity to reluctance.

The formal relation of reluctance to reluctivity with magnetic force and induction, is the same as that of resistance to resistivity with electric force and conduction current, or of elastance to elastivity with electric force and displacement.

Permeance is the reciprocal of reluctance. In this sense I have used it, though only once or twice. Prof. S. P. Thompson has also used the word in this sense in his Cantor Lectures with good effect.

If we replace our illustrative conductor by an inductor, supporting magnetic induction, and suppose it surrounded by imaginary matter of zero inductivity, and have an impressed gaussage instead of voltage at the terminals, we shall have a flux of induction which will, if the force be weak enough, vary as the force. If H be the gaussage and B the induction entering at the one and leaving at the other terminal, the ratio H/B is the reluctance, and the reciprocal B/H is the permeance. The energy stored is

$$\tfrac{1}{2}HB = \tfrac{1}{2}\left(\frac{B}{H}\right)H^2 = \tfrac{1}{2}\left(\frac{H}{B}\right)B^2. \quad . \quad . \quad (7)$$

When the relation of flux to force is not linear, we can still usefully employ the analogy with conduction current or with displacement by treating the ratio B/H as a function of H or of B; as witness the improved and simplified way of considering the dynamo in recent years. I must, however, wonder at the persistence with which the practicians have stuck to "the lines," as they usually term the flux in question.

I am aware that the use of the name induction for this flux, which I have taken from Maxwell, is in partial conflict with an older use. But it is seldom, if ever, that these uses occur together, for one thing; another thing is that the older (and often vague) use of the word induction has very largely ceased of late years. It was not without consideration that induction was adopted and, to harmonise with it, inductance and inductivity were coined.

Inductance of a Circuit.

§ 32. The meaning of inductance has sometimes been misconceived. It is not a synonym for induction, nor for self-induction, but means "the coefficient of self-induction," sometimes abbreviated to "the self-induction." It is essentially the same as permeance, the reciprocal of reluctance, but there is a practical distinction. Consider a closed conducting circuit of one turn of wire, supporting a current C_1. As will later appear, this C_1 is also the gaussage. That is, the line-integral of the magnetic force in any closed circuit (or the circuitation of the force) embracing the current once is C_1. Let also B_1 be the induction through the circuit of C_1. Then

$$B_1 = L_1 C_1, \quad . \quad . \quad . \quad . \quad . \quad (8)$$

where, by what has already been explained, L_1 is the permeance of the magnetic circuit, a function of the distribution of inductivity and of the form and position of the conducting core. The magnetic energy is

$$\tfrac{1}{2}B_1 C_1 = \tfrac{1}{2}L_1 C_1^2. \quad . \quad . \quad . \quad (9)$$

by using the first expression in (7), remembering that H there is now represented by C_1, and then using (8). This energy "of the current" resides in all parts of the field, only (usually) a small portion occupying the conductor itself.

Now substitute for the one turn of wire a bundle of wires, N in number, of the same size and form. Disregarding small differences due to the want of exact correspondence between the bundle and the one wire, everything will be the same as before if the above C_1 means the total current in the bundle. But if the same current be supported by each wire, practical convenience in respect to the external connections of the coil requires us to make the current in each wire *the* current. Let this be C, so that $C_1 = NC$. Then we shall have, by (8),

$$B_1 = (L_1 N) C \quad . \quad . \quad . \quad . \quad . \quad (8a)$$

to express the flux of induction; and by (9) and (8a),

$$\tfrac{1}{2} B_1 C_1 = \tfrac{1}{2} (L_1 N^2) C^2$$
$$= \tfrac{1}{2} L C^2, \quad . \quad . \quad . \quad . \quad . \quad (9a)$$

if $L = N^2 L_1$, to express the energy. This L is the inductance of the coil. It is N^2 times the permeance of the magnetic circuit.

Again, regarding the coil as a single circuit, $B_1 N$ is the induction through it—that is, B_1 through each winding. Calling this total B, we have, by (8a),

$$B = B_1 N = (L_1 N^2) C = LC, \quad . \quad . \quad . \quad (8b)$$

which harmonises properly with (9a).

The difference between inductance and permeance, therefore, merely depends upon the different way of reckoning the current in the coil. With one winding only, they are identical. I should here observe that I am employing at present rational units. Their connection with the Gaussian units will appear later. It would only serve to obscure the subject to bring in 4π, that arbitrary and unnecessary constant which has puzzled so many people.

It will be seen that the distinction between permeance and inductance is a practical necessity. in spite of their fundamental identity. But which should be which? On the whole, I prefer it as above stated, especially to connect with self-induction. Regarding permeability itself, it would seem that this name is more particularly suitable to express the ratio μ/μ_0 of

the inductivity of a medium to that of ether, which is, in fact,
consistent with the original meaning, I believe, as used by
Sir W. Thomson in connection with his "electromagnetic definition" of magnetic force. But to inductivity, as before-
mentioned, a wider significance should be attached. As has
been more particularly accentuated by Prof. Rücker, we really
do not know anything about the real dimensions of μ and c;
or, more strictly, we do not know the real nature of the
electromagnetic mechanism, so that μ and c are very much
what we choose to make them, by assumptions. The two principal systems are the so-called electrostatic, in which $c = 1$ in
ether, and the electromagnetic, in which $\mu = 1$ in ether. But
with these specialities we have no further concern at present.

Cross-connections of Electric and Magnetic Force. Circuital Flux. Circuitation.

§ 33. The two sets of quantities, the electric and magnetic
forces, with their corresponding fluxes and currents, and the
connected products and ratios, may be considered quite independently of one another, without any explicit connection being
stated between the electric set and the magnetic set, whether
they coexist or not. But to have a dynamical electromagnetic
theory, we require to know something more, viz., the cross-
connections or interactions between $\mathbf{E}$ and $\mathbf{H}$. Or, in another
form, we require to know how an electric field and a magnetic
field mutually influence one another.

One of these interactions has been already partially mentioned, though only incidentally, in stating the meanings of
permeance and inductance. It was observed that the electric
current in a simple conductive circuit was measured by the
gaussage in the corresponding magnetic circuit.

A word has been much wanted to express in a convenient
and concise manner the property possessed by some fluxes and
other vectors of being distributed in closed circuits. This want
has been recently supplied by Sir W. Thomson's introduction of
the word "circuital" for the purpose.[*] Thus electric current is
a circuital flux, and so is magnetic induction. The fundamental
basis of the property is that as much of a circuital flux enters

---

[*] "Mathematical and Physical Papers," Vol. III., p. 451.

any volume at some parts of its surface as leaves it at others, so that the flux has no divergence anywhere. This qualification, "anywhere," should be remembered, for a flux which may diverge locally, as, for instance, electric displacement, is not circuital in general, though even electric displacement may be circuital sometimes. Further, as a flux need not be distributed throughout a volume, but may be confined to a surface, or to a line, we have then specialised meanings of circuital and of divergence. Or a volume-distribution and a surface or line-distribution of a flux may be necessarily conjoined, without, however, any departure from the essential principle concerned.

The word "circuital," which will be often used, suggested to me the word "circuitation," to indicate the often-occurring operation of a line-integral in a closed circuit; as, for instance, in the estimation of circuital voltage or gaussage. Now, in the case of a moving fluid, Sir W. Thomson called the line-integral of the velocity in a closed circuit the "circulation." This is curiously like "circuitation." But "circulation" seems to have too specialised a meaning to be suitable for application to any vector, and I shall employ "circuitation." The operation of circuitation is applicable to any vector, whether it be circuital or not.

First Law of Circuitation.

§ 34. Now in the case of a simple conductive circuit, we have two circuital fluxes. There is a circuital conducting core supporting an electric current, and there is a circuital flux of induction through the conductive circuit. In the electric circuit we have Ohm's law,

$$E = RC, \dots \dots (1)$$

where E is the circuital voltage, C the current, and R the resistance. And in the magnetic circuit we have a formally similar relation,

$$H = L^{-1}B, \dots \dots (2)$$

where H is the circuital gaussage, B the induction, and L^{-1} the reluctance. Or,

$$B = LH, \dots \dots (3)$$

where L is the inductance (or the permeance, when there is only one turn of wire).

D

Now, the cross-connection in this special case is implied in the assertion that H and C are the same quantity, when measured in rational units The expression of the law of which this is an illustration is contained in any of the following alternative statements.

The line-integral of the magnetic force in any closed circuit measures the electric current through any surface bounded by the circuit. Or,

The circuitation of the magnetic force measures the electric current through the circuit. Or,

The electric current is measured by the magnetic circuitation, or by the circuital gaussage.

The terminology of electromagnetism is in a transitional state at present, owing to the change that is taking place in popular ideas concerning electricity, and the unsuitability of the old terminology, founded upon the fluidity of electricity, for a comprehensive view of electromagnetism. This is the excuse for so many new words and forms of expression. Some of them may find permanent acceptation.

The above law applies to any circuit of any size or shape, and irrespective of the kind of matter it passes through, meaning by "circuit" merely a closed line, along which the gaussage is reckoned. By "the current" is to be understood *the* current ; not merely the conduction current alone, or the displacement current alone, but their sum (the convection current term will be considered separately).

It is also necessary to understand that a certain convention is implied in the statement of the law, regarding positive senses of translation and rotation when taking line and surface integrals. Look at the face of a watch, and imagine its circumference to be the electric circuit. The ends of the pointers travel in this circuit in the positive sense, if you are looking through the circuit along its axis in the positive sense. Also, you are looking at the negative side of the circuit. Thus, when the current is positive in its circuit, the magnetic induction goes through it in the positive direction, from the negative side to the positive side. Otherwise, the positive sense of the current in a circuit and the induction through it are connected in the same way as the motions of rotation and translation of a nut on an ordinary right-handed screw. This is the " vine " system used

by all British writers; but some continental writers use the
"hop" system, in which the rotation is the other way, for the
same translation. It is useless trying to work both systems,
and when one comes across the left-handed system in papers,
it is, perhaps, best to marginally put the matter straight, and
then ignore the text.

Second Law of Circuitation.

§ 35. The other cross-connection required is a precisely
similar relation between voltage and magnetic current, with,
however, a change of sign. Thus :—

The negative line-integral of the electric force in any circuit
(or the electric circuitation) measures the magnetic current
through the circuit. Or,

The voltage in any circuit measures the magnetic current
through the circuit taken negatively. Or,

Magnetic current is measured by the circuital voltage reversed ; and other alternative equivalent statements.

Definition of Curl.

§ 36. In the above laws of circuitation the currents are the
concrete currents (surface-integrals), and the forces also the
concrete voltage or gaussage. When we pass to the unit
volume it is the current-density that is the flux. The circuitation of the force is then called its "curl." Thus, if J be the
electric current and G the magnetic current, the two laws are

$$\operatorname{curl} \mathbf{H}_1 = \mathbf{J}, \quad \ldots \ldots \quad (4)$$

$$-\operatorname{curl} \mathbf{E}_1 = \mathbf{G}, \quad \ldots \ldots \quad (5)$$

where $\mathbf{E}_1$ and $\mathbf{H}_1$ are the electric and magnetic force of the
field. We may now say concisely that

The electric current is the curl of the magnetic force.

The magnetic current is the negative curl of the electric force.

There is nothing transcendental about "curl." Any man
who understands the laws of circuitation also understands
what "curl" means, though he may not himself be aware of
his knowledge, being like the Frenchman who talked prose for
many years without knowing it. The concrete circuitation is
sufficient for many problems, especially those concerning linear

conductors in magnetic theory. But it does not suffice for mathematical analysis, and to go into detail we require to pass from the concrete to the specific and use curl. How to manipulate " curl " is a different matter altogether from clearly understanding what it means and the part it plays. The latter is open to everybody; for the former, vector-analysis is most suitable.

Let a unit area be chosen perpendicular to the electric current J. Its edge is then the circuit to which H_1 belongs in (4). The gaussage in this circuit measures the current-density. Similarly, regarding (5), the voltage in a unit circuit perpendicular to the magnetic current measures its density (negatively). In short, what circuitation is in general, curl is the same per unit area.

Impressed Force and Activity.

§ 37. In the statement of the laws of circuitation, I have intentionally omitted all reference to impressed forces. That there must be impressed forces is obvious enough, because a dynamical system comprehending only the electric and magnetic stored energies and the Joulean waste, is only a part of the dynamical system of Nature. We require means of showing the communication of energy to or from our electromagnetic system without having to enlarge it by making it a portion of a more complex system. Thus, taking it as it stands at present, the activity per unit volume we have seen to be

$$EJ + HG = Q + \dot{U} + \dot{T}, \quad \ldots \quad (6)$$

where the left side expresses the activity of the electric and the magnetic force on the corresponding currents, and the right side what results, viz., waste of energy, Q per second, and increase per second of the electric energy U and the magnetic T; and, as there are supposed to be no impressed forces, if we integrate through all space, we shall obtain

$$0 = \Sigma (Q + \dot{U} + \dot{T}), \quad \ldots \quad (7)$$

where Σ means summation of what follows it. Or, if Q_0 be the total waste, and similarly U_0 and T_0 the total energies,

$$0 = Q_0 + \dot{U}_0 + \dot{T}_0,$$

meaning, that whatever energy there be wasting itself is derived solely from the electric or magnetic energy, which decrease accordingly. This is the persistence of energy when there are no impressed forces.

Now, if there be impressed forces communicating energy at the rate A, the last equation must become

$$\Sigma \, \mathbf{fv} = A = Q_0 + \dot{U}_0 + \dot{T}_0, \quad \dots \quad (8)$$

and A must be the sum of the activities of the impressed forces $\mathbf{f}$ in the elements of volume, in whatever way space may be divided into elements, large or small, and however we may choose to reckon the impressed forces. There may be many ways of doing it ; $\mathbf{f}$ may sometimes, for example, be an ordinary force, and $\mathbf{v}$, the velocity to match, is then a translational velocity. But for our immediate purpose, it is naturally convenient to reckon the impressed forces electrically and magnetically ; so that the corresponding velocities are the electric and magnetic currents. We shall then have, if $\mathbf{e}$ be the impressed electric, and $\mathbf{h}$ the impressed magnetic force,

$$\mathbf{eJ} + \mathbf{hG}$$

to represent their activity per unit volume, and in all space,

$$\Sigma \, (\mathbf{eJ} + \mathbf{hG}) = \Sigma \, (Q + \dot{U} + \dot{T}) \quad \dots \quad (9)$$

instead of (7). This is the integral equation of activity. We cannot remove the sign of summation and make the same form do for the unit volume, for this would make every unit volume independent of the rest, and do away with all mutual action between contiguous elements and transfer of energy between them. This matter will be returned to in connection with the transference of energy.

Distinction between Force of the Field and Force of the Flux.

§ 38. The distinction between H_1 and H and between E_1 and E is often a matter of considerable importance. We have

$$H = h + H_1, \quad \dots \dots \quad (10)$$

$$E = e + E_1. \quad \dots \dots \quad (11)$$

Now it is **E** and **H** that are effective in producing fluxes. Thus **E** is the force of the flux **D** and also of **C** ; and **H** is the force of the flux **B**. On the other hand, in the laws of circuitation, as above expressed, the impressed forces do not count at all ; so that we have, in terms of the forces **E** and **H**,

$$\operatorname{curl} (\mathbf{H} - \mathbf{h}) = \mathbf{J}, \quad \ldots \ldots \quad (12)$$

$$-\operatorname{curl} (\mathbf{E} - \mathbf{e}) = \mathbf{G}, \quad \ldots \ldots \quad (13)$$

equivalent to (4) and (5). To distinguish from the forces of the fluxes, I sometimes call $\mathbf{E}_1$ and $\mathbf{H}_1$ the forces " of the field." Of course they only differ where there is impressed force. As the distribution of the energy, as well as of the fluxes, depends upon **E** and **H**, it is usually best to use them in the formulæ.

Classification of Impressed Forces.

§ 39. The vectors representing impressed electric and magnetic force demand consideration as to the different forms they may assume. Their line-integrals are impressed voltage and gaussage. Their activities or powers are e**J** and h**G** respectively per unit volume, and in this statement we have a sort of definition of what is to be understood by impressed force. For, **J** being the electric current anywhere, if there be an impressed force **e** acting, the amount e**J** of energy per unit volume is communicated to, or taken in by, the electromagnetic system per second ; and this should be understood to take place at the spot in question. It must then be either stored on the spot, or wasted on the spot, or be somehow transmitted away to other places, to be there stored or wasted, according to a law which will appear later on. Similarly as regards **h** and **G**.

But this concerns only the reckoning of impressed force, and is independent of its physical origin, which may be of several kinds. Thus under **e** we include—

(1.) Voltaic force.

(2.) Thermo-electric force.

(3.) The force of intrinsic electrisation.

(4.) Motional electric force.

(5.) Perhaps due to various secondary causes, especially in connection with strains.

And under **h** we include—

(1.) The force of intrinsic magnetisation.

(2.) Motional magnetic force.

(3.) Perhaps due to secondary causes.

Voltaic Force.

§ **40.** Voltaic force has its origin in chemical affinity. This is still a very obscure matter. For a rational theory of Chemistry, one of the oldest of the sciences, we may have to wait long, in spite of the activity of chemical research and of the development of the suggestive periodic law. Yet Chemistry and Electricity are so intimately connected that we cannot understand either without some explanation of the other. Electricity is, in its essentials, a far simpler matter than Chemistry, and it is possible that great light may be cast upon chemical problems (and molecular physics generally) by previous discoveries and speculations in Electricity. The very abstract nature of Electricity is, in some respects, in its favour. For there is considerable truth in the remark (which, if it has not been made before, is now originated) that the more abstract a theory is, the more likely it is to be true. For example, it may be that Maxwell's theory of displacement and induction in the ether is far more than a working theory, and is something very near the truth, though we know not what displacement and induction are. But if we try to materialise the theory by inventing a special mechanism we are almost certain to go wrong, however useful the materialisation may be for certain purposes. No one knows what matter is, any more than ether. But we do know that the properties of matter are remarkably complex. It is, therefore, a real advantage to get away from matter when possible, and think of something far more simple and uniform in its properties. We should rather explain matter in terms of ether, than go the other way to work.

However this be, we have the fact that definite chemical changes involve definite voltages, and herein lies one of the most important sources of electric current. Furthermore, there is the remarkable connection between the quantity of matter and the time-integral of the current (or quantity of electricity) produced, involved in the law of electro-chemical equivalents, which is one of the most suggestive facts in physics, and must be a necessary part of the theory of the atom which is to come. That the energy of chemical affinity may itself be partly electromagnetic is likely enough. That even conduction may be an electrolytic process is possible, in spite of the sweet simplicity of Ohm's law and that of Joule. For these laws are most probably merely laws of averages. The well-known failure of Ohm's law (apparent at any rate) when the periodicity of electromagnetic waves in a conductor amounts to billions per second may perhaps arise from the period being too short to allow of the averages concerned in Ohm's law to be established. If so, this may give a clue to the required modification.

Thermo-electric Force.

§ 41. Thermo-electric force has its origin in the heat of bodies, manifesting itself at the contact of different substances or between parts of the same substance differing in temperature. Now heat is generally supposed to consist in the energy of agitation of the molecules of bodies, and this is constantly being transferred to the ether in the form of radiant energy, *i.e.*, electromagnetic vibrations of very great frequency, but in a thoroughly irregular manner. It is this irregularity that is a general characteristic of radiation. Now the result of subjecting conductors to electric force is to dissipate energy and to heat them. This is, however, an irreversible process. But when contiguous parts of a body are at different temperatures, a differential action on the ether results, whereby a continued effect of a regular type is produced, reversible with the current, and therefore formularisable as due to an intrinsic electric force, the thermo-electric force. At the junction of different materials at the same temperature it is still the heat that is the source of energy.

The theory of thermo-electric force due to Sir W. Thomson, based upon the application of the Second Law of Thermodynamics (the First is a matter of course) to the reversible heat effects has been verified for conductive metallic circuits by the experiments of its author, and those of Prof. Tait and others. With some success the same principle has also been applied by von Helmholtz to voltaic cells, which are thermo-electric as well as voltaic cells. There are wheels within wheels, and Ohm's law is merely the crust of the pie.

Intrinsic Electrisation.

§ 42. Intrinsic electrisation is a phenomenon shown by most solid dielectrics under the continued action of electric force. It is the manifestation of a departure from perfect electric elasticity, and is probably due to a molecular rearrangement, resulting in a partial fixation of the electric displacement, whereby it is rendered independent of the " external " electrising force. Thus the displacement initially produced by a given voltage slowly increases, and upon the removal of the impressed voltage only the initial displacement will subside, if permitted, immediately. The remainder has become intrinsic, for the time, and may be considered due to an intrinsic electric force **e**. If I_1 be the intensity of intrinsic electrisation, and c the permittivity, then

$$I_1 = ce. \quad \ldots \ldots \ldots \quad (1)$$

I_1 is the full displacement the force **e** can produce elastically, all external reaction being removed by short circuiting. It is not necessarily the actual displacement. The phenomenon of " residual charge," " soakage," " absorption," &c., are accounted for by this **e** and its slow variations.

Maxwell attempted to give a physical explanation of this phenomenon by supposing the dielectric to be heterogeneously conductive. This is perhaps not the most lucidly successful of Maxwell's speculations. How far electrolysis is concerned in the matter is not thoroughly clear.

Intrinsic Magnetisation.

§ 43. Intrinsic magnetisation is, in some respects, a similar phenomenon, due to a passage from the elastic to the intrinsic form of induction externally induced in solid materials. Calling the intensity of intrinsic magnetisation I_2, we have

$$I_2 = \mu h, \quad \ldots \ldots \ldots \quad (2)$$

where **h** is the equivalent intrinsic magnetic force, and μ the inductivity (elastically reckoned).

In one important respect intrinsic induction is a less general phenomenon than intrinsic displacement. There is no magnetic

conductivity to produce similar results as regards the magnetic
current as there is electric conductivity as regards the electric
current. But if there were, then we could have a magnetic
"condenser," with a magnetically conductive external circuit,
and get our residual results to show themselves in it, quite
similarly in kind to, though varying in magnitude and perma-
nence from, what we find with an electric condenser.

The analogue of Maxwell's explanation of "absorption"
would be heterogeneous magnetic conductivity. This is infinitely more speculative than the other, which is sufficiently
doubtful.

Ewing's recent improvement of Weber's theory of magnetism
seems important. But as in static explanations of dynamical
phenomena the very vigorous molecular agitations are ignored,
it is clear that we have not got to the root of the matter. We
want another Newton, the Newton of molecular physics. Facts
there are in plenty to work upon, and perhaps another heaven-
born genius may come to make their meaning plain. Properties of matter are all very well, but what is matter, and why
their properties ? This is not a metaphysical inquiry, but concerns the construction of a physical theory.

The Motional Electric and Magnetic Forces.   Definition of a Vector-Product.

§ 44. The motional electric and magnetic forces are the
forces induced by the motion of the medium supporting the
fluxes. To express them symbolically, it will save much and
repeated circumlocution if we first define the vector-product
of a pair of vectors.

Let a and b be any vectors, and c their vector-product. This
is denoted by

$$c = Vab, \qquad \ldots \ldots \ldots (3)$$

the prefix V meaning "vector," or, more particularly here,
"vector-product." The vector c is perpendicular to the plane
of the vectors a and b, and its tensor (or magnitude) equals
the product of the tensor of a into the tensor of b into the
sine of the angle between a and b. Thus

$$c = ab \sin \theta, \qquad \ldots \ldots \ldots (4)$$

if the italic letters denote the tensors, and θ be the included

angle of the vectors. As regards the positive sense of the
vector **c**, this is reckoned in the same way as before explained
with regard to circuitation. Thus, when the tensor c is positive, a positive rotation about **c** in the plane of **a** and **b** will
carry **a** to **b**. If the time by a watch is three o'clock, and the
big hand be **a** and the little hand **b**, then the vector **c** is
directed through the watch from its face to its back. These
vector-products are of such frequent occurrence, and their
Cartesian representation is so complex, that the above concise
way of representing them should be clearly understood.

On this understanding, then, we can conveniently say that
the motional electric force is the vector-product of the velocity
and the induction, and that the motional magnetic force is the
vector-product of the displacement and the velocity. Or, in
symbols, according to (3),

$$e = VqB, \qquad \ldots \ldots \quad (5)$$

$$h = VDq, \qquad \ldots \quad \ldots \ldots \quad (6)$$

where q is the vector velocity.

Example.   A Stationary Electromagnetic Sheet.

§ 45. It should be remembered that we regard the displacement and the induction as actual states of the medium,
and therefore if the medium be moving, it carries its states
with it. Besides this, it usually happens that these states are
themselves being transferred through the medium (independently
of its translational motion), so that the resultant effect on propagation, considered with respect to fixed space, is a combination
of the natural propagation through a medium at rest, and what
we may call the convective propagation. Of course we could
not expect the two laws of circuitation for a medium at rest to
remain true when there is convective propagation.

The matter is placed in a very clear light by considering the
very simple case of an infinite plane lamina of **E** and **H** travelling at the speed of light v perpendicularly to itself through a
homogeneous dielectric. This is possible, as will appear later,
when **E** and **H** are perpendicular, and their tensors are thus
related :—

$$E = \mu vH. \qquad \ldots \ldots \quad (7)$$

Or, vectorising v to $\mathbf{v}$,

$$\mathbf{E} = \mu \mathbf{VHv}, \quad \ldots \ldots \quad (8)$$

according to the definition of a vector-product, gives the directional relations as well as the numerical.

Now, suppose we set the whole medium moving the other way at the speed of light. The travelling plane electromagnetic sheet will be brought to rest *in space*, whilst the medium pours past it. Being at rest and steady, the electric displacement and magnetic induction can only be kept up by coincident *impressed* forces, viz. :—

$$e = \mathbf{E}, \qquad h = \mathbf{H}.$$

Now compare (8) with (5) and (6); consider the directions carefully, and remember that the velocity $\mathbf{q}$ is the negative of the velocity $\mathbf{v}$, and we shall obtain the formulæ (5), (6), which are thus proved for the case of plane wave motion, by starting with a simple solution belonging to a medium at rest.

The method is, however, principally useful in showing the necessity of, and the inner meaning of the motional electric and magnetic forces. To show the general application of (5) and (6) requires a more general consideration of the motional question, to which we now proceed.

Connection between Motional Electric Force and "Electromagnetic Force."

§ 46. A second way of arriving at the motional electric force is by a consideration of the work done in moving a conducting circuit in a magnetic field. It results from Ampère's researches, and may be independently proved in a variety of ways, that the forcive (or system of forces) acting upon a conducting circuit supporting a current, may be accounted for by supposing that every element of the conductor is subject to what Maxwell termed "the electromagnetic force." This is a force perpendicular to the vector current and to the vector induction, and its magnitude equals the product of their tensors multiplied by the sine of the angle between them. In short, the electromagnetic force is the vector-product of the current and the induction. Or, by the definition of a vector-product,

$$\mathbf{F} = \mathbf{VCB}, \quad \ldots \ldots \quad (9)$$

if **F** is the force per unit volume, **C** the current, and **B** the induction. Here **F** is the force arising from the stress in the magnetic field. Its negative, say **f**, is therefore the impressed mechanical force, or ·

$$\mathbf{f} = V\mathbf{BC}, \quad \ldots \ldots \quad (10)$$

to be used when we desire to consider work done upon the electromagnetic system.

The activity of **f** is **fq**, if **q** be the velocity ; or, by (10),

$$\mathbf{fq} = \mathbf{q}V\mathbf{BC}. \quad \ldots \ldots \quad (11)$$

This is identically the same as

$$\mathbf{fq} = \mathbf{C}V\mathbf{qB}, \quad \ldots \ldots \quad (12)$$

by a fundamental formula in vector-analysis.* Here, on the left side, the activity is expressed mechanically ; on the right side, on the other hand, it is expressed electrically, as the scalar product of the current and another vector, which is the corresponding force ; it is necessarily an electric force, and necessarily impressed. So, calling it **e**, we have

$$\mathbf{e} = V\mathbf{qB} \quad \ldots \ldots \quad (13)$$

again, to express the motional electric force.

It should be observed that we are not concerned in this mode of reasoning with the explicit connection between **e** and **C**; and in this respect the process is remarkably simple. As it, however, rests upon a knowledge of the electromagnetic force, we depart from the method of deriving relations previously pursued. But, conversely, we may by (11) and (12) derive the electromagnetic force from the motional electric force.

Variation of the Induction through a Moving Circuit.

§ 47. A third method of arriving at (13) is by considering the rate of change of the amount of induction through a moving circuit. We need not think of a conducting circuit, but, more generally, of any circuit. Let it be moving in any way whatever, changing in shape and size arbitrarily. The induction through it is altering in two entirely distinct ways. First, there is the magnetic current before considered, due to the time-variation of the induction, so that, if the circuit were

---

* Proved, with other working formulæ, in the chapter on the Algebra of Vectors.

at rest in its momentary position, we should have the second
law of circuitation true in its primitive form

$$- \operatorname{curl} \mathbf{E} = \mathbf{G}, \quad . \quad . \quad . \quad . \quad . \quad (14)$$

when expressed for a unit circuit. But now, in addition, the
motion of the elements of the circuit in the magnetic field
causes, independently of the time-variation of the field, additional induction to pass through the circuit. Let its rate of
increase due to this cause be g per unit area. If, then, we
assume that the circuitation of the electric force **E** (of the flux)
equals the rate of decrease of the induction through the circuit
always, whether it be at rest or in motion, the equation (14)
becomes altered to

$$- \operatorname{curl} \mathbf{E} = \mathbf{G} + \mathbf{g}, \quad . \quad . \quad . \quad . \quad (15)$$

where the additional g may be regarded as a fictitious magnetic
current. That it is also expressible as the curl of a vector is
obvious, because it depends upon the velocity of each part of
the circuit, and is therefore a line-integral. Examination in
detail shows that

$$\mathbf{g} = - \operatorname{curl} \mathbf{VqB}, \quad . \quad . \quad . \quad . \quad (16)$$

so that we have, by inserting (16) in (15),

$$- \operatorname{curl} (\mathbf{E} - \mathbf{e}) = \mathbf{G}, \quad . \quad . \quad . \quad . \quad (17)$$

the standard form of the second law of circuitation, when we
use (13) to express the impressed force.

The method by which Maxwell deduced (13) is substantially
the same in principle ; he, however, makes use of an auxiliary
function, the vector-potential of the electric current, and this
rather complicates the matter, especially as regards the physical
meaning of the process. It is always desirable when possible
to keep as near as one can to first principles. The above may,
without any formal change, be applied to the case of assumed
magnetic conductivity, when **G** involves dissipation of energy;
the auxiliary g in (15), depending merely upon the motion of the
circuit across the induction, does not itself involve dissipation.

Modification.   Circuit Fixed.   Induction moving equivalently.

§ 48. Perhaps the matter may be put in a somewhat clearer
light by converting the case of a moving circuit into that of a

circuit at rest, and then employing the law of circuitation in
its primitive form. The moving circuit has at any instant a
definite position. Imagine it to be momentarily fixed in that
position, by stopping the motion of its parts. In order that
the relation of the circuit to the induction should be the same
as when it was moving, we must now communicate momentarily
to the lines of induction the identically opposite motion to the
(abolished) motion of the part of the circuit they touch.

We now get equation (15), on the understanding that **g** means
the additional magnetic current through a *fixed* circuit due to
a given motion of the lines of induction across its boundary,
such motion being the negative of the (abolished) motion of
the circuit. The matter, is, therefore, simplified in treatment.
For, in the former way, the process of demonstrating (15) which
I have referred to as an "examination in detail," is really
considerably complex, involving the translation, rotation, and
distortion of an elementary circuit (or equivalently for any
circuit). Fixing the circuit does away with this, and we have
merely to examine what happens at a single element of the circuit, as induction sweeps across it, in increasing the induction
through the circuit, and then apply the resulting formula to
every element.

In the consideration of a single element, it is immaterial what
the shape of the circuit may be; it may, therefore, be chosen to be

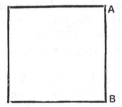

a unit square in the plane of the paper, one of whose sides, AB,
is the element of unit length. Now, suppose the induction at AB
is perpendicular to the plane of the paper, directed downwards,
and that it moves from right to left perpendicularly across
AB. Let also from A to B be the positive sense in the circuit.
It is evident, without any argumentation, that the directions
chosen for q and B are the most favourable ones possible for

increasing the induction through the circuit, and that the rate
of its increase, so far as AB alone is concerned, is simply qB,
the product of the tensors of the velocity q of transverse motion
and of the induction **B**. Further, if the velocity q be not wholly
transverse to **B** as described, but still be wholly transverse to
AB, we must take, instead of q, the effective transverse component $q \sin \theta$, if θ be the angle between q and **B**, making our
result to be qB $\sin \theta$. Now, this is the tensor of Vq**B**, whose
direction is from A to B. The motional electric force in the
element AB is therefore from B to A, and is V**B**q, because it
is the negative circuitation which measures the magnetic current through a circuit. Lastly, if the motion of **B** be not
wholly transverse to AB, we must further multiply by the
cosine of the angle between V**B**q and the element AB.
This merely amounts to taking the effective part of V**B**q along
the circuit. So, finally, we see that V**B**q fully represents the
impressed electric force per unit length in AB when it is fixed,
and the induction moves across it, or that its negative

$$e = Vq\mathbf{B}$$

represents the motional electric force when it is the element
AB that moves with velocity q through the induction **B**. Now,
apply the process of circuitation, and we see that **e** is such that
its curl represents the rate of increase of induction through the
unit circuit due to the motion alone.

This may seem rather laboured, but is perhaps quite as much
to the point as a complete analytical demonstration, where one
may get lost in the maze of differential coefficients, and have
some difficulty in interpreting the analytical steps electromagnetically.

The fictitious motion of the induction above assumed has
nothing to do with the real motion of the induction through
the medium. If there be any, its effect is fully included in the
term **G**, the real magnetic current.

The Motional Magnetic Force.

§ 49. The motional magnetic force **h** may be similarly
deduced. First we have the primitive form of the first law of
circuitation,

$$\text{curl } \mathbf{H} = \mathbf{J}, \quad \ldots \quad \ldots \quad (18)$$

when the unit circuit is at rest, where **J** is the complete electric current-density, and next

$$\text{curl } H = J + j, \quad \ldots \ldots \quad (19)$$

when the circuit moves; where the auxiliary **j** is a fictitious electric current equivalent to the increase of displacement through the circuit by its motion only. Next show that

$$j = \text{curl } h, \quad \ldots \ldots \quad (20)$$

and

$$h = VDq, \quad \ldots \ldots \quad (21)$$

by similar reasoning to that concerning **e**; so that by insertion in (19) the first law of circuitation is reduced to the standard form

$$\text{curl } (H - h) = J, \quad \ldots \ldots \quad (22)$$

with the special form of the impressed force **h** stated.

Comparing the form of **h** with that of **e** we observe that there is a reversal of direction in the vector-products, the flux being before the velocity in one and after it in the other. This arises from the opposite senses of circuitation of the electric and the magnetic force to represent the magnetic and electric currents.

The "Magneto-electric Force."

§ 50. The activity of the motional **h** is found by multiplying it by the magnetic current, and is, therefore,

$$hG = GVDq,$$
$$= qVGD,$$

by the same transformation as from (11) to (12).

We conclude that VGD is an impressed mechanical force, per unit volume, and, therefore, that VDG is a mechanical force, that is, of the Newtonian type, arising from the electric stress. By analogy with the electromagnetic force it may be termed the magnetoelectric force, acting on dielectrics supporting displacement when the induction varies with the time. Of this more hereafter.

Electrification and its Magnetic Analogue. Definition of "Divergence."

§ 51. So far nothing has been laid down about electrification. But the laws of circuitation cannot be completed without including electrification and its suggested magnetic analogue.

E

Describe a closed surface in a dielectric, and observe the net amount of displacement leaving it. This, of course, means the excess of the quantity leaving over that entering it. If the net amount be zero, there is no electrification within the region bounded by the surface. If the amount be finite, there is just that amount of electrification in the region. This is independent altogether of its distribution within the region, and of the size and shape of the region.

More formally, the surface-integral of the displacement leaving any closed surface measures the electrification within it.

This being general, if we wish to find the distribution of electrification we must break up the region into smaller regions, and in the same manner determine the electrifications in them. Carrying this on down to the infinitely small unit volume, we, by the same process of surface-integration, find the volume-density of the electrification. It is then called the divergence of the displacement.

That is, in general, the divergence of any flux is the amount of the flux leaving the unit volume.

And in particular, the divergence of the displacement measures the density of electrification.

Similarly, the divergence of the induction measures the "magnetification," if there is any to measure, which is a very doubtful matter indeed. There is no evidence that the flux induction has any divergence; it is purely a circuital flux, so far as is *certainly* known, and this is most intimately connected with the other missing link in a symmetrical electromagnetic scheme, the (unknown) magnetic conductivity.

Divergence is represented by div, thus :—

$$\operatorname{div} \mathbf{D} = \rho, \quad \cdots \cdots \quad (1)$$

$$\operatorname{div} \mathbf{B} = \sigma, \quad \cdots \cdots \quad (2)$$

if ρ and σ are the electrification and magnetification densities respectively.

In another form, electrification is the source of displacement, and magnetification the source of induction. How these fluxes are distributed after leaving their sources is a perfectly indifferent matter, so far as concerns the measure of the strength of the sources. In an isotropic uniform medium at rest, the fluxes naturally spread out uniformly and radially from point-sources of displacement or of induction. The density of the fluxes then varies as the inverse square of the distance, because the concentric spherical surfaces through which they pass vary in area directly as the square of the distance. Thus

$$D = \frac{\rho}{4\pi r^2}, \qquad B = \frac{\sigma}{4\pi r^2}, \quad \ldots \quad (3)$$

are the tensors of the displacement and induction at distance r from point-sources ρ and σ.

If the source be spread uniformly over a plane in a uniform isotropic medium to surface-density ρ or σ, then, by the mere symmetry, we see that half the flux goes one way and half the other, perpendicularly to the plane, so that

$$D = \tfrac{1}{2}\rho, \qquad B = \tfrac{1}{2}\sigma, \quad \ldots \ldots \quad (4)$$

at any distance from the plane. But if we by any means make the source send all the flux one way only, then

$$D = \rho, \qquad B = \sigma, \quad \ldots \ldots \quad (5)$$

at any distance.

A Moving Source equivalent to a Convection Current, and makes the True Current Circuital.

§ 52. The above being merely to concisely explain the essential meaning of electrification in relation to displacement, and how it is to be measured, consider a point-source or charge to be in motion through a dielectric at rest. Starting with the charge at rest at one place, the displacement is radial and stationary. When permanently at rest in another place, the displacement is the same with reference to it. In the transition, therefore, the displacement has changed its distribution. There must, therefore, be electric current. Now, the only place where the displacement diverges, however the source may be moving, is at the source itself, and therefore the only place where the displacement current diverges is at the source, because it is the time-variation of the displacement. The displacement current is therefore circuital, with the exception of a missing

E 2

link at the moving charge. If we suppose that the charge ρ moving with velocity u constitutes a current uρ, that is, in the same sense as the motion, and such that the volume-integral of the current density is uρ, then the complete system of this "convection" current, and the displacement current together form a circuital flux.

Thus, suppose the charge to be first outside a closed surface and then move across it to its inside. When outside, if the displacement goes through the surface to the inner region, it leaves it again. On the other hand, when the charge is inside, the whole displacement passes outward. Therefore, when the charge is in the very act of crossing the surface, the displacement through it outward changes from 0 to ρ, and this is the time-integral of the displacement current outward whilst the charge crosses. This is perfectly and simultaneously compensated by the convection current, making the whole current always circuital.

The electric current is, therefore, made up of three parts, the conduction current, the displacement current, and the convection current; thus,

$$ J = C + \dot{D} + \rho u, \quad \ldots \ldots \quad (6) $$

ρ being the volume-density of electrification moving through the stationary medium with the velocity u.

If the medium be also moving at velocity q referred to fixed space, we must understand by u above the velocity also referred to fixed space. The velocities q and u are only the same when the medium and the charge move together. Thus it will come to the same thing if we stop the motion of the charge altogether, and let the medium have the motion equivalent to the former relative motion.

Similarly, if there should be such a thing as diverging induction, or the "magnetification" denoted by σ above, then we shall be obliged to consider a moving magnetic charge as contributing to the magnetic current, making the complete magnetic current be expressed by

$$ G = K + \dot{B} + \sigma w, \quad \ldots \ldots \quad (7) $$

if w be the velocity of the magnetification of density σ.

Examples to illustrate Motional Forces in a Moving Medium with a Moving Source. (1.) Source and Medium with a Common Motion. Flux travels with them undisturbed.

§ 53. In order to clearly understand the sense in which motion of a charge through a medium, or motion of the medium itself, or of both together with respect to fixed space, is to be understood, and of the part played therein by the motional electric and magnetic forces, it will be desirable to give a few illustrative examples of such a nature that their meaning can be readily followed from a description, without the mathematical representation of the results. It does not, indeed, often happen that this can be done with profit and without much circumlocution. In the present case, however, it is rather easier to see the meaning of the solutions from a description, than from the formulæ.

In the first place, let us start with a single charge ρ at rest at any point A in an infinite isotropic non-conducting dielectric—ether, for example—which is also at rest. Under these circumstances the stationary condition is one of isotropic radial displacement from the charge at A according to the inverse-square law, and there is nothing to disturb this distribution.

Now, if the whole medium and the charge itself are supposed to have a common motion (referred to an assumed fixed space, in the background, as it were), no change whatever will take place in the distribution of displacement referred to the moving charge. That this should be so in a rational system we may conclude from the relativity of motion (the absolute motion of the universe being quite unknown, if not inconceivable) combined with our initial assumption that the electric flux (and the magnetic flux not here present) represent states of the medium, which may be carried with it just as states of matter are carried with matter in its motion. But as the charge, and with it the displacement, move through space as a rigid body without rotation, the changing displacement at any point constitutes an electric current, and therefore would necessitate the existence of magnetic force, if we treated the first law of circuitation in its primitive form, referred to a stationary medium. Here, however, the motional magnetic force, which is (§§ 44, 49) the vector-product of the displacement and the velocity of the medium, comes into play, and it is so constructed as to precisely annul all magnetic force under the circumstances, and leave the displacement (referred to the moving medium) unaffected ; or, in another form, it changes the law of circuitation (curl $\mathbf{H} = \mathbf{J}$) referred to fixed space, so as to refer it in the same form to the moving medium.

The result is $\mathbf{H} = 0$, and $\mathbf{D}$ moves with the medium.

Similar remarks apply to other stationary states. They are unaffected by a common motion of the whole medium and the sources (or quasi-sources), and this result is mathematically obtained by the motional electric and magnetic forces.

(2.) Source and Medium in Relative Motion. A Charge suddenly jerked into Motion at the Speed of Propagation. Generation of a Spherical Electromagnetic Sheet ; ultimately Plane. Equations of a Pure Electromagnetic Wave.

§ 54. But the case is entirely altered if the charge and the medium have a relative translational motion.

Start again with charge and medium at rest, and the displacement stationary and isotropically radial. Next, introduce the fact (the truth of which will be fully seen later) that the medium transmits all disturbances of the fluxes through itself at the speed $(\mu c)^{-\frac{1}{2}}$, which call v ; and let us suddenly set the charge moving in any direction rectilinearly through the medium at this same speed, v. The question is, what will happen ?

A part of the result can be foreseen without mathematical investigation ; the remainder is an example of the theory of the simplest spherical wave given by me in "Electromagnetic Waves." Let A (in Fig. 1) be the initial position of the charge when it first begins to move, and let AC be the direction of its subsequent motion. Describe a sphere of radius $AB = vt$; then, at the time t the charge has reached B. Now, from the mere fact that the speed of propagation is v, it follows that the displacement outside the sphere is undisturbed. It is clear that there cannot be any change to the right of B, because the charge has only just reached that place, and disturbances only

travel at the same speed as it is moving itself.   Similar considerations applied to the expanding sphere through this
charge at every moment of its passage from A to B will show
that no disturbance can have got outside the sphere.   The
radial lines, therefore, represent the actual displacement, as
well as the original displacement, though of course, in the
latter case, they extended to the point A.

We have now to complete the description of the solution.
There is no displacement whatever inside the sphere BEDF.

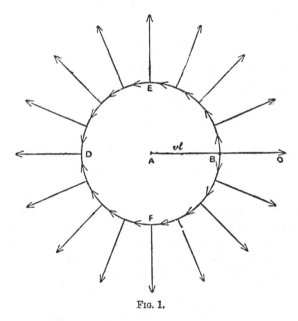

FIG. 1.

The displacement emanating from the charge at B, therefore,
joins on to the external displacement over the spherical surface.
We can say beforehand that it should do so in the simplest
conceivable manner, by the shortest paths.   On leaving the
pole B it spreads uniformly in all directions on the surface of
the sphere, and each portion goes the shortest way to the
opposite pole D.   But it leaks out externally on the way, in
such a manner that the leakages are equal from equal areas.
The displacement thus follows the lines of longitude.

This completes the case so far as the displacement is concerned. But the spherical surface constitutes an electromagnetic sheet, and corresponding to the displacement there is a distribution of coincident induction. This induction is perpendicular to the tangential displacement, and therefore follows the lines of latitude. Its direction is up through the paper above A (at E, for example), and down through the paper below A (at F, for example). The tangential displacement and induction surface-densities (or fluxes per unit area of the sheet), say, D_0 and B_0, are connected by the equation

$$B_0 = \mu v D_0,$$
or,
$$D_0 = c v B_0.$$

Or, if E_0 and H_0 be the equivalent forces got by dividing by c and by μ respectively, then, since $\mu c v^2 = 1$,

$$H_0 = c v E_0,$$
$$E_0 = \mu v H_0.$$

Or, expressing the mutual directions as well,

$$H_0 = V \mathbf{v} D_0,$$
$$E_0 = V B_0 \mathbf{v} ;$$

where $\mathbf{v}$ is the vector velocity of the electromagnetic sheet at the place considered. These last are, in fact, as we shall see later, the general equations of a wave-front or of a free wave, which though it may attenuate as it travels, does not suffer distortion by mixing up with other disturbances.

Now, as time goes on, the charge at B moves off to the right, the electromagnetic sheet simultaneously expanding. The external displacement, therefore, becomes infinitesimal ; likewise that on the D side of the shere. Practically, therefore, we are finally left with a plane electromagnetic sheet moving perpendicularly to itself at speed v, at one point of which is the moving charge, from which the displacement diverges uniformly in the sheet, following, therefore, the law of the inverse first power (instead of the original inverse square), accompanied by a distribution of induction in circles round the axis of motion, varying in density with the distance according to the same law, and connected with the displacement by the above equations. In the diagram, AB has to be very great, and the plane sheet is the portion of the spherical sheet round B, which is then of insensible curvature.

(3.) Sudden Stoppage of Charge.   Plane Sheet moves on. Spherical Sheet generated.   Final Result, the Stationary Field.

§ 55. Having thus turned the radial isotropic displacement of the stationary charge into a travelling plane distribution, let us suddenly reduce the charge to rest. We know that

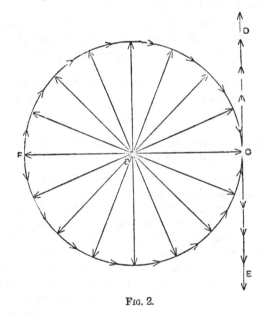

FIG. 2.

after some time has elapsed, the former isotropic distribution will be reassumed; and now the question is, how will this take place?

Let B in Fig. 2 be the position of the charge at the moment of stoppage and after. Describe a sphere of radius vt, with B for centre; then the point C is where the charge would have got at time t after the stoppage had it not been stopped, and the plane DCE would have been the position of the plane electromagnetic sheet. Now, the actual state of things is described by saying that :—

(1.) The plane sheet DCE moves on quite unaltered, except at its core C, where the charge has been taken out.

(2.) The stationary radial displacement of the charge in its new position at B is fully established within the sphere, without any induction.

(3.) The internal displacement joins itself on to the external in the plane sheet, over the spherical surface, by leaking into it and then following the shortest route to the pole C. That is, the tangential displacement follows the lines of longitude.

(4.) The induction in the spherical sheet is oppositely directed to before, still, however, following the lines of latitude, and being connected with the tangential displacement by the former relations.

In time, therefore, the plane sheet and the spherical sheet go out to infinity, and there is left behind simply the radial displacement of the stationary charge.

(4.) Medium moved instead of Charge. Or both moved with same Relative Velocity.

§ 56. Now, return to the case of § 54, and referring to Fig. 1, suppose it to be the charge that is kept at rest, whilst the medium is made to move bodily past it from right to left at speed v, so that the relative motion is the same as before. We must now suppose B to be at rest, the charge being there originally, and remaining there, whilst it is A that is travelling from right to left, and the spherical surface has a motion compounded of expansion from the centre A and translation with it. Attending to this, the former description applies exactly.

The external displacement is continuously altering, and there is electric current to correspond, but there is no magnetic force (except in the spherical sheet), and this is, as before said, accounted for by the motional magnetic force.

The final result is now a stationary plane electromagnetic sheet, as, in fact, described before in § 45, where we considered the displacement and induction in the sheet to be kept up steadily by electric and magnetic forces impressed by the motion

Now stop the motion of the medium, without altering the position of the charge, and Fig. 2 will show the growth of the radial stationary displacement, as in § 55, as it is in fact the same case precisely after the first moment. We can in a similar manner treat the cases of motion and stoppage of both charge and medium, provided the relative speed be always the speed v, however different from this may be the actual speeds.

(5.) **Meeting of a Pair of Plane Sheets with Point-Sources. Cancelment of Charges ; or else passage through one another ; different results. Spherical Sheet with two Plane Sheet Appendages.**

§ 57. From the two solutions of §§ 54, 55 (either of which may be derived from the other) we may deduce a number of other interesting cases.

Thus, let initially a pair of equal opposite charges $+\rho$ and $-\rho$ be moving towards one another, each at speed v through the medium (which for simplicity we may consider stationary), each with its accompanying plane electromagnetic sheet. When the charges meet the two sheets coincide, the two displacements cancel, leaving none, and the two inductions add, doubling the induction. We have thus, momentarily, a mere sheet of induction.

Now, if we can carry the charges through one another, without change in their motion, the two sheets will immediately reappear and separate. That is, the plane waves will pass through one another, as well as the charges.

But if the charges cancel one another continuously after their first union, a fresh case arises. It is, given a certain plane sheet of induction initially, what becomes of it, on the understanding that there is to be no electrification ?

The answer is, that the induction sheet immediately splits into two plane electromagnetic sheets, joined by a spherical sheet, as in Fig. 3. For it is the same as the problem of stoppage in § 55 with another equal charge of opposite kind moving the other way and stopped simultaneously, so that there is no electrification ever after. Touching the sphere at the point F in Fig. 2 is to be placed the additional plane wave, and the internal displacement is to be abolished. That is to say, in Fig. 3 the displacement converges uniformly to F in the plane sheet there, then flows without leakage to the opposite pole C along the lines of longitude, and there diverges uniformly in the other plane sheet. Each displacement sheet has its corresponding coincident induction, according to the former formulæ. They all move out to infinity, leaving nothing behind, as there is no source left.

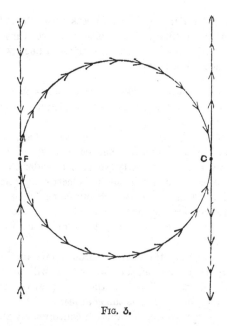

FIG. 5.

(6.) Spherical Sheet without Plane Appendages produced by sudden jerking apart of opposite Charges.

§ 58. Similarly, let there be a pair of coincident or infinitely close opposite charges, with no displacement, and let them be suddenly jerked apart, each moving at the speed of propagation of disturbances.   The result is simply a single spherical wave, without plane appendages, and without leakage of the displacement.   The charges are at opposite poles, at the ends of the axis of motion, and the displacement just flows over the

surface from one to the other symmetrically. There is the
usual induction $B_0 = \mu v D_0$ to match.

Fig. 3 also shows this case, if we leave out the plane sheets and
suppose the positive charge to be at F and the negative at C.
After a sufficient time, we have practically two widely separated plane electromagnetic sheets, although they are really
portions of a large spherical sheet.

Now, imagine the motion of the two charges to be reversed ;
if we simultaneously reverse the induction in the spherical
sheet, without altering the displacement, it will still be an
electromagnetic sheet, but will contract instead of expanding.
It will go on contracting to nothing when the charges meet.
If they are then stopped nothing more happens. But if the
charges can separate again, the result is an expanding spherical
electromagnetic sheet as before.

(7.) Collision of Equal Charges of same Name.

§ 59. If, in the case of colliding plane sheets with charges,
§ 57, they be of the same name, then, on meeting, it is the
induction that vanishes, whilst the displacement is doubled.
That is, we have momentarily a plane sheet of displacement.

If the charges be kept together thereafter, this plane
sheet splits into two plane electromagnetic sheets joined by
a spherical sheet. At the centre of the last is the (doubled)
charge 2ρ, which sends its displacement isotropically to the
surface of the sphere, where it is picked up and turned round
towards one pole or the other. The equator of the sphere is the
line of division of the oppositely flowing displacements. The
displacement gets greater and greater as the poles are neared,
the total amount reaching each pole being ρ (half the central
charge), which then diverges in the plane sheet touching the pole.

The final result, when the waves have gone out to infinity,
is, of course, merely the stationary field of the charge 2ρ.

(8.) Hemispherical Sheet. Plane, Conical and Cylindrical Boundaries.

§ 60. If a charge be initially in contact with a perfectly
conducting plane, and be then suddenly jerked away from it at
the speed v, the result is merely a hemispherical electromagnetic shell. The negative charge, corresponding to the moving

point-charge, expands in a circular ring upon the conducting plane, this ring being the equator of the (complete) sphere.

This case, in fact, merely amounts to taking one-half of the solution in § 58, and then terminating the displacement normally upon a conductor.

In Fig. 4, A is the original position of ρ on the conducting plane CAE, and when the charge has reached B the displacement terminates upon the plane in the circle D.F.

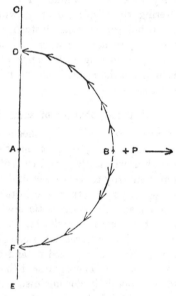

Fig. 4.

Instead of a plane conducting boundary, we may similarly have conical boundaries, internal and external (or one conical boundary alone), with portions of perfect spherical waves running along them at the speed v.

If the two conical boundaries have nearly the same angle, and this angle be small, we have a sort of concentric cable (inner and outer conductor with dielectric between), of continuously increasing thickness. The case of uniform thickness is included as an extreme case; the (portion of the) spherical wave then becomes a plane wave.

General Nature of Electrified Spherical Electromagnetic Sheet.

§ 61. The nature of a spherical electromagnetic sheet expanding or contracting at speed v, when it is charged in an arbitrary manner, may also be readily seen by the foregoing. So far, when there has been electrification on the sheet, it has been a solitary point-charge or else a pair, at opposite poles. Now, the general case of an arbitrary distribution of electrification can be followed up from the case of a pair of charges *not* at opposite ends of a diameter, and each of these may be taken by itself by means of an opposite charge at the centre or externally, so that integration does the rest of the work. When there is a pair of equal charges of opposite sign we do not need any external or internal complementary electrification, but we may make use of them argumentatively; or we may let one charge leak outward, the other inward, and have the external and internal electrification in reality. The leakage should be of the isotropic character always. But the internal electrification need not be at the central point. It may be uniformly distributed upon a concentric sphere. This, again, may be stationary, or it may itself be in motion, expanding or contracting at the speed v. The external electrification, too, may be on a concentric spherical surface, which may be in similar motion. The sheets, too, may be of finite depth, and arbitrarily electrified, so that we have any volume-distribution of electrification moving in space in radial lines to or from a centre, accompanied by electromagnetic disturbances arranged in spherical sheets.

There is thus a great variety of ways of making up problems of this character, the nature of whose solutions can be readily pictured mentally.

Two charges, q_1 and q_2, for example, on a spherical sheet. One way is to put $-q_1$ and $-q_2$ at the centre, and superpose the two solutions. Or the charges $-(q_1$ and $q_2)$ may be put externally, with isotropic leakage. Or part may be inside, and part outside. Or we may have no complementary electrification at all, but lead the displacement away into plane waves touching the sphere.

Only when the total charge on the spherical sheet is zero can we dispense with these external aids (which to use depending upon the conditions of the problem); then the displacement has sources and sinks on the sheet which balance one another. The corresponding induction is always perpendicular to the resultant tangential displacement, and is given by the above formula, or $B_0 = \mu VvD_0$, where D_0 is the tangential displacement in the sheet (volume density × depth).

One case we may notice. If the density of electrification be uniform over the surface, there is no induction at all. That is, it is not an electromagnetic sheet, but only a sheet of electrification, without tangential displacement, and therefore without induction.

So, with a condenser consisting of a pair of concentric shells uniformly electrified, either or both may expand or contract without magnetic force. This is, however, not peculiar to the case of motion at speed v. Any speed will do. But in general, if the speed be not exactly v, there result diffused disturbances. The electromagnetic waves are no longer of the same pure type.

General Remarks of the Circuital Laws. Ampere's Rule for deriving the Magnetic Force from the Current. Rational Current-element.

§ 62. The two laws of circuitation did not start into full activity all at once. On the contrary, although they express the fundamental electromagnetic principles concerned in the most concise and clear manner, it was comparatively late in the history of electromagnetism that they became clearly recognised and explicitly formularised. We have not here, however, to do the work of the electrical Todhunter, but only to notice a few points of interest.

The first law had its first beginnings in the discovery of Oersted that the electric conflict acted in a revolving manner, and in the almost simultaneous remarkable investigations of Ampère. It did not, however, receive the above used form of expression. In fact, in the long series of investigations in electro-dynamics to which Oersted's discovery, and the work of Ampère, Henry, and Faraday, gave rise, it was customary to consider an element of a conduction current as generating a

certain field of magnetic force. Natural as this course may
have seemed, it was an unfortunate one, for it left the question
of the closure of the current open; and it is quite easy to see
now that this alone constituted a great hindrance to progress.
But so far as closed currents are concerned, in a medium of
uniform inductivity, this way of regarding the relation between
current and magnetic force gives equivalent results to those
obtained from the first law of circuitation in the limited form
suitable to the circumstances stated.

If C is the density of conduction current at any place, the
corresponding field of magnetic force is given by

$$H = \frac{VCr_1}{4\pi r^2}, \quad \cdots \cdots \quad (1)$$

at distance r from the current-element C, if r_1 be a unit vector
along r from the element to the point where H is reckoned.
The intensity of H thus follows the law of the inverse square
of the distance along any radius vector proceeding from the
current-element; but, in passing from one radius vector to
another, we have to consider its inclination to the axis of the
current by means of the factor $\sin \theta$ (where θ is the angle
between r and the axis), involved in the vector product. Also,
H is perpendicular to the plane containing r and the axis of
the current-element, or the lines of H are circles about this axis.

But from the Maxwellian point of view this field of H is that
corresponding to a certain circuital distribution of electric
current, of which the current-element mentioned is only a
part; this complete current being related to the current-
element in the same way as the induction of an elementary
magnet is to the intensity of magnetisation of the latter.
Calling the complete system of electric current a rational
current-element, it may be easily seen that in a circuital distribution of rational current-elements the external portion of the
current disappears by mutual cancelling, and there is left only
the circuital current made up of the elements in the older
sense. We may, therefore, employ the formula (1) to calculate
without ambiguity the magnetic force of any circuital distribution of current. This applies not merely to conduction
current (which was all that the older electricians reckoned),
but to electric current in the wider sense introduced by

F

Maxwell. But the result will not be the real magnetic force unless the distribution of inductivity is uniform. When μ varies, we may regard the magnetic force of the current thus obtained as an impressed magnetic force, and then calculate what induction it sets up in the field of varying inductivity. This may be regarded as an independent problem.

In passing, we may remark that we can mount from magnetic current to electric force by a formula precisely similar to (1), but subject to similar reservations.

The Cardinal Feature of Maxwell's System. Advice to anti-Maxwellians.

§ 63. But this method of mounting from current to magnetic force (or equivalent methods employing potentials) is quite unsuitable to the treatment of electromagnetic waves, and is then usually of a quite unpractical nature. Besides that, the function " electric current " is then often a quite subsidiary and unimportant quantity. It is the two fluxes, induction and displacement (or equivalently the two forces to correspond), that are important and significant; and if we wish to know the electric current (which may be quite a useless piece of information) we may derive it readily from the magnetic force by differentiation ; the simplicity of the process being in striking contrast to that of the integrations by which we may mount from current to magnetic force.

To exemplify, consider the illustrations of plane and spherical electromagnetic waves of the simplest type given in §§ 53 to 61, and observe that whilst the results are rationally and simply describable in terms of the fluxes or forces, yet to describe in terms of electric current (and derive the rest from it) would introduce such complications and obscurities as would tend to anything but intelligibility.

Now, Maxwell made the first law of circuitation (not, however, in its complete form) practically the definition of electric current. This involves very important and far-reaching consequences. That it makes the electric current always circuital at once does away with a host of indeterminate and highly speculative problems relating to supposititious unclosed currents. It also necessitates the existence of electric current in perfect non-conductors or insulators. This has always been a stumbling-block to practicians who think themselves practical. But Maxwell's innovation was really the most practical improvement in electrical theory conceivable. The electric current in a nonconductor was the very thing wanted to coordinate electrostatics and electrokinetics, and consistently harmonise the equations of electromagnetism. It is the cardinal feature of Maxwell's system, and, when properly followed up, makes the insulating medium the true medium in the transmission of disturbances, and explains a multitude of phenomena that are inconceivable in any other theory (unless it be one of the same type). But let the theoretical recommendations (apart from modern experiments), which can only be appreciated after a pretty close study of theory, Maxwellian and otherwise, stand aside, and only let the tree be judged by its fruit. Is it not singular that there should be found people, the authors of works on Electricity, who are so intensely prejudiced against the Maxwellian view, which it would be quite natural not to appreciate from the theoretical standpoint, as to be apparently quite unable to recognise that the fruit has any good flavour or savour, but think it no better than Dead Sea fruit? The subject is quite sufficiently difficult to render understandable popularly, without the unnecessary obstruction evidenced by a carping and unreceptive spirit. The labours of many may be required before a satisfactory elementary presentation of the theory can be given. So much the more need, therefore, is there for the popular writer to recognise the profound significance of the remarkable experimental work of late years, a significance he appears to have so sadly missed. When that is done, then will be the time for an understanding of Maxwell's views. Let him have patience, and believe that it is not all speculative metaphysics because it is not to his present taste. Never mind the ether disturbances playing pranks with the planets. They can take care of themselves.

Changes in the Form of the First Circuital Law.

§ 64. Two or three changes I have made* in Maxwell's form of the first circuital law. One is of a formal character, the introduction of the h term to express the intrinsic force of

---

* "Electromagnetic Induction and its Propagation," *The Electrician*, 1885, January 3, and later : or reprint.

magnetisation. This somewhat simplifies the mathematics, and places the essential relations more clearly before the eye. Connected with this is a different reckoning of the energy of an intrinsic magnet, in order to get consistent results.

The second change, which is not merely one of a formal character, but is an extension of an obligatory character, is the introduction of the h term to represent the motional magnetic force. In general problems relating to electromagnetic waves it is equally important with the motional electric force.

The third change is the introduction (first done, I think, by Prof. Fitzgerald) of the term to explicitly represent the convection-current or electrification in motion as a part of the true current. Although Maxwell did not himself explicitly represent this, which was a remarkable oversight, he was strongly insistent upon the circuital nature of electric current, and would doubtless have seen the oversight the moment it was suggested to him. Now, there are spots on the sun, and I see no good reason why the many faults in Maxwell's treatise should be ignored. It is most objectionable to stereotype the work of a great man, apparently merely because it was so great an advance, and because of the great respect thereby induced. The remark applies generally ; to the science of Quaternions, for instance, which, if I understand rightly, Prof. Tait would preserve in the form given to it by Hamilton. In application to Maxwell's theory, I am sure that it is in a measure to the recognition of the faults in his treatise that a clearer view of the theory in its broader sense is due.

Introduction of the Second Circuital Law.

§ 65. The second circuital law, like the first, had an experimental origin, of course, and, like the first, was long in approximating to its present form—much longer, in fact, though in a different manner. The experimental foundation was Faraday's recognition that the voltage induced in a conducting circuit was conditioned by the variation of the number of lines of force through it.

But, rather remarkably, mathematicians did not put this straight into symbols for an elementary circuit, but went to work in a more roundabout way, and expressed it through the medium of an integration extended along a concrete circuit ;

or else, of an equation of electromotive force containing a
function called the vector potential of the current, and another
potential, the electrostatic, working together not altogether in
the most harmoniously intelligible manner—in plain English,
muddling one another. It is, I believe, a fact which has been
recognised that not even Maxwell himself quite understood how
they operated in his " general equations of propagation." We
need not wonder, then, that Maxwell's followers have not found
it a very easy task to understand what his theory really meant,
and how to work it out. I had occasion to remark, some years
since, that it was very much Maxwell's own fault that his
views obtained such slow acceptance; and, in now repeating the
remark, do not abate one jot of my appreciation of his work,
which increases daily. For he devoted the greater part of his
treatise to the working out and presentation of results which
could be equally well done in terms of other theories, and gave
only a very cursory and incomplete exposition of what were
peculiarly his own views and their consequences, which are
of the utmost importance. At the same time, it is easily to be
recognised that he was himself fully aware of their importance,
by the tone of quiet confidence in which he wrote concerning
them.

Finding these equations of propagation containing the two
potentials unmanageable, and also not sufficiently comprehensive, I was obliged to dispense with them ; and, going back to
first principles, introduced* what I term the second circuital law
as a fundamental equation, the natural companion to the first.
The change is, I believe, a practical one, and enables us to considerably simplify and clarify the treatment of general questions, whilst bringing to light interesting relations which were
formerly hidden from view by the intervention of the vector
potential **A**, and its parasites J and Ψ.

Another rather curious point relates to the old German
electro-dynamic investigations and their extensions to endeavour
to include, supersede, or generalise Maxwell by anti-Maxwellian
methods. It would be slaying the slain to attack them ; but
one point about them deserves notice. The main causes of the
variety of formulæ, and the great complexity of the investigations, were—first, the indefiniteness produced by the want of

* See footnote, p. 67.

circuitality in the current, and next the potential methods
employed. [J. J. Thomson's "Report on Electrical Theories"
contains an account of many. As a very full example, that
most astoundingly complex investigation of Clausius, in the
second volume of his "Mechanische Wärmetheorie," may be
referred to.] But if the critical reader will look through these
investigations, and eliminate the potentials, he will find, as a
useful residuum, the second circuital law; and, bringing it
into full view, will see that many of these investigations are
purely artificial elaborations, devoid of physical significance;
gropings after mares' nests, so to speak. Now, using this law
in the investigations, it will be seen to involve, merely as a
matter of the mathematical fitness of things, the use of another,
viz., the first circuital law, and so to justify Maxwell in his
doctrine of the circuitality of the current. The useful moral
to be deduced is, I think, that in the choice of variables to express physical phenomena, one should keep as close as possible
to those with which we are experimentally acquainted, and
which are of dynamical significance, and be on one's guard
against being led away from the straight and narrow path in
the pursuit of the Will-o' the-wisp.

As regards the terms in the expression of the second law
which stand for unknown properties, they may be regarded
merely as mathematical extensions which, by symmetrizing the
equations, render the correct electric and magnetic analogies
plainer, and sometimes assist working out. But some other
extensions of meaning, yet to be considered, have a more substantial foundation.

Meaning of True Current. Criterion.

§ 66. One of these extensions refers to the meaning to be
attached to the term current, electric or magnetic respectively.
As we have seen, electric current, which was originally conduction current only, had a second part, the displacement current,
added to it by Maxwell, to produce a circuital flux; and further,
when there is electrification in motion, we must, working to the
same end, add a third term, the convection current, to preserve
circuitality.

A fourth term may now be added to make up the "true"
current, when the medium supporting the fluxes is in motion.

Thus, let us separate the motional electric and magnetic forces from all other intrinsic forces that go with them in the two circuital laws (voltaic, thermo-electric, &c., § 39); denoting the former by e and h, and the latter by e_0 and h_0. The two equations of circuitation [(12), (13) § 38, and (6), (7) § 52)] now read

$$\operatorname{curl} (H - h_0 - h) = J = C + \dot{D} + u\rho, \quad . \quad . \quad (1)$$
$$- \operatorname{curl} (E - e_0 - e) = G = K + \dot{B} + w\sigma. \quad . \quad . \quad (2)$$

Now transfer the e and h terms to the right side, producing

$$\operatorname{curl} (H - h_0) = J_0 = C + \dot{D} + u\rho + j, \quad . \quad . \quad (3)$$
$$- \operatorname{curl} (E - e_0) = G_0 = K + \dot{B} + w\sigma + g, \quad . \quad . \quad (4)$$

where j and g are the auxiliaries before used, given by

$$j = \operatorname{curl} h, \qquad g = - \operatorname{curl} e.$$

It is the new vectors J_0 and G_0 which should be regarded as the true currents when the medium moves.

As the auxiliaries j and g are themselves circuital vectors, there may not at first sight appear to be any reason for further complicating the meaning of true current when separated into component parts, for the current J is circuital, and so is J_0, which is of course the same as J when the medium is stationary. The extension would appear to be an unnecessary one, of a merely formal character.

On the other hand, it is to be observed that from the Maxwellian method of regarding the current as a function of the magnetic force, the extended meaning of true current is not a further complication, but is a simplification. For whereas in the equation (2) we deduct from the force E of the flux not only the intrinsic force e_0 but also the motional force e to obtain the effective force whose curl measures the current; on the other hand, in (4) we deduct only the intrinsic force. Away, therefore, from the sources of energy which are independent of the motion of the medium, the force whose curl is taken in (4) is the force of the flux, which specifies the electric state of the medium, whether it be stationary or moving.

To show the effect of the change, consider the example of § 56, in which the medium moves past the charge, and there is continuously changing displacement outside a certain sphere. If we consider J the true current, we should say there is electric current, although there is no magnetic force. But according to the other way there is no true current, and the absence of magnetic force implies the absence of true electric current. But still this question remains, so far, somewhat of a conventional one. Is there any test to be applied which shall effectually discriminate between J and J_0 as the true current? There would appear to be one and only one; viz., that e_0 being an intrinsic force, its activity, whatever it be, say e_0x, expresses the rate of communication of energy to the electromagnetic system from a source not included therein, and not connected with the motional force e. When the medium is stationary x is J, and the question is, is x to be J or J_0 (or anything else) when the medium moves?

Now this question can only be answered by making it a part of a much larger and more important one; that is to say, by a comprehensive examination of all the fluxes of energy concerned in the equations (1), (2), or (3), (4), and their mutual harmonisation. This being done, the result is that e_0J_0 is the activity of e_0, so that it is J_0 that is the measure of the true current, with the simpler relation to the force of the flux; and it is naturally suggested that in case of further possible extensions, we should follow in the same track, and consider the true current to be always the curl of $(H - h_0)$, independently of the make up of its component parts.

This is not the place for a full investigation of this complex question, but the main steps can be given; not for the exhibition of the mathematical working, which may either be taken for granted, or filled in by those who can do it, but especially with a view to the understanding in a broad manner of the course of the argument. Up to the present I have used the notation of vectors for the concise and plain presentation of principles and results, but not for working purposes. This course will be continued now. How to work vectors may form the subject of a future chapter. It is not so hard when you know how to do it.

The Persistence of Energy. Continuity in Time and Space and Flux of Energy.

§ 67. The principle of the conservation or persistence of energy is certainly as old as Newton, when viewed from the standpoint of theoretical dynamics.   But only (roughly speaking) about the middle of the present century did it become recognised by scientific men as a universal truth, extending to all the phenomena of Nature.   This arose from the experimental demonstrations given (principally those of Joule), that in various cases in which energy disappeared from one form, it was not lost, but made its appearance in other forms, and to the same amount.   The principle is now-a-days a sure article of scientific faith.   Its ultimate basis is probably the conviction (a thoroughly reasonable one) that the laws of motion hold good in the invisible world as well as in the visible, confirmed by the repeated experimental verification.   But the principle is now believed in quite generally; by the ordinary unscientific man, for example, who could not tell you correctly what the principle meant, or energy either, to save his life, although he might tell you that he could not conceive the possibility of energy being destroyed. He goes by faith, having taken it in when young.

So it will be with the modern view of the ether as the medium through which energy is being sent when a wire supports a current.   Only train up the young to believe this, and they will afterwards look upon the notion of its going through the conductor as perfectly absurd, and will wonder how anyone ever could have believed it.   But before this inevitable state of things comes to pass, an intermediate period must be passed through, in which scientific men themselves are learning to believe the modern view thoroughly from the evidence of its truth, and to teach it to others.   This period is certainly not yet come to its end ; for almost weekly we may read about "the velocity of electricity in wires," such language emanating from scientific men who are engaged in repeating and extending Hertz's experiments.

The principle of the continuity of energy is a special form of that of its conservation.   In the ordinary understanding of the conservation principle it is the integral amount of energy that is conserved, and nothing is said about its distribution or its motion.   This involves continuity of existence in time, but not necessarily in space also.

But if we can localise energy definitely in space, then we are bound to ask how energy gets from place to place.   If it

possessed continuity in time only, it might go out of existence
at one place and come into existence simultaneously at
another. This is sufficient for its conservation. This view,
however, does not recommend itself. The alternative is to
assert continuity of existence in space also, and to enunciate
the principle thus :—

When energy goes from place to place, it traverses the
intermediate space.

This is so intelligible and practical a form of the principle,
that we should do our utmost to carry it out.

The idea that energy has position, therefore, naturally
involves the idea of a flux of energy. Let **A** be the vector flux
of energy, or the amount transferred per unit of time per unit
of area normal to the direction of its transfer ; and let T be
the density of the energy. We may then write

$$\text{conv } \mathbf{A} = \dot{\mathrm{T}}, \quad \ldots \ldots (5)$$

or
$$\text{div } \mathbf{A} = -\dot{\mathrm{T}}. \quad \ldots \ldots (6)$$

That is, the convergence of the flux of energy is accounted for
by increased energy in the unit volume where it converges ;
or, the divergence of the flux of energy, or the rate at which
it leaves the unit volume, is accompanied by simultaneous and
corresponding decrease in the density of the energy.

Examples. Convection of Energy and Flux of Energy due to an active Stress. Gravitational difficulty.

§ 68. Now, there are numerous cases in which the flux of
energy is perfectly plain, and only needs to be pointed out to
be recognised. The simplest of all is the mere convection of
energy by motion of the matter with which it is associated. A
body in translational motion, for instance, carries its energy
with it; not merely the kinetic energy of its translational
motion, but also its rotational energy if it rotates, and likewise the energy of any internal vibratory or rotatory motions
it may possess, and other energy not known to be kinetic,
and therefore included under "potential energy," which we
may have good reason to localise in the body. Even
if part of the energy be outside the body, yet, if it be
associated with the body, it will travel with it, on the whole.

and so (at least sometimes) may be considered to be rigidly attached to it. This case may be regarded as the convection of energy, if we do not go too closely into detail.

When energy is thus conveyed, the energy flux is qT, where q is the velocity and T the density of the energy conveyed (of any kind). If, then, *all* energy were conveyed, equation (5) would become

$$\operatorname{conv} qT = \dot{T}. \qquad \ldots \ldots \quad (7)$$

Now this equation, if we regard T as the density of matter, is the well known equation of continuity of matter used in hydrodynamics and elsewhere.

This brings us to Prof. Lodge's theory of the identity of energy. (*Phil. Mag.*, 1885.) Has energy personal identity, like matter? I cannot see it, for one ; and think it is pushing the principle of continuity of energy, which Prof. Lodge was writing about, too far. It is difficult to endow energy with objectivity, or thinginess, or personal identity, like matter. The relativity of motion seems to be entirely against the idea. Nor are we able to write the equation (7) in all cases. Energy may be transferred in other ways than by convection of associated matter. If we atomise the energy we can then imagine the q above to be the velocity of the energy, not of the matter. But as the science of dynamics is at present understood, we cannot make use of this idea profitably, I think.

The chief reason why the founders of the modern science of energy did not explicitly make use of the idea of the continuity of energy was probably the very obscure nature of gravitational energy. Where was it before it became localised as the kinetic energy of a mass? It is of no use to call it potential energy if that is to explain anything, which it does not. It would seem that the energy must have been in the ether, somewhere, and was transferred into the body, somehow. This makes the ether the great store-house of all gravitational energy. But we are entirely ignorant of its distribution in the ether, and of its mode of transference.

Observe here that ether must be regarded as a form of matter, because it is the recipient of energy, and that is the characteristic of ordinary matter. There is an unceasing enormous flux of energy through the ether from the Sun, for instance, and we know that it takes several minutes to come; it comes through the ether, without bringing the ether with it; it is not a convection of energy, therefore. Of course, to avoid confusion, it is well to distinguish ether from ordinary matter by its separate name; but it is important to note that it *has* some of the characteristics of ordinary matter. It need not be gravitating matter; it is, perhaps, more likely to be the medium of gravitational action than to gravitate itself.

In default of ability to represent the flux of gravitational energy its entry into a body must be otherwise represented than by the convergence of a flux. Turn (7) into

$$fq + \text{conv } \mathbf{A} = \dot{T}. \quad \ldots \quad (8)$$

Here **fq** is the activity of the force **f**, and indicates energy communicated to the unit volume in an unstated manner, generated within the unit volume, so to speak. The force **f** is thus an impressed force. By this device we can allow for unknown fluxes of energy. We should also explicitly represent the convective flux. Thus,

$$fq + \text{conv} (qT + \mathbf{A}) = \dot{T}, \quad \ldots \quad (9)$$

where **A** is the flux of energy other than convective, not associated with **f**.

When a stress works, for example, we shall have a flux of energy, **A**, which can hardly be considered to be convective, in the sense explained, a flux of energy through matter or ether, as in wave propagation through an elastic solid, or radiation through the ether. We can, however, sometimes reduce it to the form qT_1, and then it would *appear* to be convection of energy.

This occurs in a moving frictionless fluid, for example, when the stress is an isotropic pressure p. The flux of energy other than convective is $p\mathbf{q}$. But when we pass to a less ideal case, as an elastic solid, in which the stress is of a more general character, the energy flux expressive of the activity of the stress takes the form

$$\mathbf{A} = -\mathbf{P}_q \, q, \quad \ldots \quad (10)$$

where $\mathbf{P}_q$ is the vector expressing the pull per unit area on the plane perpendicular to the velocity **q**. In this equation q is the tensor of **q**.

Specialised form of expression of the Continuity of Energy.

§ 69. Passing to more practical and specialised forms of
expression of the continuity of energy, it is to be observed
first that we are not usually concerned with a resultant flux
of energy from all causes, but only with the particular forms
relating to the dynamical question that may be under consideration.   Secondly, that it is convenient to divide the
energy above denoted by T into different parts, denoting
potential energy, and kinetic energy, and wasted energy.
Thus we have the following as a practical form,

$$fq + \text{conv} \left[A + q(U + T) \right] = Q + \dot{U} + \dot{T}, \quad . \quad (11)$$

where the terms in the square brackets indicate the energy
flux, partly convective, with the factor q, U being the density
of potential energy, and T that of kinetic energy, and partly
non-convective, viz., **A**, which may be due to a working stress.
Also **fq** is the activity of impressed force (here merely a translational force). Thus on the left side of equation (11) we
have a statement of the supply of energy to the unit volume
fixed in space.   On the right side we account for it by the rates
of increase of the stored potential and kinetic energy, and by
Q, which means the rate of waste of energy in the unit volume.
The wasted energy is also stored, at least temporarily ; but not
recoverably, so that we may ignore energy altogether after it is
once wasted.

In place of the term **fq**, in which the impressed force is
translational, we may have other terms possessing a similar
meaning, indicating a supply of energy from certain sources.
These sources, too, may be not external, but internal or intrinsic, as for instance when there is a thermo-electric or voltaic
source of energy within the unit volume considered.

Also, the terms **A**, U, T, and Q may have to be split up
into different parts, according to the nature of the dynamical
connections.   The important thing to be grasped is, that whenever we definitely localise energy we can obtain an equation
showing its continuity in space and time, and that when we
can only partially localise it, we can still, by proper devices,
allow for the absence of definiteness. The equation is simply
the equation of activity of the dynamical system, suitably

arranged to show the flux of energy, and can always be obtained when the equations of motion are known and also the nature of the stored and wasted energies.

Electromagnetic Application. Medium at Rest. The Poynting Flux.

§ 70. Now, in the electromagnetic case, the "equations of motion" are the two circuital laws, and to form the equation of activity, we may multiply (1) by $(\mathbf{H} - \mathbf{h}_0 - \mathbf{h})$ and (2) by $(\mathbf{E} - \mathbf{e}_0 - \mathbf{e})$ and add the results; or else multiply (3) by $(\mathbf{H} - \mathbf{h}_0)$ and (4) by $(\mathbf{E} - \mathbf{e}_0)$ and add the results. The equation of activity thus obtained has then to be dynamically interpreted in accordance with the principle of continuity of energy.

When the medium is stationary, there is no difficulty with the interpretation. We obtain

$$\mathbf{e}_0 \mathbf{J} + \mathbf{h}_0 \mathbf{G} = Q + \dot{U} + \dot{T} + \operatorname{div} \mathbf{W}, \quad . \quad . \quad (12)$$

where $\mathbf{W}$ is a new vector given by

$$\mathbf{W} = V(\mathbf{E} - \mathbf{e}_0)(\mathbf{H} - \mathbf{h}_0), \quad . \quad . \quad . \quad (13)$$

and U is the electric energy, T the magnetic energy, and Q the waste per unit volume.

On the left side of (12) is exhibited the rate of supply of energy from intrinsic sources. On the right side it is accounted for partly by the waste Q, and by increase of the stored energy U and T. The rest is exhibited as the divergence of the flux $\mathbf{W}$ given by (13). We conclude, therefore, that $\mathbf{W}$ expresses the flux of energy in the electromagnetic field when it is stationary.

This remarkable formula was first discovered and interpreted by Prof. Poynting [*Phil. Trans.*, 1884, Pt. 2], and independently by myself a little later. It was this discovery that brought the principle of continuity of energy into prominence. But it should be remembered that there is nothing peculiarly electromagnetic about a flux of energy. It is here made distinct, because the energy is distinctly localised in Maxwell's theory.

The flux of energy takes place in the direction perpendicular to the plane containing the electric and magnetic forces (of the field), say

$$\mathbf{W} = V\mathbf{E}_1\mathbf{H}_1, \quad . \quad . \quad . \quad . \quad . \quad (14)$$

if $\mathbf{E} = \mathbf{e}_0 + \mathbf{E}_1 \; ; \quad \mathbf{H} = \mathbf{h}_0 + \mathbf{H}_1.$

The formula is not a working formula, in general, for to know **W** we must first know the distribution of the electric and magnetic forces. But it is a valuable and instructive formula for all that. Its discovery furnished the first *proof* that Maxwell's theory implied that the insulating medium outside a conducting wire supporting an electric current was the medium through which energy is transferred from its source at a distance, provided that we admit the postulated storage of energy. For inside a wire the electric force is axial, and the magnetic force circular about the axis ; the flux of energy is therefore radial. When the current is steady, it comes from the boundary of the wire, and ceases at its axis. It delivers up energy on the way, which is wasted in the Joule-heating. But when the current is not steady, the magnetic and electric energy will be also varying.

If we work out the distributions of electric and magnetic force outside the wire, according to the conditions to which it is subjected and its environment, we can similarly fully trace how the energy is supplied to the wire from its source. It is passed out from the source into the dielectric medium, and then converges upon the wire where it is wasted. The flux of energy usually takes place nearly parallel to the wire (because the electric force is nearly perpendicular to its boundary) ; its slight slope towards the wire indicates its convergence thereupon. If the wire had no resistance, there would be no convergence of energy upon it, the flux. of energy would be quite parallel to it. Details are best studied in the concrete application, and it is only by the consideration of variable states, and the propagation of electromagnetic waves, that we can obtain a full understanding of the meaning of **W** considered as a flux of energy.

In the case of a simple progressive plane wave disturbance, in which a distribution of **E** and **H** (mutually perpendicular) in the plane of the wave is propagated unchanged through a medium at constant speed, it is a self-evident result that the energy of the disturbance travels with it. The flux of energy is, therefore (since $e_0 = 0$, $h_0 = 0$),

$$\mathbf{W} = \mathbf{v}\,(U + T) = V\mathbf{EH}, \quad . \quad . \quad . \quad . \quad (15)$$

where **v** is the velocity of the wave, and U, T are the densities (which are here equal) of the electric and magnetic energies. In this example, which is approximately that of radiant energy from the sun, the idea of a flux of energy, and the conclusion that its proper measure is the density of the energy multiplied by the wave velocity, are perfectly plain and reasonable. Now, many cases of the propagation of waves along wires can be reduced to this simple case, with a correction for the resistance of the wire, and other cases can be represented by two or more oppositely travelling plane disturbances. In a very complex electromagnetic field, the flux of energy is necessarily also very complex, and hard to follow ; but the fundamental principles concerned are the same throughout.

The flux of energy **W** arises from the internal structure of the ether. It is somewhat analogous to the activity of a stress. But the only dynamical analogy that is satisfactory in this respect is that furnished by Sir W. Thomson's rotational ether, when interpreted in a certain manner, so that 2**E** shall represent a torque, and **H** the velocity of the medium, with the result that (on this understanding) V**EH** is the flux of energy, whilst U is the potential energy of the rotation, and T the kinetic translational energy. But it is very difficult to extend this analogy to include electromagnetic phenomena more comprehensively. [*See* Appendix at the end of this chapter.]

Extension to a Moving Medium. Full interpretation of the Equation of Activity and derivation of the Flux of Energy.

§ 71. Passing now to the case of a moving medium, we shall obtain, from equations (1), (2), § 66, the equation of activity,

$$(e_0 + e)J + (h_0 + h)G = EJ + HG$$
$$+ \operatorname{div} V(E - e_0 - e)(H - h_0 - h). \quad (16)$$

Or from equations (3), (4), § 66, we may get the form

$$e_0 J_0 + h_0 G_0 = EJ_0 + HG_0 + \operatorname{div} V(E - e_0)(H - h_0), \quad (17)$$

and from these a variety of other forms may be derived. The dynamical interpretation in accordance with the principle of continuity of energy is not so easy as in the former case. I have recently given a full discussion of these equations in another place.   The following gives an outline of the results :—

Since there is a flux of energy when the medium is stationary, there will still be a flux of this kind (*i.e.*, independent of the motion) when it is moving, whatever other flux of energy there may then be. I conclude that the Poynting flux **W** still preserves the form

$$\mathbf{W} = V(\mathbf{E} - \mathbf{e}_0)\,(\mathbf{H} - \mathbf{h}_0), \quad \cdots \quad (18)$$

from the fact that disturbances are propagated through a medium endowed with a uniform translational motion in the same manner as when it is at rest. Otherwise we could only know that it must reduce to this form when at rest.

Next, there is the convective flux of energy

$$\mathbf{q}(U + T), \quad \cdots \quad (19)$$

where **q** is the velocity of the medium.

Thirdly, there is a flux of energy representing the activity, **A**, of the electromagnetic stress. It is given by

$$\mathbf{A} = -\,[V\mathbf{e}\mathbf{H} + V\mathbf{E}\mathbf{h} + \mathbf{q}(U + T)]. \quad \cdots \quad (20)$$

Fourthly, there is, in association with this stress, a convective flux of other energy, say,

$$\mathbf{q}(U_0 + T_0). \quad \cdots \quad (21)$$

The complete flux of energy is the sum **X**, of these four vectors, *i.e.* :—

$$\mathbf{X} = \mathbf{W} + \mathbf{q}(U + T) + \mathbf{A} + \mathbf{q}(U_0 + T_0), \quad \cdots \quad (22)$$

W and **A** being given by (18) and (20).

Finally, the activity of the intrinsic forces is

$$\mathbf{e}_0 \mathbf{J}_0 + \mathbf{h}_0 \mathbf{G}_0, \quad \cdots \quad (23)$$

so that $\mathbf{J}_0$ is the true electric current.

We, therefore, have the equation of activity brought to the standard form

$$\mathbf{e}_0 \mathbf{J}_0 + \mathbf{h}_0 \mathbf{G}_0 = (Q + \dot{U} + \dot{T}) + (Q_0 + \dot{U}_0 + \dot{T}_0) + \text{div.}\ \mathbf{X}, \quad (24)$$

which is a special form of equation (11), with the convergence of the energy flux in it replaced by its divergence (the negative of convergence) on the other side of the equation.

But the unknown terms, on the right side of (24), with the
zero suffix, may be entirely eliminated, including those in **X**, by
making use of the secondary equation of translational activity

$$\mathbf{Fq} = Q_0 + \dot{U}_0 + \dot{T}_0 + \text{div.}\,\mathbf{q}(U_0 + T_0), \quad . \quad . \quad (25)$$

where **F** is the translational force due to the stress.

When this is done, equation (24) takes the form

$$e_0 J_0 + h_0 G_0 = Q + \dot{U} + \dot{T} + \mathbf{Fq} + \text{div.}[\mathbf{W} + \mathbf{A} + \mathbf{q}\,(U + T)] \quad (26)$$

where the energy flux represented includes the Poynting flux,
the stress flux, and the convective flux of electric and magnetic
energy.

We may observe that in the expression (20) for **A** occurs the
term $-\mathbf{q}\,(U + T)$, so that this term may be eliminated from (26),
making the energy flux in it become

$$V(\mathbf{E} - e_0)(\mathbf{H} - h_0) - V e \mathbf{H} - V \mathbf{E} h. \quad . \quad . \quad (27)$$

But these changes, with a view to the simplification of expression, cause us to altogether lose sight of the dynamical significance of the equation of activity, and of the stress function.
Equation (26) is, therefore, the best form.

It should be understood that it is an identity, subject to the
two laws of circuitation and the distribution of energy according to that of **E** and **H** in the field. But it should be also
mentioned that in the establishment of (26) it has been assumed
that the medium, as it moves, carries its intrinsic properties of
permittivity and inductivity with it unchanged. That is, these
properties do not alter for the same portion of the medium,
irrespective of its position, although within the same unit
volume these properties may be changing by the exit of one
and entry of another part of the medium of different permittivity and inductivity ; understanding by " medium " whatever
is supporting the fluxes, whether matter and ether together, or
ether alone ; and it is also to be understood that the three
velocities **q**, **u**, and **w** are identical, or that electrification,
which is always found associated with matter, moves with the
medium, which is then the matter and ether, moving together.
In other respects equation (26) is unrestricted as regards either
homogeneity and isotropy in respect to permittivity, inductivity,
and conductivity (electric, and fictitious magnetic).

Whether, when matter moves, it carries the immediately surrounding ether with it, or moves through the ether, or only partially carries it forward, and what is the nature of the motion produced in the ether by moving matter, are questions which cannot be answered at present. Optical evidence is difficult of interpretation, and conclusions therefrom are conflicting.

But in ordinary large scale electromagnetic phenomena, it can make very little difference whether the ether moves or stands stock-still in space. For the speed with which it propagates disturbances through itself is so enormous that if the ether round a magnet were stirred up, artificially, like water in a basin, with any not excessive velocity, the distortion in the magnetic field produced by the stirring would be next to nothing.

Strictly speaking, when matter is strained its elastic and other constants must be somewhat altered by the distortion of the matter. The assumption, therefore, that the permittivity and inductivity of the same part of the medium remain the same as it moves is not strictly correct. The dependence of the permittivity and inductivity on the strain can be allowed for in the reckoning of the stress function. This matter has been lately considered by Prof. Hertz. But the constants may also vary in other ways. It is unnecessary to consider here these small corrections. As usual in such cases, the magnitude of the expressions for the corrections is out of all proportion to their importance, in relation to the primary formula to which they are added.

Derivation of the Electromagnetic Stress from the Flux of Energy. Division into an Electric and a Magnetic Stress.

§ 72. From the form of the expression for **A** in equation (20), viz., the flux of energy due to the stress, we may derive the expression for the electromagnetic stress itself. If the stress were of the irrotational type considered in works on Elasticity, we could do this by means of the formula (10), § 68. But for a stress of the most general type the corresponding formula is

$$A = -Q_q q, \qquad \qquad (28)$$

where Q_q is the stress vector conjugate to P_q; these are identically the same when the stress is irrotational. This gives

$$Q_N = D.EN + B.HN - N(U + T), \quad . \quad . \quad (29)$$

from which we obtain P_N by merely exchanging E and D, and
H and B; thus

$$P_N = E.DN + H.BN - N(U + T). \quad . \quad . \quad (30)$$

This is the stress vector for any plane defined by N, a unit
vector normal to the plane.

But it is only in eolotropic bodies that we have to distinguish
between the directions of a force and of the corresponding
flux. Putting these on one side, and considering only ordinary isotropy, the interpretation is simple enough. It will be
observed that the electromagnetic stress (30) divides into an
electric stress

$$E.DN - N.\tfrac{1}{2}ED, \quad . \quad . \quad . \quad . \quad (31)$$

and a magnetic stress

$$H.BN - N.\tfrac{1}{2}HB. \quad . \quad . \quad . \quad . \quad (32)$$

To find their meaning, take N in turns parallel to and perpendicular to the force E (or to D, since its direction is the same).

In the first case, the stress (31) becomes

$$N(ED - \tfrac{1}{2}ED) = N.\tfrac{1}{2}ED = NU,$$

indicating a tension parallel to the electric force of amount U
per unit area.

In the second case, when N is perpendicular to E, we have
$DN = 0$, so that the stress is

$$- NU,$$

that is, a pressure of amount U. This applies to any direction
perpendicular to E, so that the electric stress consists of a
tension U parallel to the electric force, combined with an equal
lateral pressure.

Similarly the magnetic stress consists of a tension T parallel
to the magnetic force, combined with an equal lateral pressure.

It will also be found that the tensor of the electric stress
vector is always U, and that of the magnetic stress vector is
always T. The following construction (Fig. 5) is also useful:—
Let ABC be the plane on which the electric stress is required,
BN the unit normal, BE the electric force, BP the stress.

Then **N**, **E**, and **P** are in the same plane, and the angle between **N** and **E** equals that between **E** and **P**. Or, the same
operation which turns **N** to **E** also turns **E** to **P**, except as
regards the tensor of **P**.

To show the transition from a tension to an equal pressure,
imagine the plane ABC to be turned round, and with it the
normal **N**. Of course **E** remains fixed, being the electric force
at the point B. Start with coincidence of **N** and **E**. Then **P**
also coincides with them, and represents a normal pull on the
surface ABC. As **N** and **E** separate, so do **E** and **P** equally,
so that when **E** makes an angle of 45° with **N** the normal pull
is turned into a tangential pull, or a shearing stress, **P** being

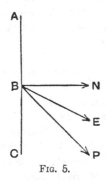

FIG. 5.

now at right angles to **N**. Further increase in the angle **E**
makes with **N** brings BP to the other side of BC; and when **E**
is at right angles to **N**, we have **P** and **N** in the same line, but
oppositely directed. That is, the tension has become converted
into an equal pressure.

Uncertainty regarding the General Application of the Electromagnetic Stress.

§ 73. We may now consider the practical meaning of the
stress whose relation to the electric and magnetic forces has,
under certain suppositions, been formularised. Go back to the
foundation of electromagnetic theory, viz., the mechanical
forces experienced by electrically charged bodies, by conductors
supporting currents, and by magnets, intrinsic or induced. It

is by observation of these forces in the first place, followed by
the induction of the laws they obey, and then by deductive
work, that the carving out of space into tubes of force follows;
and now, further, we see that the localisation of the stored
energies, according to the square of the electric and magnetic
force respectively, combined with the two circuital laws, leads
definitely to a stress existing in the electromagnetic field,
which is the natural concomitant of the stored energy, and
which is the immediate cause of the mechanical forces observed
in certain cases. But the theory of the stress goes so far beyond experimental knowledge in some respects, although agreeing with it in others, that we could only expect it to be true if
the theoretical foundations were also rigidly true in all respects.
Such is not the case, however. To begin with, the way of expressing the action of ordinary matter merely by altering the
values of the two ether constants, and by a fresh property,
that of conductivity, is extremely bald. It is, indeed, surprising what a variety of phenomena is explained by so crude a
method.

The objection is sometimes made against some modern theoretical developments that they are complicated. Considered as
an argument, the objection is valueless, and only worthy of
superficial minds. Whatever do they expect ? Do they not
know that experimental knowledge, even as at present existent,
shows that the theory of electromagnetism, when matter is
present, must, to be comprehensive, be something far more (instead of less) complicated than theory as now developed ? The
latter is, as it were, merely a rough sketch of a most elaborate subject, only small parts of which can be seen at one
time.

In the next place, even if we take the stated influence of
matter on the ether as sufficient for the purposes of a rough
sketch, the theory of the stress should, except in certain relatively simple cases, be received with much caution. Why this
should be so will be apparent on examining the manner in which
the stress has been obtained from the circuital laws. If we investigate the subject statically, and, starting from certain mechanical
forces regarded as known, endeavour to arrive at a stress which
shall explain those forces, we shall find that the problem is essentially an indeterminate one. All sorts of stress functions may

be made up which are precisely equivalent in their effects in the
gross, that is, as regards translating or rotating solid bodies
placed in an electric or magnetic field. To remove this indeterminateness a dynamical method must be adopted, wherein what
goes on in the unit volume whilst its electric and magnetic
states are changing, and the matter concerned is itself in motion,
are considered. If our system of connections is dynamically
complete and consistent, and is such that the flux of energy can
be traced, then a determinate stress comes out, as we have
found. The method is, at any rate, a correct one, however the
results may require to be modified by alterations in the data.
Besides that, the distribution of energy (electric and magnetic) in
bodies is in some cases open to question ; and a really speculative
datum is that concerning the motion of the ether as controlled
by the motion of matter. Now, this datum appears to be one
which is essential to the dynamical method ; the only alternative
is the statical and quite indeterminate method. Our attitude
towards the general application of the special form of the
stress theory obtained should, therefore, be one of scientific
scepticism. This should, however, be carefully distinguished
from an obstinate prejudice founded upon ignorance, such as
is displayed by some anti-Maxwellians, even towards parts of
Maxwell's theory which have received experimental demonstration.

The stress theory can, nevertheless, sometimes be received
with considerable confidence, if not absolute certainty. The
simplest case is that of ordinary electrostatics.

The Electrostatic Stress in Air.

§ 74. Let there be no magnetic force at all, and the electric
force be quite steady, and the medium be at rest, and there be
no impressed forces. These limitations bring the circuital laws
down to

$$0 = k\mathbf{E}, \quad \ldots \ldots \quad (1)$$

$$-\operatorname{curl}\mathbf{E} = 0 ; \quad \ldots \ldots \quad (2)$$

that is, there must be no conduction current anywhere, and
the voltage in any circuit must be zero. The first condition
(1) implies that there is no electric force in conductors. We
may, therefore, divide space into conducting and non-conducting

regions, and our electric field is entirely confined to the latter.
The second condition (2) implies tangential continuity of **E**
at the boundary of the non-conductor, so that as there is no **E**
in the other or conducting side, there must be no tangential
E on the non-conducting side. The lines of force, therefore,
terminate perpendicularly on the conducting matter. Whether
the conductors are also dielectrics or not is quite immaterial.
The displacement also terminates normally on the conducting
surface in the usual case of isotropy. Thus D, the tensor of **D**,
measures the surface density of electrification, when the positive
direction of **D** is from the conductor to the insulator. But in
general it is the normal component of **D**, that is **DN**, where
N is the unit normal vector, that measures the density of the
electrification. Besides this, there may be interior electrification
of the non-conductor, its volume density being measured by the
divergence of the displacement. The arrangement of the
electric force, so that the circuital voltage shall be zero
throughout the non-conductor, and give the proper internal
electrification, and the charges on the conductors, is uniquely
determinate.

We have thus the ordinary case of a number of charged conductors in air, with the difference that the air, or parts thereof,
may be replaced by matter of different permittivity. It is also
to be noted that one non-conducting region which is entirely
separated from another by conducting matter may be taken by
itself, and all the rest ignored.

Now, first without replacing the air by matter of different
permittivity, we see that there are two entirely different ways
of considering the mutual actions of the conductors. The old
way is analogous to Newton's way of expressing the fact of
gravitation. We may say that any element of electrification ρ
repels any other ρ' with a force

$$\rho\rho' / 4\pi c r^2,$$

if r be the distance between the two charges, and that the
resultant of all such forces makes up the real forcive.

In the other way, appropriate to the philosophy of Faraday,
as developed by Maxwell, these forces acting at a distance are
mathematical abstractions only, and have no real existence.
What is real is a stress in the electric field, of a peculiar nature,

being a tension of amount U parallel to **E**, combined with an equal lateral pressure, and it is the action of this stress that causes the electrified conductors to move, or strains them, according to the way they are supported, when by constraints they cannot appreciably move.

Since the electric force is normal to the conducting surfaces, the stress vector is entirely a normal pull of amount U per unit area, and the motions or tendencies to move of the conductors are perfectly accounted for by this pull. They do not move because of electrical forces acting at a distance across the air, but because they are subjected to moving force on the spot by the stress terminating upon them.

Thus a charged soap-bubble is subjected to an external radial tension, and therefore expands; and so, no doubt, does a charged metal sphere to some small extent. The parallel plates of a condenser are pulled together. When they are very large compared with their distance apart, the force on either is

$$\tfrac{1}{2} \text{ED} \times \text{area},$$

$$= \tfrac{1}{2} \text{ E} \times \text{charge}.$$

Here E is the transverse voltage divided by the distance between the plates, so that, if the plates be connected to a constant source of voltage, the attraction varies inversely as the square of the distance between them; whereas, if the plates be insulated and their charges constant, the attraction is the same at any distance sufficiently small compared with the size of the plates.

But by sufficiently separating the plates, or by using smaller plates, the displacement, which was formerly almost entirely between them, will spread out, and will terminate in appreciable amount upon the sides remote from one another. By the pull on the remote sides thus produced the attraction will be lessened, and the further the plates are separated the more displacement goes to their backs, and the less is the attraction. When the distance is great enough it tends to be simply the attraction between two point charges. Thus the attraction between two distant oppositely charged conducting spheres, which varies closely as the inverse square of the distance, depends entirely upon the slight departure from uniformity of distribution of the electrification over their surfaces, whereby the normal pull on either is made a little greater on the side next the other than on the remote side. Also, the inverse square law itself, which is exactly true for point charges, is merely the ultimate limit of this operation.

Some attacks have been made on the law of inverse squares, especially in its magnetic aspect. But these attacks appear to have been founded upon misapprehension. The law is true, and always will be.

The moving Force on Electrification, bodily and superficial. Harmonisation.

§ 75. In the above electrostatic application of the stress, it will be observed that the tension alone comes into play, at least explicitly, owing to the tubes of displacement terminating perpendicularly on the conductors. Thus each tube may be compared with a rope in a state of tension, pulling whatever its ends may be attached to. But the lateral pressure is needed to keep the medium itself in equilibrium, so that the only places where translational force arises from the stress is where there is electrification. The mechanical force is

$$\mathbf{F} = \mathbf{E} \operatorname{div} \mathbf{D} = \mathbf{E}\rho \quad \dots \quad (3)$$

per unit volume. This is the force on volume electrification, and is the result of the differential action of the stress round about the electrification, as in the case of the inverse square law between point charges, lately mentioned. The corresponding surface force is

$$\mathbf{F} = \tfrac{1}{2}\mathbf{E}.\mathbf{DN} = \mathbf{NU} \quad \dots \quad (4)$$

per unit area. Now, here $\mathbf{DN}$ is the surface equivalent of div $\mathbf{D}$, so there is at first sight a discrepancy between the expressions for the force per unit volume (3), and per unit area (4), on bodily and surface electrification. How the coefficient $\tfrac{1}{2}$ comes in may be seen by taking the limiting form of the previous expression. Let there be a thin skin of electrification, of amount σ per unit area ; $\mathbf{E}$ falling off from $\mathbf{E}$ outside to 0 inside the skin. Evidently the mean $\mathbf{E}$ is $\tfrac{1}{2}\mathbf{E}$, so that the total force on unit area of the skin obtained by summation of the forces on the volume electrification in the skin, is not $\mathbf{E}\sigma$, but $\tfrac{1}{2}\mathbf{E}\sigma$. This is merely a mathematical harmonisation.

From the point of view of the stress the difficulty does not present itself.

The harmonisation is simply evident when the layer is of uniform density, for the electric force will then fall off in intensity uniformly. It might, however, be suspected that, perhaps, the result would not come out quite the same if we assumed any other law of distribution, and kept to it in proceeding to the limit by making the skin infinitely thin. But a cursory examination will show that it is all right ; for if E is the electric force within the layer, the electrification density will be $c \, (dE/dx)$ and the translational force will be $cE \, (dE/dx)$ per unit volume, if x is measured perpendicularly to the skin's surfaces. Integrate through the skin, and the result is

$$F = \tfrac{1}{2} cE_1{}^2 - \tfrac{1}{2} cE_2{}^2 = U_1 - U_2,$$

where the suffixes refer to the value just outside the skin, on its two sides. In the present case the second term U_2 is zero (within the conductor), so that the result is the single normal pull of the tension on the non-conducting side.

Depth of Electrified Layer on a Conductor.

§ 76. In this connection the old question of the depth of the layer of electrification on a conducting surface crops up. Has it any depth at all, and, if so, how much ? The question is not so superficial as it looks, and the answer thereto lies in the application thereof. If a powerful mental microscope be applied to magnify the molecules and produce evident heterogeneity, the surface of a conductor would become indefinite ; and unless the molecules were found to be very closely packed, it is evident that the displacement in the ether outside the conductor would not terminate entirely upon those molecules which happen to be most superficially situated, but that a portion of the displacement would go deeper and in sensible amount reach molecules beneath the first set, and an insensible amount might penetrate through many layers. Thus in a molecular theory the depth of the layer of electrification has meaning, and could be roughly estimated.

But the case is entirely different in a theory which deliberately ignores molecules, and assumes continuity of structure.

A conductor is then a conductor all through, and not a heterogeneous mixture; and the surface of a conductor is an unbroken surface. The electrification on it is therefore surface electrification, and has no depth. For it to be otherwise is simply to make nonsense. It is desirable to be consistent in working out a theory, for the sake of distinctness of ideas; if, then, we wish to give depth to surface electrification, and still keep in harmony with Maxwell's theory, we must change our way of regarding a conductor, and bring in heterogeneity. Each view is true, in its own way; but as in the mathematical theory continuity of structure is tacitly assumed, we have a simultaneous evanescence of one dimension in the distribution of electrification.

The same question occurs in another form in the estimation of bodily electrification, when the meaning of volume density of electrification is considered. When air is electrified, it is probable that the electrification is carried upon the foreign particles suspended in the air, and it may be partly upon the air molecules themselves. In either case it is ultimately surface electrification, and quite discontinuous. But, merely for the sake of facility of working, it is desirable to ignore all the discontinuity, and assume a continuous and practically equivalent distribution of bodily electrification. Thus, as previously we saw surface density to be a kind of volume density, so now we see that volume density is a kind of surface density. When, therefore, we say that the translational force per unit volume is $E\rho$, where E is the electric force and ρ the volume density of electrification, we really mean that $E\rho$ is the average, obtained by summation, of the translational forces on the multitudinous electrified particles, every one of these forces being itself a differential effect, as before seen, viz., the resultant of the unequal pulls on different parts of a particle exerted by the electric stress.

As ether has some of the properties of matter, and as electrification is found in association with matter, it is possible, however improbable, that ether itself may become electrified. But of this nothing is known. Nor, more importantly, is it understood why the electric stress appears to act differently on positively and on negatively electrified matter. But, if we begin to talk about what is not understood, we enter illimitable regions. Men who are engaged in expounding practical problems sometimes make the boast that they take and discuss things as they are, not as they might be. There is a sound of specious plausibility here, which is grateful and comforting; but, as a matter of plain fact, questions of physics never are theorised upon as they are, but always as they might be. It is a necessity to limit the field of inquiry, for to take things as they are, or as they seem to be, would lead at once to a complete tangle. For the problem, as it presents itself in reality, there is always substituted a far simpler one, containing certain features of the real one emphasised, as it were, and others altogether omitted. The juveniles, who take things as they are, do not do it; they only think so. They may strain out a few gnats successfully, but swallow, quite unawares, all the camels in Arabia. But the principle and practice of limitation and substitution is the same all over; in politics, for instance, where a fictitious British Constitution does brave duty, as a scarecrow, and in other useful ways.

Electric Field disturbed by Foreign Body.   Effect of a Spherical Non-conductor.

§ 77. To further exemplify the significance of the electric stress, let us introduce a foreign body into a stationary electric field. The field will be disturbed by its introduction, and will settle down to a new state; the change depending upon the nature of the foreign body, whether conducting or non-conducting, in substance or superficially, and upon whether it has a charge itself, or contains any other source of displacement. If it be a good conductor, either charged or uncharged, the final state, reached very quickly, will be such that the displacement will terminate normally upon its surface, thus reproducing the previous case (§§ 74 to 76). But if it be a non-conductor, the result is somewhat different. If superficially conducting, we may indeed have an ultimate electrification of the surface, so as to come wholly or partly under the same case : but if there be no superficial or internal conduction, or only so little that a long time must elapse for it to become fully effective, what we do is simply to replace the dielectric air in a certain region by another dielectric of different permittivity, usually greater.

Then, supposing the external field to be due to charges upon insulated conductors or to internal electrification in the air (kept at rest for the purposes of the argument), no change will occur in their amounts; but there will be merely an alteration in the distribution of the electrification on the conductors, caused by the displacement becoming denser within the foreign body than before (or less dense, if its permittivity can be less than that of the air). The intrinsic electrification of the body itself, if any, must also be allowed for; but should it have none previously, it will remain unelectrified when introduced into the field, and the displacement will pass freely through it, and out again in a solenoidal manner. This is expressed by the surface condition

$$\mathbf{D}_1\mathbf{N}_1 + \mathbf{D}_2\mathbf{N}_2 = 0,$$

where $\mathbf{D}_1$ and $\mathbf{D}_2$ are the displacements in the air and foreign body respectively at their interface, and $\mathbf{N}_1$, $\mathbf{N}_2$ are the unit normals from the interface to the two media.

The only quite simple case (excepting that of infinite plane sheets) is that of a sphere of uniform permittivity brought into a previously uniform field. The ultimate displacement in the sphere is then parallel to the original displacement in the air. It may vary between zero and three times the original displacement, as the permittivity of the sphere varies from zero up to infinity. It is certainly a little surprising that the ultimate displacement with infinite permittivity should be only three times the original (and it is not much less when the permittivity is only 10 times that of the air); whilst, on the other hand, the zero displacement when the permittivity is zero (a quite ideal case) is obvious enough, because the displacement never enters the sphere at all, but goes round it. The original uniform field may be conveniently that between the parallel plates of a very large air condenser. There is, however, a double action taking place. When the transverse voltage of the condenser is maintained constant by connection with a suitable source, the insertion of the foreign body, which increases or reduces the permittance of the condenser, will increase or reduce its charge under the action of the constant source, besides concentrating the displacement within the body, or the reverse. With constant charges, however, when the plates are insulated, the insertion

of the foreign body will reduce the transverse voltage when its
permittivity exceeds that of the air ; and conversely.   Increase
of permittivity also increases the stored energy when the transverse voltage is constant, but reduces it when the charges are
constant.

Now, since the electric force does not terminate on the boundary of the foreign body, but extends all through it, so does the
electric stress. So far, however, as the resultant force and
torque on the body, when solid, is concerned, we may ignore
the internal stress altogether, and consider only the external, or
stress in the air. This is a particular case of a somewhat important and wide property in abstract dynamics, which we may
state separately thus.

Dynamical Principle.   Any Stress Self-equilibrating.

§ 78. The resultant force and torque due to any stress in any
region is zero.   Or, any stress in any region forms a self-equili-
brating system.

Imagine any distribution of stress to exist in a region A, and
to terminate abruptly on its boundary.   Or, equivalently,
imagine a piece of a stressed solid to be removed from its place
without altering the stress.   The stress-variation, when estimated in a certain way, constitutes mechanical force tending to
move the body. This will be, in the case of an ordinary irrotational stress, entirely translational force. But there are two kinds.
First, there is internal force, reckoned per unit volume, due to
the continuous variation of the stress in the body. Next,
there is superficial force, reckoned per unit area, due to the
abrupt cessation of the stress. This surface traction is represented simply by the stress vector itself, acting on the inner
side of the surface of the body.   Now, the resultant effect of
these two forcives, over the surface and throughout the volume
of the solid respectively, in tending to translate and rotate it,
is zero. Or, in other words, the force and torque equivalent to the
surface forcive are the negatives of those due to the internal
forcive. If it were otherwise, the differential action would
cause indefinite increase in the translational and rotational
energy to arise out of the internal mutual forces only of a
body.

If the stress be of the rotational type, there will be an internal torque (per unit volume) as well as a translational force. Still, however, the resultant force and torque due to the surface tractions will cancel those due to the internal forces and torques.

In case there be any difficulty in conceiving the traction exerted on the surface of a body by the stress within itself, we may replace the sudden cessation of the stress by a gradual cessation through a thin skin. The solid is then under the influence of continuous bodily force only (and torque also, if the stress be rotational) conveniently divisible into the internal force all over, and the force in the skin. Otherwise it is the same.

Now put the solid piece back into its place again. Since its own stress balances itself, we see that whatever the forcive on the piece may be it must be statically equivalent to the action upon its boundary of the external stress only, constituting an external surface traction. There need be no connection between the external and the internal stress. The latter may be anything we like, so far as the resultant force and torque on the piece are concerned. The difference will arise in the strains produced, or in the relative internal motions, when for one stress another is substituted.

Electric Application of the Principle. Resultant Action on Solid Body independent of the Internal Stress, which is statically indeterminate. Real Surface Traction is the Stress Difference.

§ 79. Returning to the electric field, we see that whatever be the nature of the reaction of the foreign body on the original state of the field, the resultant mechanical action on the body as a whole is fully represented by the stress in the air just outside it, in its actual state, as modified by the presence of the body, and that we need not concern ourselves with the internal state of stress. Nor are we limited, in this respect, by any assumed proportionality of electric force to displacement in the body, or assumed absence of absorption, or other irregularities and complications. That is, we need not have any theory to explicitly account for the change made in the electric field by the body. Nor do we gain any information regarding the internal stress from merely a knowledge of the external stress, although that involves the reaction of the body on the electric field. This is the meaning of the statical indeterminateness of the stress before referred to, and the principle applies generally.

The air stress vector **P** will usually have both a normal and a tangential component at the surface of a body, viz.:—

$$\text{U} \cos 2\theta \quad \text{normal,}$$
$$\text{U} \sin 2\theta \quad \text{tangential,}$$

if θ be the angle between the normal **N** to the surface (drawn from the body to the air) and the electric force **E**, and U the density of the electric energy, or the tensor of the stress vector.

In only one case, however, will this external surface traction represent the real forcive in detail (as well as in the lump), viz., when there is no stress at all on the other side of the boundary, that is, in the case of a conductor in static equilibrium. In general, the real surface force is represented by the vector $\mathbf{P}_1 - \mathbf{P}_2$, the stress difference at the boundary, $\mathbf{P}_1$ being the external and $\mathbf{P}_2$ the internal stress vector, which two stresses may, if we please, be imagined to be united continuously by a gradually changing intermediate stress existing in a thin skin, an idea appropriate to molecular theories. (It may be remembered that **P** when positive means a pull.) Each of these may be split into a normal and a tangential component. Now, the tangential components are

$$\text{U}_1 \sin 2\theta_1 \quad \text{and} \quad \text{U}_2 \sin 2\theta_2,$$

where the suffix $_1$ relates to the air, and $_2$ to the other medium.

Or $\qquad \text{E}_1\text{D}_1 \sin \theta_1 \cos \theta_1 \quad \text{and} \quad \text{E}_2\text{D}_2 \sin \theta_2 \cos \theta_2.$

But here we have normal continuity of the flux **D**, and tangential continuity of the force **E**; (otherwise the surface would be electrified and covered with a magnetic current sheet); that is

$$\left.\begin{array}{l} \text{D}_1 \cos \theta_1 = \text{D}_2 \cos \theta_2, \\ \text{E}_1 \sin \theta_1 = \text{E}_2 \sin \theta_2. \end{array}\right\} \quad \cdots \cdots \quad (5)$$

and

These relations make the tangential tractions equal and opposite, so that there is no resultant tangential traction, and the actual traction is entirely normal, being the difference of the normal components of P_1 and P_2, or,

$$U_1 \cos 2\theta_1 - U_2 \cos 2\theta_2, \quad . \quad . \quad . \quad (6)$$

which is the same as (being subject to (5),)

$$\tfrac{1}{2}D_1 \cos \theta_1 \, (E_1 \cos \theta_1 - E_2 \cos \theta_2)$$

$$- \tfrac{1}{2}E_1 \sin \theta_1 \, (D_1 \sin \theta_1 - D_2 \sin \theta_2) \quad . \quad . \quad . \quad (7)$$

The coefficient $\tfrac{1}{2}$ comes in for a similar reason to before, § 75.

This formula (6) or (7) being the real surface traction when E varies as D in the body, and the stress is of the same type as in air, furnishes a second way of calculating the resultant force and torque on the body, when its permittivity is uniform ; and it is noteworthy that the surface traction is, as in the case of an electrified conductor in equilibrium, entirely normal, although it may now be either a pull or a push.

Translational Force due to Variation of Permittivity. Harmonisation with Surface Traction.

§ 80. Noting that the mechanical force on the elastically electrizable body is situated where the change of permittivity occurs, and is in the direction of this change, it may be inferred that when the permittivity varies continuously there is a bodily translational force due to the stress variation which is in the direction of the most rapid change of permittivity. This is, in fact, what the stress vector indicates when c varies continuously, viz., the force represented by

$$- \tfrac{1}{2}E^2 \nabla c = - \nabla_c U, \quad . \quad . \quad . \quad . \quad (8)$$

where ∇c means the vector rate of fastest increase of c round-about the point considered. Since $U = \tfrac{1}{2}cE^2$, the second form in (8) will be understood, meaning the vector slope of U as dependent upon the variation of c only.

This bodily force, and the previous surface force may be harmonised by letting c vary not abruptly, but continuously from the value c_1 on one side to c_2 on the other side of the surface of discontinuity, through a thin skin, and summing up the translational forces in the skin by the formula (8). Thus, if x be measured normal to the skin, the translational force per unit volume is $-\frac{1}{2}E^2 (dc/dx)$ in the direction of the normal; or

$$-\frac{1}{2}E_n^2 \frac{dc}{dx} - \frac{1}{2}E_t^2 \frac{dc}{dx}$$

$$= -\frac{1}{2}D_n^2 \frac{1}{c^2}\frac{dc}{dx} - \frac{1}{2}E_t^2 \frac{dc}{dx}, \quad \cdot \quad \cdot \quad (9)$$

where E_n and D_n are the normal components of E and D, and E_t the tangential component of E. Now D_n and E_t are constant ultimately, for the reason before given, so we can integrate (9) with respect to x immediately, giving

$$\left[\frac{1}{2}\frac{D_n^2}{c} - \frac{1}{2}cE_t^2 \right] = \left[\frac{1}{2}cE_n^2 - \frac{1}{2}cE_t^2 \right] \quad \cdot \quad \cdot \quad (10)$$

between the limits; or, since E_n and E_t are proportional to the cosine and the sine of θ respectively.

$$[\frac{1}{2}cE^2 \cos 2\theta] \quad \cdot \quad \cdot \quad \cdot \quad \cdot \quad (11)$$

between the limits, which is the same as (6), which was to be verified.

Movement of Insulators in Electric Field. Effect on the Stored Energy.

§ 81. Since a sphere of uniform permittivity placed in a uniform field causes the external lines of electric force to be symmetrically distorted fore and aft, it has no tendency to move, but is merely strained. But if the body be not in an initially uniform field, or be not spherical, complex calculations are usually needed to determine the effect. If, however, it be only a small piece, the tendency is for it to move in the direction in which the energy, or the stress, in the field increases most rapidly, independent of the direction of the electric force, when its permittivity exceeds that of the gaseous medium. The total electric energy will be diminished by permitting the motion when the charges are constant ; but increased should the field be kept up by constant sources of voltage.

These properties are rendered particularly evident by taking the extreme cases of infinite and zero permittivity of a small body placed in a widely varying field, that surrounding a charged sphere, for example, the electric force varying in intensity as

H 2

the inverse square of the distance from its centre. Here the
non-permittive body is repelled, for the lines of force go round
it, and the lateral pressure comes fully into play, and is greater
on the side next the charged sphere. On the other hand, with
the infinitely permittive body, concentrating the displacement,
it is the tension that comes fully into play, and this being
greater on the side next the charged sphere, the result is an
attraction. The permittance of the sphere, also, is increased in
the latter case, and decreased in the former—that is, when the
natural motion of the body to or from the sphere is allowed;
so, since the total electric energy is $\frac{1}{2}SV^2$ where S is the permittance and V the voltage, or, equivalently, $\frac{1}{2}VQ$, if Q is the
charge, we have always a diminution of energy when the
natural motion is allowed, whether resulting from attraction or
repulsion, if the charge is constant; but an increase of energy
if the voltage is constant.

In intermediate cases the tension is dominant when the c of
the body exceeds, and the pressure is dominant when it is less
than that of the air, there being perfect equilibrium of a piece
of any shape in any field if there be equality of permittivity,
and therefore no disturbance of the field. Here we see the part
played by the lateral pressure in the case of conductors in
equilibrium. It has no influence on them immediately, and
might be thought wholly unnecessary, but it is equally
important with the tension in the non-conducting dielectric
itself.

Magnetic Stress. Force due to Abrupt or Gradual Change of Inductivity. Movement of Elastically Magnetised Bodies.

§ 82. Passing now to the corresponding magnetic side of the
stress question, we may observe that the analogy is an imperfect
one. Thus, proceeding as at the beginning of § 74, to have a
stationary magnetic field without impressed forces, we shall
find that there must first be no magnetic conduction current;
and next, that the gaussage in any circuit must be zero. The
magnetic force we then conclude to be confined entirely to the
magnetically non-conducting regions, and to terminate perpendicularly upon their boundaries. Thus we come to the
conception of a number of detached magnetic conductors im-

mersed in a magnetically non-conductive medium, these conductors having magnetic charges on them measured by the
amount of induction leaving or terminating upon them, with
possible associated volume magnetification in the non-conducting medium.

The magnetic stress will exert a normal traction

$$\mathbf{NT} = \tfrac{1}{2}\mu \mathbf{H}^2 \mathbf{N} \quad \cdots \cdots \quad (12)$$

per unit area on the conductors, and a force

$$\mathbf{H}\sigma = \mathbf{H} \operatorname{div.} \mathbf{B} \quad \cdots \cdots \quad (13)$$

per unit volume on tridimensional magnetification, of density σ,
measured by the divergence of the induction. These are analogous to the forces on surface and volume electrification.

Also, when the inductivity μ varies, we shall have a normal
surface traction of amount,

$$T_1 \cos 2\theta_1 - T_2 \cos 2\theta_2 \quad \cdots \cdots \quad (14)$$

per unit area, analogous to (6), when μ changes value abruptly
at the interface of two media ; and a force

$$-\tfrac{1}{2}\mathbf{H}^2 \nabla \mu \quad \cdots \cdots \cdots \quad (15)$$

per unit volume, when μ varies continuously.

The forces (12), (13), however, are absent, because of the absence of magnetic conductivity, and, in connection therewith,
the absence of "magnetification." But (12) may be sometimes
used, nevertheless, when it is the stress across any surface that
is in question, and we create surface magnetification by regarding one side only.

We are, therefore, left with the forces (14), (15), depending
upon variation of inductivity, abrupt or gradual. These explain the mechanical action upon elastically magnetised
media, e.g., the motions of bodies to or from a magnet pole,
according as they are paramagnetic or diamagnetic, which, it
should be remembered, depends fundamentally upon the variation in the intensity of magnetic force near the pole ; and the
axial equilibrium of a paramagnetic bar, and equatorial equilibrium of a diamagnetic bar. Faraday's remarkable sagacity
led him to the essence of the explanation of these and other
allied phenomena, as was later mathematically demonstrated by

Sir W. Thomson. Questions relating to diamagnetic polarity
are, in comparison, mere trifling.

**Force on Electric Current Conductors.   The Lateral Pressure
becomes prominent, but no Stress Discontinuity in general.**

§ 82a. But the magnetic stress has other work to do than to
move elastically magnetised matter under the circumstances
stated. Wholly independent of magnetisation, it produces the
very important moving force on conductors supporting electric
current, first mathematically investigated by Ampère.

The lateral pressure of the stress here comes prominently
into view, when we ignore the stress in the interior of the conductors, so that the stress vector in the air at the boundary of
a conductor represents the moving force on it per unit area.
Thus, two parallel conducting wires supporting similar currents
attract one another, because their magnetic forces are additive
on the sides remote from one another, rendering the lateral
pressure on them greater there than on the sides in proximity.
But when the currents are dissimilar the magnetic force is
greater on the sides in proximity, and, therefore, the lateral
pressure of the stress is greater there, producing repulsion.

Proceeding further, and considering the stress within the
conductor also, according to the same law, we find this peculiarity. In the case of unmagnetisable conductors (typified
practically by copper), there is no superficial discontinuity in
the stress, and therefore no surface forcive of the kind stated.
This may be easily seen from § 79, translating the results from
the electric to the magnetic stress.   There is no tangential
discontinuity in the stress because the normal induction and
tangential magnetic force are continuous ; and there is no discontinuity in the normal component of the stress because (since
there is no difference of inductivity) the normal magnetic force
and the tangential induction are also continuous.

In (5), (6), (7) turn E to H, and D to B, and U to T, and
note that (5), as transformed, are true when the force and flux
are exchanged, so that the transformed expressions (6) or (7)
vanish.

In the case of a real conductor, therefore, with finite volume
density of electric current, the moving force is distributed

throughout its substance ; and the variation of the stress indicates that the translational force per unit volume is expressed by

$$\mathbf{F} = \mathbf{VCB}, \quad \ldots \ldots \quad (16)$$

where **C** is the current density. This is what Maxwell termed "the electromagnetic force," and it is what is so extensively made use of by engineers in their dynamos, motors, and things of that sort. It is probable, I think, that in grinding away at the ether they also stir it about a good deal, though not fast enough to produce sensible disturbances due to etherial displacement.

Force on Intrinsically Magnetised Matter. Difficulty. Maxwell's Solution probably wrong. Special Estimation of Energy of a Magnet and the Moving Force it leads to.

§ 83. There is next the force on intrinsically magnetised matter to be considered in connection with the magnetic stress. This is, perhaps, the most difficult part of magnetic science. Although, so far as the resultant effect on a magnet is concerned, we need not trouble about its internal state, but, as before, merely regard it as being pushed or pulled by the stress in the surrounding air, such stress being calculable from the distribution of magnetic force immediately outside it, as modified by the magnet itself, yet it is impossible that the real forcive can be represented merely by the surface traction $\mathbf{P_N}$.

Now, it is possible to find a distribution of magnetification over the surface, which shall be externally equivalent to the interior magnetisation, or to whatever other source of induction there may be. Then we may substitute for the surface traction $\mathbf{P_N}$, another traction, namely, upon the surface magnetification.

Or, we may find a distribution of fictitious electric current upon the boundary of the magnet, which shall be externally equivalent to the interior sources, and then represent the forcive by means of fictitious electromagnetic force on this current, § 82a.

Or we may combine these methods in various ways. Evidently, however, such methods are purely artificial, and that to obtain the real forcive we must go inside the magnet. This can only be done hypothetically, and with precarious validity.

One way of exhibiting the mechanical action on a magnet, or
on magnetised matter generally, is that given by Maxwell in his
chapter on the stresses (Vol. II.). This I believe to be quite
erroneous for many reasons, the principal being that it does
not harmonise with his scheme generally, and that it lumps together intrinsic and induced magnetisation, which have essential differences and are physically distinct. There are many
other ways of exhibiting the resultant force and torque as made
up of elementary forces, acting upon magnetisation, or on free
magnetism, or on the variation of magnetism estimated in
different ways. It is unnecessary to enter into detail regarding
them. Nobody would read it. It will be sufficient to point out
the particular way which harmonises with Maxwell's scheme in
general, in the form in which I display it, with a special estimation of magnetic energy. Proportionality of force and flux
is assumed. The want of this proportionality is quite a separate
question. Given a definite relation between force and flux, the
accompanying change in the stress vector, in accordance with
the continuity of energy, can be estimated. This I have recently shown how to do in another place, which shall not be
more explicitly referred to.

Intrinsic magnetisation possesses the peculiarity that it is, in
a manner, outside the dynamical system formulated in the
electromagnetic equations, inasmuch as it needs to be exhibited in them through the medium of an impressed force,
although this is disguised in the ordinary mode of representation. Calling this intrinsic magnetic force h_0 (any distribution),
as before, the induction due to it in a medium of any inductivity
(varying continuously or abruptly, if required) is found in the
same way as the displacement due to intrinsic electric force in
a non-conducting medium of similarly distributed permittivity ;
or as the conduction current due to the same in a medium of
similar conductivity ; or, to make a fourfold analogy, as the
magnetic conduction current due to h_0 in a (fictitious) medium
of similar magnetic conductivity.

I may here point out that a clear recognition of the correct
analogies between the electric and magnetic sides of electromagnetism is essential to permanently useful work. Many
have been misled in this respect, especially in comparing
Maxwell's displacement with magnetic polarisation. The true

analogue of **D** is **B**. Investigations based upon the false foundation mentioned can lead to nothing but confusion. There are enough sources of error without bringing in gratuitous ones.

Now, presuming we have induction set up by $\mathbf{h}_0$, how is the energy to be reckoned? I reckon its amount per unit volume to be $\frac{1}{2}\mathbf{HB}$ generally, whether outside or within the magnet; or, in the usual isotropic case, $\frac{1}{2}\mu H^2$ or $\frac{1}{2}\mu^{-1}B^2$. This makes the total work done by $\mathbf{h}_0$ to be $\mathbf{h}_0\mathbf{B}$ per unit volume, if $\mathbf{h}_0$ is suddenly established; of which one half is wasted, and the rest remains as stored magnetic energy. That is,

$$\Sigma T = \Sigma\tfrac{1}{2}\mathbf{h}_0\mathbf{B} = \Sigma\tfrac{1}{2}\mathbf{HB}, \quad \ldots \quad (17)$$

if the Σ indicates space-summation. If $\mathbf{h}_0$ be suddenly destroyed, the energy ΣT is set free and is dissipated, mainly by the heat of currents induced within the magnet itself and surrounding conductors. This is the meaning of $\frac{1}{2}\mathbf{HB}$ being the stored energy per unit volume. But it may not be immediately available. To take an extreme case, if we have a complete magnetic circuit, so magnetised intrinsically that there is no external field, the energy, as above reckoned, is the greatest possible, since $\mathbf{H} = \mathbf{h}_0$. But it is now not at all available, unless $\mathbf{h}_0$ be destroyed. On the other hand, Maxwell would appear to have considered the energy of a magnet to be $\Sigma\frac{1}{2}\mu(\mathbf{H} - \mathbf{h}_0)^2$, which is zero in the just mentioned case. This reckoning does not harmonise with the continuity of energy, although it has significance, considered as energy more or less immediately available without destruction of $\mathbf{h}_0$. The connection of the two reckonings is shown by

$$\Sigma\tfrac{1}{2}\mathbf{h}_0\mathbf{B} = \Sigma\tfrac{1}{2}\mathbf{h}_0\mu\mathbf{h}_0 - \Sigma\tfrac{1}{2}(\mathbf{H} - \mathbf{h}_0)\mu(\mathbf{H} - \mathbf{h}_0); \quad . \quad (18);$$

or, when **H** and **B** are parallel,

$$\Sigma\tfrac{1}{2}\mathbf{h}_0\mathbf{B} = \Sigma\tfrac{1}{2}\mu\mathbf{h}_0^2 - \Sigma\tfrac{1}{2}\mu(\mathbf{H} - \mathbf{h}_0)^2 \quad . \quad . \quad (19).$$

Now, according to the reckoning (17) of the stored energy, and the consequent flux of energy, the stress vector derived therefrom indicates that the moving force per unit volume is

$$\mathbf{F} = V\mathbf{j}_0\mathbf{B} . \quad . \quad . \quad . \quad . \quad (20),$$

where $\qquad\qquad \mathbf{j}_0 = \text{curl } \mathbf{h}_0 . \quad . \quad . \quad . \quad . \quad (21).$

The interpretation is that the vector j_0 represents the distribution of (fictitious) electric current, which would, under the same circumstances as regards inductivity, set up, or be associated with, the same distribution of induction B as the intrinsic magnet is. It may be remarked that the induction due to an intrinsic magnetic force does not depend upon *its* distribution primarily, but solely upon that of its curl.

Substituting this current system for the intrinsic magnetisation, equation (20) indicates that the moving force is "the electromagnetic force" corresponding thereto, according to (16), § 82. The accompanying surface distribution of current, representing the abrupt cessation of h_0, must not be forgotten; or we may let it cease gradually, and have a current layer in the skin of the magnet. It may be far more important than the internal current, which may, indeed, be non-existent. For instance, in the case of a uniformly longitudinally magnetised bar, the equivalent current forms a cylindrical sheet round the magnet. In general, the surface representative of j_0 is

$$VNh_0, \qquad \ldots \ldots \ldots \quad (22)$$

where N is a unit normal drawn from the boundary into the magnet. It should also not be forgotten that if the inductivity changes, there is also the moving force (15) or (14) to be reckoned, besides that dependent upon the intrinsic magnetisation.

Force on Intrinsically Electrized Matter.

§ 84. The electric analogue of intrinsic magnetisation is intrinsic electrisation, represented in a solid dielectric in which "absorption" has occurred, and perhaps in pyroelectric crystals. It seems very probable that there is a true electrisation, quite apart from complications due to conduction, surface actions, and electrolysis. When formulated in a similar manner to intrinsic magnetisation, by means of intrinsic electric force e_0 producing displacement according to the permittivity of the medium (which displacement, however, is now also affected by the presence of conducting matter), and with a similar reckoning of the stored energy, viz., $\frac{1}{2}cE^2$ per unit volume, where E

includes e_0, we may expect to find, and do find, a moving force analogous to (20), viz.,

$$\mathbf{F} = \mathbf{VD}g_0, \quad \ldots \ldots \ldots \quad (23)$$

where $$g_0 = -\text{curl } e_0 \ldots \ldots \quad (24).$$

That is, magnetoelectric force on the fictitious magnetic current g_0 which is equivalent to the intrinsic force e_0. And, similarly to before, we may remark that the flux due to e_0 depends solely upon its curl.

Summary of the Forces. Extension to include varying States in a Moving Medium.

§ 85. Now bring together the different moving forces we have gone over. On the electric side we have

$$\mathbf{F}_e = \mathbf{E}\rho - \tfrac{1}{2}\mathbf{E}^2\nabla c \left[+ \mathbf{VDK} \right] + \mathbf{VD}g_0, \quad \ldots \quad (25),$$

and on the magnetic side

$$\mathbf{F}_m = [\mathbf{H}\sigma] - \tfrac{1}{2}\mathbf{H}^2\nabla\mu + \mathbf{VCB} + \mathbf{V}j_0\mathbf{B}, \quad \ldots \quad (26),$$

where the third term on the right of (25), and the first on the right of (26), in square brackets, are zero; $\mathbf{H}\sigma$ being force on magnetification and $\mathbf{VDK}$ the magnetic analogue of $\mathbf{VCB}$. The other component forces can all separately exist, and in stationary states.

Passing to unrestricted variable states, with motion of the flux-supporting media also, the electric and magnetic stress vectors indicate that the moving forces arising therefrom are obtainable from those exhibited in (25), (26), by simply changing the meaning of the electric current and the magnetic current symbols in the third terms on the right. Above, they stand for conduction current only, and one of them is fictitious. They must be altered to $\mathbf{G}_0$ and $\mathbf{J}_0$, the "true" currents, as explained in § 66. Thus

$$\mathbf{F}_e = \mathbf{E}\rho - \tfrac{1}{2}\mathbf{E}^2\nabla c + \mathbf{VD}(\mathbf{G}_0 + g_0), \quad \ldots \quad (27)$$

$$\mathbf{F}_m = [\mathbf{H}\sigma] - \tfrac{1}{2}\mathbf{H}^2\nabla\mu + \mathbf{V}(\mathbf{J}_0 + j_0)\mathbf{B}, \quad \ldots \quad (28)$$

express the complete translational forces due to the electric and magnetic stresses (31), (32), § 72.

The division of the resultant translational force into a
number of distinct forces is sometimes useless and artificial.
But as many of them can be isolated, and studied separately,
it is not desirable to overlook the division.

The equilibrium of a dielectric medium free from electrification and intrinsic forces, which obtains when the electric and
magnetic forces are steady, is upset when they vary. There is
then the electromagnetic force in virtue of the displacement
(electric) current and the magnetoelectric force in virtue of the
magnetic current ; that is,

$$\mathbf{F} = \mathbf{VDB} + \mathbf{VDB} \quad \cdots \quad (29)$$

$$= \frac{d}{dt}\mathbf{VDB} = \frac{1}{v^2}\frac{d\mathbf{W}}{dt}, \quad \cdots \quad (30),$$

where **W** is the flux of energy. Here we neglect possible small
terms depending on the motion of the medium.

That there should be, in a material dielectric, moving force
brought into play under the action of varying displacement and
induction does not present any improbability. But it is less
easy to grasp the idea when it is the ether itself that is the
dielectric concerned. Perhaps this is, for some people, because
of old associations—the elastic solid theory of light, for instance, wherein displacement of the ether represents the
disturbance.

But if we take an all round view of the electromagnetic connections and their consequences, the idea of moving force on
the ether when its electromagnetic state is changing will be
found to be quite natural, if not imperatively necessary. We
do know something about how disturbances are propagated
through the ether, and we can, on the same principles, allow
for bodily motion of the ether itself. Further, reactions on the
ether, tending to move it, are indicated. But here we are
stopped. We have no knowledge of the density of the ether,
nor of its mechanical properties in bulk, so, from default of real
data, are unable to say, except upon speculative data, what
motions actually result, and whether they make any sensible
difference in phenomena calculated on the supposition that the
ether is fixed.

Union of Electric and Magnetic to produce Electromagnetic Stress. Principal Axes.

§ 86. From the last formula we see that to have moving force in a non-conductor (free from electrification, &c.) requires not merely the coexistence of electric and magnetic force, but also that one or other of them, or both, should be varying with the time. That is, when the energy flux is steady, there is no moving force, but when it varies, its vector time-variation, divided by v^2, expresses the moving force.

The direction of W is a natural one to choose as one of the axes of reference of the stress, being perpendicular to both E and H, which indicate the axes of symmetry of the electric and magnetic stresses. The two lateral pressures combine together to produce a stress on the W plane

$$P_{W1} = -W_1(U + T), \quad \ldots \quad (31)$$

where W_1 is a unit vector parallel to W. (Take $N = W_1$ in the general formula (30) § 72 for P_N).

If, further, E and H are perpendicular to one another, the stresses on the planes perpendicular to them are

$$P_{E1} = E_1(U - T), \quad \ldots \quad (32)$$

$$P_{H1} = H_1(T - U), \quad \ldots \quad (33)$$

which are also entirely normal. Here E_1 and H_1 are unit vectors. Thus, W_1 is always a principal stress axis, while E_1 and H_1 become the other pair of principal axes when they are perpendicular, as in various cases of electromagnetic waves.

Thus when a long straight wire supports a steady current, or else is transmitting waves, the principal axes of the stress at a point near the wire are respectively parallel to it and perpendicular to it, radially and circularly. The first one has a pressure $(U + T)$ acting along it, the second (parallel to E) a pressure $(T - U)$, and the third (parallel to H) a pressure $(U - T)$. (This legitimate use of pressure must not be confounded with the utterly vicious misuse of pressure to indicate E.M.F. or voltage, by men who are old enough to know oetter, and do.) In general, U and T are unequal. But in the case of a solitary wave or train of waves with negligible distortion U and T are equal, and there is but one principal stress, viz., that with axis parallel to **W**, a pressure 2U or 2T. It is this pressure (or its mean value) that is referred to as the pressure exerted by solar radiation, and its space-variation constitutes the moving force before mentioned. This matter is still in a somewhat speculative stage.

It is natural to ask what part do the stresses play in the propagation of disturbances ? The stresses and accompanying strains in an elastic body are materially concerned in the transmission of motion through them, and it might be thought that it would be the same here. But it does not *appear* to be so from the electromagnetic equations and their dynamical consequences—that is to say, we represent the propagation of disturbances by particular relations between the space- and the time-variations of **E** and **H**; and the electromagnetic stress and possible bodily motions *seem* to be accompaniments rather than the main theme.

Dependence of the Fluxes due to an Impressed Forcive upon its Curl only. General Demonstration of this Property.

§ 87. In § 83 it was remarked that the flux induction due to an intrinsic magnetic forcive depends not upon itself directly, but upon its curl, and in § 84 a similar property was pointed out connecting the flux displacement and intrinsic electric force. That is to say, the fluxes depend upon the vectors j_0 and g_0, not upon e_0 and h_0. This remarkable and, at first sight, strange property, which is general, admits of being demonstrated in a manner which shall make its truth evident in a wider sense, and lead to a connected property of considerable importance in the theory of electromagnetic waves.

Let there be any impressed forcive e_0 in a stationary medium. Its activity is $e_0 J$ per unit volume, where **J** is the electric current, and the total activity is $\Sigma e_0 J$, where the Σ indicates summation through all space, or at any rate so far as to include every place where e_0 exists. Its equivalent is Q_0, the total rate of waste of energy, and the rate of increase of the total stored energies, say U_0 and T_0. Thus,

$$\Sigma e_0 J = Q_0 + \dot{U}_0 + \dot{T}_0 \quad . \quad . \quad . \quad . \quad (34).$$

Now, the value of the summation is zero if e_0 has no curl, or is irrotational—that is, an irrotational forcive does no work upon a circuital flux. This proposition may be rendered evident by employing a particular method of effecting the space summation, viz., instead of the Cartesian method of cubic subdivision of space, or any method employing co-ordinates, divide space into the elementary circuital tubes belonging to the flux. (Here it is J.) Fixing the attention upon any one of these circuits, in which the flux is a constant quantity, we see at once that the part of the summation belonging to it is the product of the impressed voltage in the circuit and the flux therein. But there is no impressed voltage, because e_0 has no curl ; hence the circuit contributes nothing to the summation. Further, since this is true for every one of the elementary circuits, and inclusion of them all includes all space, we see that the summation $\Sigma e_0 J$ necessarily vanishes under the circumstance stated of e_0 being irrotational.

Now return to equation (34), and suppose that the initial state of things is absence of **E** and **H** everywhere, so that Q_0, U_0 and T_0 are all zero. Next start any irrotational e_0, and see what will happen. The left side of (34) being zero, the right side must also be and continue zero. But Q_0, U_0 and T_0 when not zero are essentially positive. Now if $(U_0 + T_0)$ becomes positive, Q_0 should become negative, in order to keep the right number of (34) at zero. This negativity of Q_0 being impossible, these quantities Q_0, U_0 and T_0 must all remain zero. Consequently **E** and **H** must remain zero. That is, an irrotational forcive can produce no fluxes at all, if the flux corresponding to the force is restricted to be circuital.

It should be observed that this proposition applies not merely to the steady distribution appropriate to the given forcive, but to all intermediate stages, involving both electric and magnetic force, and flux of energy. Nothing happens, in fact, when any distribution of impressed force is made to vary in time and in space, provided it be restricted to be of the irrotational type, so that the voltage (or the gaussage) in every circuit is nil.

Notice further the dependence of the property upon circuitality of the factor with which the impressed force is associated (thus **J** with e_0, and similarly **G** with h_0), and the positivity of energy ; and, more strikingly, the independence of such details as are not concerned in equation (34), such as the distribution of conductivity, permittivity, &c., or of the forces being linear functions of the fluxes.

Identity of the Disturbances due to Impressed Forcives having the same Curl. Example:—A Single Circuital Source of Disturbance.

§ 88. Returning to (34) again, we see that any two impressed forcives produce the same results in every particular, as regards the varying states of **E** and **H** gone through, if their curl is the same. For the difference of the two forcives is an irrotational forcive, and is inoperative. Here it is desirable to take

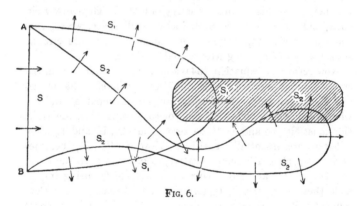

Fig. 6.

an explicit example for illustration of the meaning and effect. Describe a linear circuit in space, and a surface bounded by the circuit (Fig 6). Over this surface, which call S, let an impressed force V act normally, V being the same all over the surface. This system of force is irrotational everywhere except at the bounding circuit, where there is a circuital distribution of g_0 the curl of e_0, of strength V.

Now, our present proposition asserts that the disturbances due to V over the surface S are in every respect the same as those due to V (the same normal impressed force), spread over any other surface bounded by the same circuit. The comprehensiveness of this property may be illustrated by supposing that one surface is wholly in a non-conductor, whilst the other passes through a conductor ; or that one surface is wholly within a conductor, whilst the other passes out of it and through another conductor insulated from the first. Thus, in the diagram, let S be the first surface (in section), and S_1 the second, passing through a conductor represented by the shaded region. The points A, B are where the common boundary of the two surfaces cuts the paper, whilst the arrows serve to show the direction of action of the impressed force. S_2 is another surface of impressed force, also reaching into the conductor. Now all these forcives (each by itself) will produce the same final state of displacement in the dielectric and electrification of the conductor, and will do so in the same manner ; that is, the electromagnetic disturbances generated will be the same when expressed in terms of E and H. The distribution of energy, for example, and the stresses, will be the same.

But there must be some difference made by thus shifting the source of energy. Obviously the nature of the flux of energy is changed. This being

$$W = V(E - e_0)(H - h_0),$$

where we deduct the intrinsic forces to obtain the forces effective in transferring energy, we see that every change made in the distribution of the intrinsic sources affects the flux of energy, in spite of the independence of E and H of their distributions (subject to the limitation mentioned). In our example, however, the only change is in the sheet of impressed force itself.

Production of Steady State due to Impressed Forcive by crossing of Electromagnetic Waves. Example of a Circular Source. Distinction between Source of Energy and of Disturbance.

§ 89. It may be readily suspected from the preceding, that, as far as the production of electromagnetic disturbances goes, we may ignore e_0 and h_0 altogether, and regard the circuital vectors g_0 and j_0 as the real sources. This is, in fact, the case when ultimately analysed. In the example just taken the circuit ABA is the source of the disturbances. That is, they emanate from this line. If disturbances were propagated

I

infinitely rapidly there might be some difficulty in recognising
the property, because the steady state appropriate to the
instantaneous state of the impressed force would exist
(if the conductor were away); but, as we shall see later,
the speed of propagation is always finite, depending upon the
values of c and μ in the medium; and conduction does not alter
this property, although by its attenuating and distorting effects
it may profoundly alter the nature of the resultant phenomena.
With, then, a finite speed of propagation, we have merely to
cause the impressed force to vary or fluctuate sufficiently
rapidly to obtain distinct evidence of the emanation of waves
from the real sources of disturbance. Thus, let the source be

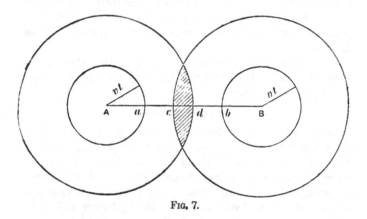

Fig. 7.

circular in a plane perpendicular to the paper (Fig. 7), and A, B
the points where it cuts the paper. When the source is suddenly
started, the circle ABA is the first line of magnetic force. At
any time t later, such that vt is less than the distance $\frac{1}{2}$AB,
the electromagnetic disturbance will be confined wholly within
a ring-shaped region having the circular source for core and of
radius vt round A or round B. But when the distance vt increases to $\frac{1}{2}$AB overlapping commences, and a little later there
is (as in the shaded part of the figure) a region occupied by two
coincident waves crossing one another. Now, the union of
these waves produces the steady state of displacement without
magnetic force—that is, within the shaded region the magnetic

force vanishes, and the displacement is that which belongs to the final steady state. As time goes on, of course the shaded region enlarges itself indefinitely, although outside it is still a region occupied by electromagnetic disturbances going out to infinity. This supposes that there is no conductor in the way. Should there be one, then, as soon as it is struck by the initial electromagnetic wave, it becomes and continues to be a secondary source of disturbance, and the final steady state, different from the former, now arises from the superimposition of the primary waves and the secondary.

It is, of course, impossible to go into detail at present, the object being merely to point out the difference between sources of energy and sources of disturbance. The source of energy only works when there is electric current at the spot, and this comes to it from the vortex lines (the circuital sources of disturbance). Thus, in the last figure, at time vt, when the ring of disturbance is of thickness $2vt$, if the impressed force be in a circular disc whose trace is the straight line AB, the force is working in Aa and in Bb, but inoperative elsewhere. Again, after overlapping has begun it is still working in Aac and Bbd, but is inoperative in cd, having done its work. Similarly when the sheet of impressed force has any other shape. The impressed force only works when it is allowed to work by the electromagnetic wave reaching it.

To emphasize the matter, take another case. Let the impressed force in any telegraph circuit be confined to a single plane section across the conductor, so that the vortex line is a line on its surface, going round it. If this be at Valencia, we may shift the "seat of the E.M.F." to Newfoundland, provided we preserve continuity of connexion with the vortex line, as before explained ; for instance, by extending the sheet over the whole surface of the conductor between the two places. The sheet at Newfoundland and the surface sheet will together produce the same effects as the original sheet at Valencia. Or, we may have the sheet entirely outside the circuit, provided only that it is bounded by the original vortex line, in touch with the conductor.

What the practical interpretation of this extraordinary property is in connection with the "seat of E.M.F." of galvanic batteries and in electrolysis generally still remains obscure.

All impressed voltages and gaussages are more or less difficult to understand. But there need be no doubt as to the general truth of the property if the circuitality of the current can be trusted.

The Eruption of "4π"s.

§ 90. It may have been observed that the equations employed by me in the preceding differ from those in use in all mathematical treatises on the subject in a certain respect (amongst others), inasmuch as the constant 4π, which is usually so obtrusively prominent, has been conspicuous by its absence. This constant 4π was formerly supposed to be an essential part of all electric and magnetic theories. One of the earliest results to which a student of the mathematics of electricity was introduced in pre-Maxwellian days was Coulomb's law of the relation between the density of the electric layer on a conductor and the intensity of electric force just outside it, say

$$E = 4\pi\sigma ; \quad . \quad . \quad . \quad . \quad . \quad (1)$$

and, since this was proved by mathematics, it seemed that the 4π was an essential ratio between two physical quantities, viz., electricity and the force it exerted on other electricity. Never a hint was given that this 4π was purely conventional; it was not, indeed, even recognised to be conventional, and is not at the present day in some quarters. Then, again, at the beginning of magnetism, was Gauss's celebrated theorem proving mathematically that the total flux of magnetic force outward through a closed surface equalled precisely 4π times the total amount of magnetism enclosed within the surface ; and, for all that might appear to the contrary, this remarkable result flowed out necessarily from the properties of the potential function and its derivatives, and of the three direction cosines of the normal to an element of the surface. It was funny—very funny. How ever the 4π managed to find its way in was the puzzle in these and similar results; for instance, in the well-known

$$\mu = 1 + 4\pi\kappa, \quad . \quad . \quad . \quad . \quad . \quad (2)$$

where μ is the permeability and κ another physical property, the susceptibility to magnetisation of a substance. The dark mystery was carefully covered up by the mathematics. It was as hard to understand as the monarch found it to explain the presence of the apple in the dumpling—how did the goodwife manage to get it in ? Nor was the matter rendered plainer by Maxwell's great treatise. Maxwell thought his theory of electric displacement explained the meaning of the 4π (in the corresponding electric theorem), as if it were a matter of physics, instead, as is the fact, of irrationally chosen units.

As the present chapter has been mainly devoted to a general outline of electromagnetic theory expressed in formulæ involving rational units, it will be fitting, in bringing it to a conclusion, to explain here the relation these rational units bear to the ordinary irrational units.

The Origin and Spread of the Eruption.

§ 91. The origin of the 4π absurdity lay in the wisdom of our ancestors,—literally. The inverse square law being recognised, say that two charges q_1 and q_2 repelled one another with a force varying inversely as the square of the distance between them ; thus,

$$F = aq_1q_2/r^2, \quad \cdots \quad \cdots \quad (3)$$

where a is a constant ; what was more natural than to make the expression of the law as simple as possible by giving the constant a the value unity, if, indeed, it were thought of at all ? Our ancestors could not see into the future,—that is to say, beyond their noses,—and perceive that this system would work out absurdly. They were sufficiently wise in their generation, and were not to blame.

But, after learning that certain physical quantities bear to one another invariable relations, we should, in forming a systematic representation of the same, endeavour to avoid the introduction of arbitrary and unnecessary constants. This valuable principle was recognised to a small extent by our ancestors, as above ; it was emphasised by Maxwell and Jenkin in their little treatise on units in one of the Reports of the B. A. Committee on Electrical Standards (1863, Appendix C ; p. 59 of Spon's Reprint). Thus, referring to the magnetic law of inverse squares, we have the following :—

" The strength of the pole is necessarily defined as proportional to the force it is capable of exerting on any other pole. Hence the force f exerted between two poles of the strengths m and m_1, must be proportional to the product mm_1. The force f is also found to be inversely proportional to the square of the distance, D, separating the poles, and to depend on no other quantity ; hence, we have, *unless an absurd and useless coefficient be introduced,*

$$f = mm_1/D^2."\quad\quad\quad (4)$$

Observe the words which I have italicised. When it is considered what Maxwell had then done in the way of framing a broad theory of electromagnetism, it is marvellous that he should have written in that way. By mere force of habit one might, indeed, not consider there to be anything anomalous about the 4π in Coulomb's and Gauss's theorems. But did not Maxwell's electrostatic energy $K\mathfrak{E}^2/8\pi$ and his magnetic energy $\mu\mathfrak{H}^2/8\pi$ per unit volume loudly proclaim that there was something radically wrong in the system to lead to such a mode of expression, which fault should be attended to and set right *ab-initio*, especially in framing a permanent system of practical units, which was what Maxwell and his colleagues were about?

It would seem that the proclamation fell upon deaf ears, for not only were the units irrationally constructed, but in his Treatise, which followed some years later, we find the following statement (p. 155, second edition). After an account of his theory of electric displacement, we are told that "the theory completely accounts for the theorem of Art. 77, that the total induction through a closed surface is equal to the total quantity of electricity within the surface multiplied by 4π. For what we have called the induction through the surface is simply the electric displacement multiplied by 4π, and the total displacement outward is necessarily equal to the total charge within the surface." That is, his theory of electric displacement accounts for the 4π. So it seems to do; and yet the 4π has no essential connection with his theory, or with any one else's. Though by no means evident until it is pointed out, it is entirely a question of the proper choice of units, and is independent of all theories of electricity. It depends upon something much more fundamental.

The Cure of the Disease by Proper Measure of the Strength of Sources.

§ 92. When looked into carefully, the question is simply this : What is the natural measure of the strength of a source? Suppose, for instance, we have a source of heat in a medium which does not absorb heat, how should we measure the intensity of the source? Plainly by the amount of heat emitted per second, passing out through any surface enclosing the source. If the flux of heat be isotropically regular, its density will vary inversely as the square of the distance from a point source, giving

$$C = S/4\pi r^2, \quad \ldots \ldots \quad (5)$$

if C be the heat flux (per unit area) at distance r from the source of strength S. If we knock out the 4π we shall obviously have an unnatural measure of the strength of the source.

Similarly, if we send water along a pipe and let it flow from its end, which may therefore be regarded as a source, we should naturally measure its strength in a similar manner, viz., by the current in the pipe, or by the *total* outflow.

Now in an electric field, or in a magnetic field, or in the field of any vector magnitude, we have everywhere mathematically analogous cases. We find, for instance, that electrostatic force is distributed like velocity in an incompressible liquid, except at certain places, where it is, by analogy, generated, or has its source. If, then, we observe that the flux of force through a closed surface is not zero, there must be sources within the region enclosed, and the natural measure of the total strength of the sources is the total flux of force itself. I have put this in terms of electric force rather than of electric displacement, merely to exemplify that the matter has no particular connexion with electric displacement. In the former case it is a source of " electric force " that is considered; in the latter, it would be of displacement; and the principle concerned in a rational reckoning of the strength of a source is the same in either case. It is a part of the fitness of things, and holds good in the abstract theory of the space-variation of vector magnitudes, apart from all physical application.

Using, temporarily, the language of lines of force, or of tubes of force (which, however, as here, sometimes works out rather nonsensically), we may say that a *unit* pole sends out *one* line of force, or one tube, when rationally estimated.

Next there is the proper measure of the strength of circuital fluxes to be considered; electric current, for instance. The universal property here is that the circuitation of the magnetic force is proportional to the current through the circuit; and the natural measure of the strength of the current is the circuitation itself, without, as usual, dividing by 4π. Now this division by 4π arises out of the irrational reckoning of the strength of point sources. We may therefore expect that when the point sources are measured rationally, the 4π will disappear from the reckoning of electric current, making the circuitation of magnetic force the proper reckoning. This is so, as may be easily seen by substituting for a linear electric current an equivalent magnetic shell, and so bringing in point sources distributed over its two faces.

Thus, in rational units, if we have a point source q of displacement, and a point source m of induction, we have

$$D = q/4\pi r^2, \qquad B = m/4\pi r^2, \quad . \quad . \quad . \quad (6)$$

to express the displacement and induction at distance r, when the fluxes emanate isotropically; and

$$E = (q/c)/4\pi r^2, \qquad H = (m/\mu)/4\pi r^2, \quad . \quad . \quad (7)$$

if q/c and m/μ are the measures of the sources of electric and magnetic force respectively. In the magnetic case m may represent the strength of a pole, on the understanding that (since the induction is really circuital) we ignore the flux coming to the pole (as along a filamentary magnet), and consider only the diverging induction. Also,

$$H = C/2\pi r \quad . \quad . \quad . \quad . \quad (8)$$

expresses the intensity of magnetic force at distance r from a long solitary straight current of strength C.

Obnoxious Effects of the Eruption.

§ 93. Considering merely the formulæ belonging to point sources with uniform divergence, we see that the effect of changing from irrational to rational units is to introduce 4π. If this were all, we might overlook the fundamental irrationality and use irrational units for practical convenience. But, as a matter of fact, it works out quite differently. For the unnatural suppression of the 4π in the formulæ of central force, where it has a right to be, drives it into the blood, there to multiply itself, and afterwards break out all over the body of electromagnetic theory. The few formulæ where 4π should be are principally scholastic formulæ and little used; the many formulæ where it is forced out are, on the contrary, useful formulæ of actual practice, and of the practice of theory. A practical theorist would knock them out merely from the trouble they give, let alone the desire to see things in their right places. Furthermore, it should be remarked that the irrationality of the formulæ is a great impediment in the way of a clear understanding of electromagnetic theory. The interpretation of equation (2) above, for instance, or of the similar well-known equation

$$\mathfrak{B} = \mathfrak{H} + 4\pi\mathfrak{I},$$

presents some difficulty even to a student of ability, unless he be given beforehand a hint or two to assist him. For if κ is a rational physical quantity, then μ cannot be ; or if μ is right, then κ must be wrong. Or $\mathfrak{B}$, $\mathfrak{H}$, and $\mathfrak{I}$ must be incongruous. The 4π is also particularly inconvenient in descriptive matter relating to tubes of force or flux, and in everything connected with them.

This difficulty in the way of understanding the inner meaning of theory is still further increased by the 4π not entering into the magnetic formulæ in the same way as into the electric. Thus, it is $\mathfrak{D}$ and $\mathfrak{B}/4\pi$ which are analogues. Again, in many of the irrational formulæ the irrationality *appears* to disappear. For instance, in $\frac{1}{2}\Psi\rho$, in $\frac{1}{2}\mathfrak{A}\mathfrak{C}$, in $\frac{1}{2}\mathfrak{E}\mathfrak{D}$, in $\mathfrak{E}\mathfrak{C}$, and in some others. This is because both the factors are irrational, and the two irrationalities cancel. It looks as if $\mathfrak{D}$ were the flux belonging to $\mathfrak{E}$, but it is not. In reality, we have, if $x = (4\pi)^{\frac{1}{2}}$,

$$\mathfrak{E} = x\mathbf{E}, \qquad c\mathfrak{E} = x\mathbf{D}, \qquad \mathfrak{D} = x^{-1}\mathbf{D} ;$$

therefore, $$\mathbf{ED} = \mathfrak{E}\mathfrak{D},$$

whilst c is the same in both irrational and rational units.

A Plea for the Removal of the Eruption by the Radical Cure.

§ 94. The question now is, what is to be done? Are we modern pigmies, who by looking over the shoulders of the giants can see somewhat further than they did, to go on perpetrating and perpetuating their errors for ever and ever, and even legalising them? If they are to be enforced, it is to be hoped that it will not be made a penal offence not to use the legal and imperial units.

The "brain-wasting perversity" of the British nation in submitting year after year to be ruled by such a heterogeneous and incongruous collection of units as the yard, foot, inch, mile, knot, pound, ounce, pint, quart, gallon, acre, pole, horse-power, etc., etc., has been repeatedly lamented by would-be reformers, who would introduce the common-sense decimal system; and amongst them have been prominent electricians who hoped to insert the thin end of the wedge by means of the decimal subdivision of the electrical units, and their connection with the metre and gramme, and thus lead to the abolition of the present British system of weights and measures with its absurd and useless arbitrary connecting constants. But what a satire it is upon their labours that they should have fallen into the very pit they were professedly avoiding! The perverse British nation—practically the British engineers—have surely a right to expect that the electricians will first set their own house in order.

The ohm and the volt, etc., are now legalised, so that, as I am informed, it is too late to alter them. This is a *non seq.*, however, for the yard and the gallon are legalised ; and if it is not too late to alter *them*, it cannot be too late to put the newfangled ohm and its companions right. It is *never* too late to mend. No new physical laboratory determinations will be needed. The value of π has been calculated to hundreds of places of decimals ; so that rational ohms, volts, etc., can be now fixed with the same degree of accuracy as the irrational ones, by any calculator.

When I first brought up this matter in *The Electrician* in 1882-3, explaining the origin of the 4π absurdity and its cure, I did not go further than to use rational units in explaining the theory of potentials, scalar and vector, and similar matters; then returning to irrational units in order to preserve harmony with the formulæ in Maxwell's treatise, and I did not think a change was practicable, on account of the B. A. Committee's work, and the general ignorance and want of interest in the subject. But much has happened since then. The spread of electrical knowledge has been immense, concurrent with the development of electrical industries, to say nothing of theoretical and experimental developments. I therefore now think the change is perfectly practicable. At any rate, some one must set the example, if the change is to occur. I have, therefore, in the preceding, wholly avoided the irrational units *ab initio* ; and shall continue to use rational units in the remainder of this work.

So far as theoretical papers and treatises are concerned, there is no difficulty. Every treatise on Electricity should be done in rational formulæ, their connexion with the irrational (so long as they exist) being explained separately (in a chapter at the end of the book, for instance), along with the method of converting into volts, amperes, etc., the present practical units. At present you have to first settle whether to use the electrostatic or the electromagnetic units, and then introduce the appropriate powers of 10. If rational formulæ are used, then, in addition, you must first insert the constant 4π in certain places, so long as the irrational units last.

When, however, the real advantages of the rational system become widely recognized and thoroughly assimilated, then will come a demand for the rationalisation of the practical units. Even at present the poor practician is complaining that he cannot even pass from magnetomotive force to ampere-turns without a "stupid" 4π coefficient getting in the way. That the practical electrical units should be reformed as a preliminary to the general reform of the British units requires no argument to maintain. That this general reform is coming I have not the least doubt. Even the perversity of the British engineer has its limits.

Rational v. Irrational Electric Poles.

§ 95. We may now briefly compare some of the more important formulæ in the two systems. Let us denote quantities in

rational units as in the preceding (except that when vector relations are not in question we need not employ special type), and
the corresponding quantities in irrational units by the same
symbols with the suffix $_i$; thus, q and q_i.   Also denote $(4\pi)^{\frac{1}{2}}$ by x.

Let q_i be the irrational charge which repels an equal
irrational charge at distance r with the *same* force F as a
rational charge q repels an equal charge at the same distance.
Then,

$$F = \frac{1}{c}\left(\frac{q}{xr}\right)^2 = \frac{1}{c}\left(\frac{q_i}{r}\right)^2, \quad \ldots \quad (9)$$

Therefore

$$q = xq_i. \quad \ldots \ldots \quad (10)$$

As regards the ratio of the units, it is sufficient merely to
observe that the magnitude of the number expressing a
quantity varies inversely as the size of the unit.   This
applies throughout, so that we need not bring in units at all,
but keep to the concretes appearing in formulæ.

By (10) we shall have

$$x = \frac{q}{q_i} = \frac{\rho}{\rho_i} = \frac{\sigma}{\sigma_i} = \frac{D}{D_i} = \frac{C}{C_i}, \quad \ldots \quad (11)$$

if ρ is volume-density of electrification, σ surface density,
D displacement, C electric current density (or else the total
displacement and current).

Since $D = cE$,   and $D_i = cE_i/4\pi$, $\quad \ldots \quad (12)$

whilst $D = xD_i$ by (11), we have also

$$x = \frac{E_i}{E} = \frac{A_i}{A} = \frac{V_i}{V} = \frac{e_i}{e} = \frac{P_i}{P}, \quad \ldots \quad (13)$$

where A is the time-integral of E, V the line-integral of E, or
voltage, e impressed electric force, P electric potential or potential difference.

The permittivity c is the same in both systems.   So is the
ratio c/c_0 of the permittivity to that of ether, or the specific
inductive capacity (electric).   The rational permittance of a
unit cube condenser is c, and the irrational is $c_i/4\pi$.   If S is
the permittance of any condenser

$$S = x^2 S_i, \quad \ldots \ldots \quad (14)$$

and its energy is

$$\tfrac{1}{2}VQ = \tfrac{1}{2}V_i Q_i = \tfrac{1}{2}SV^2 = \tfrac{1}{2}S_i V_i^2, \quad \dots \quad (15)$$

if Q is its charge and V its voltage. We also have

$$\tfrac{1}{2}ED = \tfrac{1}{2}E_i D_i = \tfrac{1}{2}cE^2 = \tfrac{1}{2}cE_i^2/4\pi, \quad \dots \quad (16)$$

$$\tfrac{1}{2}P\rho = \tfrac{1}{2}P_i \rho_i, \quad \dots \quad \dots \quad (17)$$

$$\tfrac{1}{2}AC = \tfrac{1}{2}A_i C_i, \quad \dots \quad \dots \quad (18)$$

the first and second sets relating to electric energy, the third to magnetic energy.

By Ohm's law

$$V = RC, \quad \text{and} \quad V_i = R\,C_i ; \quad \dots \quad (19)$$

therefore by (11) and (13)

$$x^2 = \frac{R_i}{R} = \frac{r_i}{r}, \quad \dots \quad \dots \quad (20)$$

where r is resistivity, so that

$$EC = E_i C_i = RC^2 = R_i C_i^2. \quad \dots \quad (21)$$

Rational v. Irrational Magnetic Poles.

§ 96. If the repulsion between two magnetic poles m and m is F at distance r, and this is also the repulsion between irrational poles m_i and m_i, we have

$$F = \frac{1}{\mu}\left(\frac{m}{xr}\right)^2 = \frac{1}{\mu}\left(\frac{m_i}{r}\right)^2, \quad \dots \quad (22)$$

where μ is the inductivity of the medium, common to the two systems, as is likewise μ/μ_0 the ratio of μ to that of the ether, or the permeability. So

$$m = xm_i. \quad \dots \quad \dots \quad (23)$$

Here some discrepancies come in. For

$$B = \mu H, \quad \text{and} \quad B_i = \mu H_i ; \quad \dots \quad (24)$$

so we have

$$x = \frac{m}{m_i} = \frac{H_i}{H} = \frac{B_i}{B} = \frac{\Omega_i}{\Omega} = \frac{h_i}{h}, \quad \dots \quad (25)$$

where Ω is magnetic potential or gaussage. If I be intrinsic magnetisation (intensity of), then

$$I = \mu h, \quad \text{and} \quad I_i = \mu h_i/x^2, \quad \dots \quad (26)$$

from which, and by (25), we have

$$I = xI_i. \qquad \qquad (27)$$

The equation $\mathbf{B} = \mu\mathbf{H}$, in the sense used by me, expands to

$$\mathbf{B} = \mu(\mathbf{h} + \mathbf{F}), \qquad \qquad (28)$$

where $\mathbf{h}$ is intrinsic, and $\mathbf{F}$ is the magnetic force of the field. Also

$$\mu = \mu_0(1 + \kappa) = \mu_0(1 + 4\pi\kappa_i), \qquad \qquad (29)$$

so we have, by (28) and (26),

$$\mathbf{B} = \mathbf{I} + \mu_0\mathbf{F} + \mu_0\kappa\mathbf{F}, \qquad \qquad (30)$$

where $\mu_0\kappa\mathbf{F}$ is the induced magnetisation.

If L is inductance, $\frac{1}{2}LC^2 = \frac{1}{2}L_iC_i^2$, whence $x^2 = \dfrac{L_i}{L} = \dfrac{M_i}{M}$, if M is mutual inductance.

In the common equation,

$$\mathbf{B}_i = \mathbf{F}_i + 4\pi\mathbf{I}_i \qquad \qquad (31),$$

the intrinsic and induced magnetisations are lumped together for one thing, and it is assumed that $\mu_0 = 1$. The quantity κ is thus essentially a numeric, whilst μ is only a numeric by assumption. But whilst $\mu = \mu_i$, we have $\kappa = x^2\kappa_i$. Although magnetisation, whether intrinsic or induced, are of the same kind as induction, yet the reckoning is discrepant. Compare (27) with (25) for B and $\mathbf{B}_i$. In (27) also, I may be either intrinsic or induced, so far as the ratio x goes.

The common equation, div $\mathbf{D}_i = 4\pi\rho_i$, becomes

$$\text{div } \mathbf{D} = \rho, \qquad \qquad (32)$$

and the characteristic equation of Poisson becomes

$$\nabla^2 P = -\rho/c, \qquad \qquad (33)$$

the 4π going out by the rationalisation. But P itself is given by

$$P = \Sigma\rho/4\pi rc. \qquad \qquad (34)$$

The definition of current density,

$$\text{curl } \mathbf{H}_i = 4\pi\mathbf{C}_i,$$

becomes

$$\text{curl } \mathbf{H} = \mathbf{C}. \qquad \qquad (35)$$

Other changes readily follow.   But now that I have explicitly stated the relation between my rational formulæ and the ordinary, I leave the irrationals—for good, I hope—and return to the rational and simplified formulæ, which are so much superior.

APPENDIX.

THE ROTATIONAL ETHER IN ITS APPLICATION TO ELECTROMAGNETISM.

According to Maxwell's theory of electric displacement, disturbances in the electric displacement and magnetic induction are propagated in a non-conducting dielectric after the manner of motions in an incompressible solid.   The subject is somewhat obscured in Maxwell's treatise by his equations of propagation containing **A**, Ψ, **J**, all of which are functions considerably remote from the vectors which represent the state of the field, viz., the electric and magnetic forces, and by some dubious reasoning concerning Ψ and **J**.   There is, however, no doubt about the statement with which I commenced, as it becomes immediately evident when we ignore the potentials and use **E** or **H** instead, the electric or the magnetic force.

The analogy has been made use of in more ways than one, and can be used in very many ways.   The easiest of all is to assume that the magnetic force is the velocity of the medium, magnetic induction the momentum, and so on, as is done by Prof. Lodge (Appendix to "Modern Views of Electricity"). I have also used this method for private purposes, on account of the facility with which electromagnetic problems may be made elastic-solid problems.   I have shown that when impressed electric force acts it is the curl or rotation of the electric force which is to be considered as the source of the resulting disturbances.   Now, on the assumption that the magnetic force is the velocity in the elastic solid, we find that the curl of the impressed electric force is represented simply by impressed mechanical force of the ordinary Newtonian type.   This is very convenient.

But the difficulties in the way of a complete and satisfactory representation of electromagnetic phenomena by an elastic-solid ether are insuperable. Recognising this, Sir W. Thomson has recently brought out a new ether; a rotational ether. It is incompressible, and has no true rigidity, but possesses a quasi-rigidity arising from elastic resistance to absolute rotation.

The stress consists partly of a hydrostatic pressure (which I shall ignore later), but there is no distorting stress, and its place is taken by a rotating stress. It gives rise to a translational force and a torque. If $\mathbf{E}$ be the torque, the stress on any plane $\mathbf{N}$ (unit normal) is simply $V\mathbf{EN}$, the vector product of the torque and the normal vector.

The force is $-\operatorname{curl} \mathbf{E}$. We have therefore the equation of motion

$$-\operatorname{curl} \mathbf{E} = \mu \dot{\mathbf{H}},$$

if $\mathbf{H}$ is the velocity and μ the density. But, alas, the torque is proportional to the rotation. This gives

$$\operatorname{curl} \mathbf{H} = c \dot{\mathbf{E}},$$

where c is the compliancy, the reciprocal of the quasi-rigidity.

Now these are the equations connecting electric and magnetic force in a non-conducting dielectric, when μ is the inductivity and c the permittancy. We have a parallelism in detail, not merely in some particulars. The kinetic energy $\frac{1}{2}\mu \mathbf{H}^2$ represents the magnetic energy, and the potential energy $\frac{1}{2}c\mathbf{E}^2$ the electric energy. The vector-flux of energy is $V\mathbf{EH}$, the activity of the stress.

This mode of representation differs from that of Sir W. Thomson, who represents magnetic force by rotation. This system makes electric energy kinetic, and magnetic energy potential, which I do not find so easy to follow.

Now let us, if possible, extend our analogy to conductors. Let the translational and the rotational motions be both frictionally resisted, and let the above equations become

$$-\operatorname{curl} \mathbf{E} = g\mathbf{H} + \mu \dot{\mathbf{H}},$$

$$\operatorname{curl} \mathbf{H} = k\mathbf{E} + c\dot{\mathbf{E}},$$

where g is the translational frictionality; k will be considered later. We have now the equations of electric and magnetic force in a dielectric with duplex conductivity, k being the electric and g the magnetic conductivity (by analogy with electric force, but a frictionality in our present dynamical analogy). We have, therefore, still a parallelism in every detail. We have waste of energy by friction $g\mathrm{H}^2$ (translational) and $k\mathrm{E}^2$ (rotational). If $g/\mu = k/c$ the propagation of disturbances will take place precisely as in a non-conducting dielectric, though with attenuation caused by the loss of energy.

To show how this analogy works out in practice, consider a telegraph circuit, which is most simply taken to be three co-axial tubes. Let, A, B, and C be the tubes; A the innermost, C the outermost, B between them; all closely fitted. Let their material be the rotational ether. In the first place, suppose that there is perfect slip between B and its neighbours. Then, when a torque is applied to the end of B (the axis of torque to be that of the tubes), and circular motion thus given to B, the motion is (in virtue of the perfect slip) transmitted along B, without change of type, at constant speed, and without affecting A and C.

This is the analogue of a concentric cable, if the conductors A and C be perfect conductors, and the dielectric B a perfect insulator. The terminal torque corresponds to the impressed voltage. It should be so distributed over the end of B that the applied force there is circular tangential traction, varying inversely as the distance from the axis; like the distribution of magnetic force, in fact.

Now, if we introduce translational and rotational resistance in B, in the above manner, still keeping the slip perfect, we make the dielectric not only conducting electrically but also magnetically. This will not do. Abolish the translational resistance in B altogether, and let there be no slip at all between B and A, and B and C. Let also there be rotational resistance in A and C.

We have now the analogue of a real cable : two conductors separated by a third. All are dielectrics, but the middle one should have practically very slight conductivity, so that it is pre-eminently a dielectric; whilst the other two should have very high conductivity, so that they are pre-eminently conductors. The three constants, μ, c, k, may have any value in the three tubes, but practically k should be in the middle tube a very small fraction of what it is in the others.

K

It is remarkable that the *quasi-rotational* resistance in A and C should tend to counteract the distorting effect on waves of the *quasi-rotational* resistance in B. But the two rotations, it should be observed, are practically perpendicular, being axial or longitudinal (now) in A and C, and transverse or radial in B; due to the relative smallness of k in the middle tube.

To make this neutralising property work exactly we must transfer the resistance in the tubes A and C to the tube B, at the same time making it translational resistance. Also restore the slip. Then we can have perfect annihilation of distortion in the propagation of disturbances, viz., when k and g are so proportioned as to make the two wastes of energy equal. In the passage of a disturbance along B there is partial absorption, but no reflection.

But as regards the meaning of the above k there is a difficulty. In the original rotational ether the torque varies as the rotation. If we superadd a real frictional resistance to rotation we get an equation of the form

$$\dot{\mathbf{E}} = \left(a + b\frac{d}{dt} \right) \text{curl } \mathbf{H},$$

$\mathbf{E}$ being (as before) the torque, and $\mathbf{H}$ the velocity. But this is not of the right form, which is (as above)

$$\text{curl } \mathbf{H} = \left(k + c\frac{d}{dt} \right)\mathbf{E};$$

therefore some special arrangement is required (to produce the dissipation of energy $k\mathbf{E}^2$), which does not obviously present itself in the mechanics of the rotational ether.

On the other hand, if we follow up the other system, in which magnetic force is allied with rotation, we may put $g = 0$, let $-\mathbf{E}$ be the velocity and $\mathbf{H}$ the torque; μ the compliancy, c the density, and k the translational frictionality. This gives

$$-\text{curl } \mathbf{E} = \mu\dot{\mathbf{H}}$$

$$\text{curl } \mathbf{H} = k\mathbf{E} + c\dot{\mathbf{E}}.$$

We thus represent a homogeneous conducting dielectric, with a translational resistance to cause the Joulean waste of energy.

But it is now seemingly impossible to properly satisfy the conditions of continuity at the interface of different media. For instance, the velocity − **E** should be continuous, but we do not have normal continuity of electric force at an interface. In the case of the tubes we avoided this difficulty by having the velocity tangential.

Either way, then, the matter is left, for the present, in an imperfect state.

In the general case, the d/dt of our equations should receive an extended meaning, on account of the translational motion of the medium. The analogy will, therefore, work out less satisfactorily. And it must be remembered that it is only an analogy in virtue of similitude of relations. We cannot, for instance, deduce the Maxwellian stresses and mechanical forces on charged or currented bodies. The similitude does not extend so far. But certainly the new ether goes somewhat further than anything known to me that has been yet proposed in the way of a stressed solid.

[P.S.—The special reckonings of torque and rotation in the above are merely designed to facilitate the elastic-solid and electromagnetic comparisons without unnecessary constants.]

CHAPTER III.

———

THE ELEMENTS OF VECTORIAL ALGEBRA AND ANALYSIS.

Scalars and Vectors.

§ 97. Ordinary algebra, as is well known, treats of quantities and their relations. If, however, we examine geometry, we shall soon find that the fundamental entity concerned, namely a straight line, when regarded as an entity, cannot be treated simply as a quantity in the algebraical sense. It has, indeed, size, viz., its length; but with this is conjoined another important property, its direction. Taken as a whole, it is a Vector. In contrast with this, an ordinary quantity, having size only, is a Scalar.

Again, if we consider the mathematics of physical questions, we find two distinct kinds of magnitudes prominently present. All such magnitudes as mass, density, energy, temperature, are evidently quantities in the simple algebraical sense; that is, scalar magnitudes, or simply scalars. A certain (it may be an unstated) unit of density, for instance, being implied, any density is expressed by a simple number. (The question of the "dimensions" of physical magnitudes is not in question.) All magnitudes whatever which have no directional peculiarity and which are each specified by a single number are scalars, and subject to scalar algebra.

But such magnitudes as displacement, velocity, acceleration, force, momentum, electric current, &c., which have direction as well as size, and which are fully specified by statement of the size and direction, are vector magnitudes, or simply vectors.

Now, just as there is an algebra and analysis for scalars, so is there a vector algebra and analysis appropriate to vectors; and it is the object of the present chapter to give a brief account of the latter, especially in respect to its application to electromagnetism.

Characteristics of Cartesian and Vectorial Analysis.

§ 98. Algebraical or analytical geometry in the usual Cartesian form, though dealing ultimately with vectors, is not vectorial algebra. It is, in fact, a reduction to scalar algebra by resolution of every vector into three rectangular components, which are manipulated as scalars. Similarly, in the usual treatment of physical vectors, there is an avoidance of the vectors themselves by their resolution into components. That this is a highly artificial process is obvious, but it is often convenient. More often, however, the Cartesian mathematics is ill-adapted to the work it has to do, being lengthy and cumbrous, and frequently calculated to conceal rather than to furnish and exhibit useful results and relations in a ready manner. When we work directly with vectors, we have our attention fixed upon them, and on their mutual relations; and these are usually exhibited in a neat, compact, and expressive form, whose inner meaning is evident at a glance to the practised eye. Put the same formula, however, into the Cartesian form, and—what a difference ! The formula which was expressed by a few letters and symbols in a single line, readable at once, sometimes swells out and covers a whole page ! A very close study of the complex array of symbols is then required to find out what it means ; and, even though the notation be thoroughly symmetrical, it becomes a work of time and great patience. In this interpretation we shall, either consciously or unconsciously, be endeavouring to translate the Cartesian formulæ into the language of vectors.

Again, in the Cartesian method, we are led away from the physical relations that it is so desirable to bear in mind, to the working out of mathematical exercises upon the components. It becomes, or tends to become, blind mathematics. It was once told as a good joke upon a mathematician that the poor man went mad and mistook his symbols for realities ; as M for the moon and S for the sun. There is another side to the story, however. If our object be ultimately physical, rather than mathematical, then the more closely we can identify the symbols with their physical representatives the more usefully can we work, with avoidance of useless—though equally true— mathematical exercises. The mere sight of the arrangement of symbols should call up an immediate picture of the physics symbolised, so that our formulæ may become *alive*, as it were. Now this is possible, and indeed, comparatively easy, in vectorial analysis ; but is very difficult in Cartesian analysis, beyond a certain point, owing to the geometrically progressive complexity of the expressions to be interpreted and manipulated. Vectorial algebra is the natural language of vectors, and no one who has ever learnt it (not too late in life, however) will ever care to go back from the vitality of vectors to the bulky inanimateness of the Cartesian system.

Abstrusity of Quaternions and Comparative Simplicity gained by ignoring them.

§ 99. But supposing, as is generally supposed, vector algebra is something "awfully difficult," involving metaphysical considerations of an abstruse nature, only to be thoroughly understood by consummately profound metaphysicomathematicians, such as Prof. Tait, for example. Well, if so, there would not be the slightest chance for vector algebra and analysis to ever become generally useful; and I should not be writing this, nor should I have, for several years past, persisted in using vector algebra in electromagnetic theory—a prophet howling in the wilderness. It will readily be concluded, then, that I believe that the vector analysis is going to become generally used in scientific work, and that what is needed is not "awfully difficult." There was a time, indeed, when I, although recognising the appropriateness of vector analysis in electromagnetic theory (and in mathematical physics generally), did think it was harder to understand and to work than the Cartesian analysis. But that was before I had thrown off the quaternionic old-man-of-the-sea who fastened himself on my shoulders when reading the only accessible treatise on the subject—Prof. Tait's Quaternions. But I came later to see that, so far as the vector analysis I required was concerned, the quaternion was not only not required, but was a positive evil of no inconsiderable magnitude; and that by its avoidance the establishment of vector analysis was made quite simple and its working also simplified, and that it could be conveniently harmonised with ordinary Cartesian work. There is not a ghost of a quaternion in any of my papers (except in one, for a special purpose). The vector analysis I use may be described either as a convenient and systematic abbreviation of Cartesian analysis; or else, as Quaternions without the quaternions, and with a simplified notation harmonising with Cartesians. In this form, it is not more difficult, but easier to work than Cartesians. Of course you must learn how to work it. Initially, unfamiliarity may make it difficult. But no amount of familiarity will make Quaternions an easy subject.

Maxwell, in his great treatise on Electricity and Magnetism, whilst pointing out the suitability of vectorial methods to the treatment of his subject, did not go any further than to freely make use of the idea of a vector, in the first place, and to occasionally express his results in vectorial form. In this way his readers became familiarised with the idea of a vector, and also with the appearance of certain formulæ when exhibited in the quaternionic notation. They did not, however, derive any information how to work vectors. On the whole, I am inclined to think that the omission of this information has not tended to impede the diffusion of a knowledge of vector analysis. For, had he given an account of the theory, he would certainly have followed the Hamilton-Tait system; and this would probably, for reasons I shall shortly mention, have violently prejudiced his readers against the whole thing.

But the diffusion of vector analysis has, undoubtedly, in my opinion, been impeded by the absence of sufficiently elementary works on the subject, with a method of establishment of principles adapted to ordinary minds, and with a conveniently workable notation. For the reader of Maxwell's treatise who desired to learn to work vectors in analysis had either to go to Hamilton's ponderous volumes, or else to Prof. Tait's treatise. The former are out of the question for initiatory purposes. But the latter is excessively difficult, although described as " an elementary treatise "—not the same thing as " a treatise on the elements." The difficulty arises in a great measure from the quaternionic basis.

Elementary Vector Analysis independent of the Quaternion.

§ 100. Suppose a sufficiently competent mathematician desired to find out from the Cartesian mathematics what vector algebra was like, and its laws. He could do so by careful inspection and comparison of the Cartesian formulæ. He would find certain combinations of symbols and quantities occurring again and again, usually in systems of threes. He might introduce tentatively an abbreviated notation for these combinations. After a little practice he would perceive the laws according to which these combinations arose and how they operated. Finally, he would come to a very compact system in which vectors themselves and certain simple functions of vectors appeared, and would be delighted to find that the rules for the multiplication and general manipulation of these vectors were, considering the complexity of the Cartesian mathematics out of which he had discovered them, of an almost incredible simplicity. But there would be no sign of a quaternion in his results, for one thing; and, for another, there would be no metaphysics or abstruse reasoning required to establish the rules of manipulation of his vectors. Vector analysis is, in its elements, entirely independent of the exceedingly difficult theory of quaternions; that is, when the latter is treated quaternionically *ab initio*.

"Quaternion" was, I think, defined by an American school-girl to be "an ancient religious ceremony." This was, however, a complete mistake. The ancients—unlike Prof. Tait—knew not, and did not worship Quaternions. The quaternion and its laws were discovered by that extraordinary genius Sir W. Hamilton. A quaternion is neither a scalar, nor a vector, but a sort of combination of both. It has no physical representa-tives, but is a highly abstract mathematical concept. It is the "operator" which turns one vector into another. It has a stretching faculty first, to make the one vector become as long as the other; and a rotating faculty, to bring the one into parallelism with the other.

Now in Quaternions the quaternion is the master, and lays down the law to the vector and scalar. Everything revolves round the quaternion. The laws of vector algebra themselves are established through quaternions, assisted by the imaginary $\sqrt{-1}$. But I am not sure that any one has ever quite understood this establishment. It is done in the second chapter of Tait's treatise. I never understood it, but had to pass on. That the establishment is not demonstrative may be the reason of the important changes made therein in the third edition. But it is still undemonstrative to me, though much improved. Now this relates to the very elements of the subject, viz., the scalar and vector products of a pair of vectors, the laws of which are quite plain in the Cartesian mathematics. Clearly, then, the quaternionic is an undesirable way of beginning the subject, and impedes the diffusion of vectorial analysis in a way which is as vexatious and brain-wasting as it is unnecessary.

Tait v. Gibbs and Gibbs v. Tait.

§ 101. Considering the obligations I am personally under to Prof. Tait (in spite of that lamentable second chapter), it does seem ungrateful that I should complain. But I have at heart the spread of a working knowledge of elementary vector analysis quite as much as Prof. Tait has the extension of the theory of quaternions. Besides, Prof. Tait has assumed a very conservative attitude in relation to Hamilton's grand system. For instance, to "more than one correspondent" who had written for explanation of something they found obscure—and the same thing occurred to me—described in his treatise by "It is evident, . . .", he, "on full consideration," decides not to modify it, but to italicise the words ! He also told them that if they did not see it, in the light of certain preceding parts of the treatise, then they had "begun the study of Quaternions too soon" (Third edition, p. 110). This is as characteristic of the sardonic philosopher as a certain heavy kind of "flippancy" is of the Cockney. Again, in his Preface he states one cause of the little progress made in the development of Quaternions to be that workers have (especially in France) been more intent on modifying the notation or the mode of presentation of the elementary principles, than in extending the application of the calculus. "Even Prof. Willard Gibbs must be ranked as one of the retarders of quaternionic progress, in virtue of his pamphlet on Vector

Analysis, a sort of hermaphrodite monster, compounded of the notations of Hamilton and Grassmann." Grassmann, I may observe, established, *inter alia*, a calculus of vectors, but not of quaternions. Prof. W. Gibbs is well able to take care of himself. I may, however, remark that the modifications referred to are evidence of modifications being felt to be needed ; and that Prof. Gibbs's pamphlet (NOT PUBLISHED, Newhaven, 1881-4, pp. 83), is not a quaternionic treatise, but an able and in some respects original little treatise on vector analysis, though too condensed and also too advanced for learners' use ; and that Prof. Gibbs, being no doubt a little touched by Prof. Tait's condemnation, has recently (in the pages of *Nature*) made a powerful defence of his position. He has by a long way the best of the argument, unless Prof. Tait's rejoinder has still to appear. Prof. Gibbs clearly separates the quaternionic question from the question of a suitable notation, and argues strongly against the quaternionic establishment of vector analysis. I am able (and am happy) to express a general concurrence of opinion with him about the quaternion, and its comparative uselessness in practical vector analysis. As regards his notation, however, I do not like it. Mine is Tait's, but simplified, and made to harmonise with Cartesians.

Abolition of the Minus Sign of Quaternions.

§ 102. In Quaternions, the square of a unit vector is -1. This singular convention is quaternionically convenient. But in the practical vector analysis of physics it is particularly inconvenient, being indeed, an obtrusive stumbling-block. All positive scalar products have the *minus* sign prefixed ; there is thus a want of harmony with scalar investigations, and a difficulty in readily passing from Cartesians to Vectors and conversely. My notation, on the other hand, is expressly arranged to facilitate this mutual conversion.

As regards the establishment of the elementary vector algebra, that is quite simple (freed from the quaternion) ; it all follows from the definitions of a vector and of the scalar and vector products of a pair of vectors.

Now I can imagine a quaternionist (unless prejudiced) admitting the simplicity of establishment and of operation, and the convenience of the notation, and its sufficiency for practical requirements up to a certain point ; and yet adding the inquiry whether there is not, over and above this vector analysis, a theory of Quaternions which is overlooked. To this I would reply, Certainly, but it is not food for the average mathematician, and can, therefore, never be generally used by him, his practical requirements being more suitably satisfied by the rudimentary vector analysis divested of the mysterious quaternion. This does not exclude the important theory of ∇ and its applications. Prof. Gibbs would, I think, go further, and maintain that the anti- or ex-quaternionic vectorial analysis was far superior to the quaternionic, which is uniquely adapted to three dimensions, whilst the other admits of appropriate extension to more generalised cases. I, however, find it sufficient to take my stand upon the superior simplicity and practical utility of the ex-quaternionic system.

We may, however, if we wish to go further, after ex-quaternionic establishment of vector algebra, conjoin the scalar and vector, and make the quaternion, and so deduce the whole body of Quaternions.   But sufficient for the day is the labour thereof ; and we shall now be concerned with scalars and vectors only.   The reader should entirely divest his mind of any idea that we are concerned with the imaginary $\sqrt{-1}$ in vector analysis.   Also, he should remember that unfamiliarity with notation and processes may give an appearance of difficulty that is entirely fictitious, even to an intrinsically easy matter ; so that it is necessary to thoroughly master the notation and ideas involved.   The best plan is to sit down and work ; all that books can do is to show the way.

Type for Vectors.   Greek, German, and Roman Letters unsuitable.   Clarendon Type suitable.   Typographical Backsliding in the Present Generation.

§ 103. We should, in the first place, fix how to represent vectors, although in reality this is the outcome of experience. A vector may obviously be denoted by a single letter ; and, having defined certain letters to stand for scalars and others for vectors, it is certainly unnecessary that the vectors should be distinguished from the scalars by the use of different kinds of types.   But practical experience shows that it is **very** desirable that this should be done, in order to facilitate the reading of vectorial work, by showing at a glance which letters are vectors and which are scalars, and thus easing the stress and strain on the memory. This is all the more important because the manipulation of vectors sometimes differs from that of scalars.

Now Prof. Tait usually indicates vectors by Greek letters. But it is well known that a considerable familiarity with the Greek letters—such as is acquired by studying the literature of ancient Greece—is required before they can be read and manipulated with facility. On the other hand, few are Greek scholars, and in fact many people think it is about time the dead languages were buried. Greek letters are, at any rate, not very well adapted to a vector analysis which aims at practicality.

Maxwell employed German or Gothic type. This was an unfortunate choice, being by itself sufficient to prejudice readers against vectorial analysis. Perhaps some few readers who were educated at a commercial academy where the writing of German letters was taught might be able to manage the German vectors without much difficulty ; but to others it is a work of great pains to form German letters legibly. Nor is the reading of the printed letters an easy matter. Some of them are so much alike that a close scrutiny with a glass is needed to distinguish them, unless one is lynx-eyed. This is a fatal objection. But, irrespective of this, the flourishing ornamental character of the letters is against legibility. In fact, the German type is so thoroughly unpractical that the Germans themselves are giving it up in favour of the plain Roman characters, which he who runs may read. It is a relic of mediæval monkery, and is quite unsuited to the present day. Besides, there can be little doubt that the prevalent shortsightedness of the German nation has (in a great measure) arisen from the character of the printed and written letters employed for so many generations, by inheritance and accumulation. It became racial; cultivated in youth, it was intensified in the adult, and again transmitted to posterity. German letters must go.

Rejecting Germans and Greeks, I formerly used ordinary Roman letters to mean the same as Maxwell's corresponding

Germans. They are plain enough, of course; but, as before mentioned, are open to objection. Finally, I found salvation in **Clarendons,** and introduced the use of the kind of type so called, I believe, for vectors (*Phil. Mag.*, August, 1886), and have found it thoroughly suitable. It is always in stock; it is very neat; it is perfectly legible (sometimes alarmingly so), and is suitable for use in formulæ along with other types, Roman or italic, as the case may be, contrasting and also harmonising well with them.

Sometimes block letters have been used; but it is sufficient merely to look at a mixed formula containing them to see that they are not quite suitable. I should mention here, however, that it is not the mere use of special types that converts scalar to vector algebra. For instance, engineers have often to deal with vector magnitudes in their calculations, but not (save exceptionally) in their vectorial signification. That is, it is merely their size that is in question, and when this is the case there is no particular reason why a special kind of type should be used, whether blocks or another sort.

In connection with Clarendon type, a remark may be made on a subject which is important to the community in general. I refer to the retrograde movement in typography which has been going on for the last 20 years or so. Many people, who possess fairly good and normal eyesight, find a difficulty in reading printed books without straining the eyes, and do not know the reason. They may think the print is too small, or that the light is not good, or that their eyes are not right. But the size of type *per se* has little to do with its legibility; a far more important factor is the style of type, especially as regards the fineness of the marks printed. The "old style," revived a generation since, and now largely in use, differs from the more legible "modern style" (of this page, for instance) in two respects. It has certain eccentricities of shape, which make it somewhat less easy to read; but, more importantly, the letters on the types are cut a good deal finer, which results in a pale impression, as if the ink were watery. This is the main cause of the strain upon the eyes, and the good light wanted. Even very small print is easy to read if it be bold and black, not thin and pale. It would be a public benefit if the retrograde step were reversed, and the revived old style, which threatens

to drive out the modern style, discarded. Not that the latter
is perfect. A still better style would be arrived at by thicken-
ing the fine lines in it, producing something intermediate
between it and the Clarendon style (which last, of course, would
be too much of a good thing). As everyone who has had to
read MSS. knows, the most legible handwritings—irrespective
of the proper formation of the letters—are those in which the
writing-master's fine upstroke is discarded. Now it is the same
in print. Thicken the fine lines, and the effect is magical.

**Notation.   Tensor and Components of a Vector.   Unit
Vectors of Reference.**

§ 104. The tensor of a vector is its size, or magnitude apart
from direction. Other important connected quantities are its
three rectangular scalar components—the Cartesian components—which are the tensors of the three rectangular vector
components. It is usual, in Cartesian work, to use three separate letters for these components, as F, G, H; or u, v, w; or
ξ, η, ζ, &c. One objection to this practice is that when there
is a large number of vectors the memory is strongly taxed to
remember their proper constituents. Another is the prodigal
waste of useful letters. One alphabet is soon exhausted, and
others have to be drawn upon.

In my notation the same letter serves for the vector itself,
and for its tensor and components. Thus, **E** denoting any
vector, its tensor is E, and its components are E_1, E_2, E_3. Similarly, the tensor of **a** is a, and the components are a_1, a_2, a_3. A
large stock of letters is thus set free for other use.

But a remark must be made concerning MS. work, as distinguished from printed work. In MS. work it is inconvenient
to be at the trouble of writing two kinds of letters; ordinary
letters will suffice for both scalars and vectors. Or the ordinary letters representing vectors may receive some conventional
mark to vectorise them. But, presuming ordinary letters are
written, something is required to distinguish between the vector
and its tensor. This may be satisfied by calling E_0 the tensor
of **E**. Only when an investigation is to be written out for the
printers is it necessary to bring in special letters, and this is
most simply done by a conventional mark affixed to every
(to be) vector. Compositors are very intelligent, read mathe-

matics like winking, and carry out all instructions made by the author.

If a vector **a** be multiplied by a scalar x, the result, written x**a** or **a**x, is a vector x times as big as **a**, and having the same direction. Thus, if **a**$_1$ be a unit vector parallel to **a** (or, more strictly, parallel to and concurrent with **a**), of unit length, we have $\mathbf{a} = a\mathbf{a}_1$. It is sometimes useful to separately represent the direction and length of a vector, and the above is a convenient way of doing it without introducing new letters. This applies to any vector. But it need not be an absolute rule. For instance, the Cartesian co-ordinates x, y, z of a point may be retained. Thus, let **r** be the vector from the origin to any point, and let **i**, **j**, **k** be unit vectors from the origin along the three rectangular axes. We shall then have

$$\mathbf{x} = x\mathbf{i}, \qquad \mathbf{y} = y\mathbf{j}, \qquad \mathbf{z} = z\mathbf{k}, \qquad . \quad . \quad (1)$$

where **x** is the vector projection of **r** on the **i** axis, as we know by the elementary geometry of a rectangular parallelepiped or brick. Also, the direction-cosines of **r** are x/r, y/r, z/r, and

$$r^2 = x^2 + y^2 + z^2, \quad . \quad . \quad . \quad . \quad . \quad (2)$$

by Euclid I., 47.

The Addition of Vectors. Circuital Property.

§ 105. This brings us to vector addition. The vector **x** signifies translation through the distance x in the direction **i**. If, now, after performing this operation, we carry out the operation indicated by **y**, viz., translation through the distance y in the direction **j** ; and, lastly, carry out the operation **z**, or translation through the distance z in the **k** direction, we shall arrive at the end of the vector **r**. That is, starting from one corner of a brick, we may reach the opposite corner by three mutually perpendicular journeys along three edges of the brick. The final result is the same as if we went straight across from corner to corner, that is, by carrying out the operation indicated by the vector **r**. This equivalence is expressed by

$$\mathbf{r} = \mathbf{x} + \mathbf{y} + \mathbf{z}, \quad . \quad . \quad . \quad . \quad . \quad (3)$$

Or, by (1), $\qquad\qquad \mathbf{r} = x\mathbf{i} + y\mathbf{j} + z\mathbf{k}. \quad . \quad . \quad . \quad . \quad (4)$

The meaning of addition of vectors in this example is simply the carrying out of the operations implied by the individual vectors added, the geometrical vector meaning a displacement in space, or translation from one point to another.   The order of addition is indifferent, since there are six ways of going from one corner to the opposite one of a brick along its edges.

We see that any vector may be expressed as the sum of three mutually perpendicular vectors, viz., its vector projections on the axes.   Furthermore, by the use of a skew parallelepiped instead of a brick we see that any vector may be expressed as the sum of three other vectors having any directions we please, provided they are independent, or not all in the same plane. For in the latter case the parallelepiped degenerates to a plane figure.

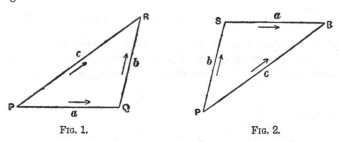

FIG. 1.   FIG. 2.

But it is perhaps best to explain vector addition in general without any reference to axes.   Thus let **a** and **b** be vectors to be added, **a** meaning translation from P to Q in the first figure, or through an equal distance along any parallel line, as from S to R in the second figure, whilst **b** means translation from Q to R in the first figure, or through an equal distance along any parallel line, as from P to S in the second figure.   In the first case, performing the operation **a** first, and then **b**, we go from P to R *viâ* Q ; in the second case, with **b** first and then **a**, we go from P to R *viâ* S. The final result in either case is equivalent to direct translation from P to R, symbolised by the vector **c**.

Thus, $c = a + b = b + a.$ . . . . . . (5)

The above may be extended to any number of vectors. Or we may reason thus :—Let **A** be any given vector, translating, say,

from P to Q. We need not go from P to Q direct, but may
follow any one of an infinite number of paths, as for example,

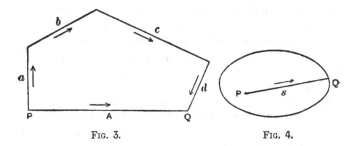

FIG. 3.   FIG. 4.

$a + b + c + d$ in the figure. The final result of the successive
translations is always the same, viz., the direct translation **A**.

Or $$A = a + b + c + d. \quad \ldots \ldots \quad (6)$$

Thus any vector **A** may be split up into the sum of any
number n of vectors, of which $n - 1$ are perfectly arbitrary, for
instance a, b, c in the figure. The remaining one d is, of
course, not arbitrary. It is the vector required to complete
the circuit of vectors.

The vectors need not be in one plane. Nor need the path
followed consist of finite straight portions. It may be wholly
or partly curved. The curved portions are then made up of
infinitesimal vectors. Each curved portion may be replaced
by the vector joining its terminals.

Since the original vector and the substituted vectors form a
circuit, if the positive direction in the circuit be the same for
all the vectors, we may express vector addition thus :—The
sum of any number of vectors which make a circuit is zero.
That is, $\Sigma a = 0$, if a is the type of the vectors summed. For
a curved circuit we shall have

$$\int d\mathbf{s} = 0, \quad \ldots \ldots \quad (7)$$

where $d\mathbf{s}$ is the vector element of the circuit. Here s itself
may be taken to be the vector from *any* fixed point P to a
point Q on the circuit, Fig. 4. Then $d\mathbf{s}$ is the infinitesimal

L

change in s made in an infinitesimal step along the circuit, that is, it is the vector element of the circuit itself.

In a vector equation every term is a vector, of course, however the individual terms may be made up, and every vector equation expresses the fact symbolised by (7), or by $\Sigma a = 0$ when the vectors are finite. In the latter case, a may be also regarded as Δs, the *finite* change in the vector s from any fixed origin produced by passing from beginning to end of the vector a.

The $-$ sign prefixed to a vector is the same as multiplication by -1, and its effect is simply to reverse the direction of translation. Or it may be regarded as reversing the tensor, without altering the direction: thus

$$-\mathbf{a} = -a\mathbf{a}_1 = a \times (-\mathbf{a}_1) = (-a) \times \mathbf{a}_1 \quad . \quad . \quad (8).$$

Thus, in any vector equation, for example

$$\mathbf{a} + \mathbf{b} + \mathbf{c} + \mathbf{d} + \mathbf{e} + \mathbf{f} = 0, \quad . \quad . \quad . \quad (9)$$

we may transfer any terms to the other side by prefixing the $-$ sign. Thus,

$$\mathbf{a} + \mathbf{b} + \mathbf{c} = -\mathbf{d} - \mathbf{e} - \mathbf{f} = -(\mathbf{d} + \mathbf{e} + \mathbf{f}). \quad . \quad . \quad (10)$$

In the form (9) we express the fact that translation in a circuit is equivalent to no translation. In (10), however, we express the equivalence of two translations by different paths from one point to another.

If any trouble be experienced in seeing the necessary truth of (9) for a circuit in whatever order the addition be made, the matter may be clinched by means of i, j, k, the unit rectangular vectors ; as we saw before, the resolution of vectors into rectangular component vectors depends only upon the properties of the right angled triangle. Thus split up the vectors a, b, c into components, we have

$$\left. \begin{array}{l} \mathbf{a} = \mathbf{i}a_1 + \mathbf{j}a_2 + \mathbf{k}a_3, \\ \mathbf{b} = \mathbf{i}b_1 + \mathbf{j}b_2 + \mathbf{k}b_3, \\ \mathbf{c} = \mathbf{i}c_1 + \mathbf{j}c_2 + \mathbf{k}c_3, \end{array} \right\} . \quad . \quad . \quad (11)$$

Now add up. All vectors parallel to i add in scalar fashion, for there is no change of direction. Similarly for j and k; so we have

$$(\mathbf{a} + \mathbf{b} + \mathbf{c}) = \mathbf{i}(a_1 + b_1 + c_1) + \mathbf{j}(a_2 + b_2 + c_2) + \mathbf{k}(a_3 + b_3 + c_3).$$

Now the scalar additions may be done in any order we please. It follows that the same is true in vector addition. The addition and subtraction of vectors and transfer from one side of an equation to the other are thus done identically as in the algebra of scalars. Addition has, it is true, not the same meaning, but the vector meaning is not inconsistent with the scalar meaning, and, in fact, includes the latter as a particular case. There is never any conflict between + put between vectors and + put between scalars.

Application to Physical Vectors. Futility of Popular Demonstrations. Barbarity of Euclid.

§ 106. In the above we have referred entirely to the geometrical vector. But a very important step further can be made referring to physical vectors, a step which immediately does away with piles of ingeniously constructed and brain-wasting " demonstrations," especially compiled for the torturing of students and discouragement of learning. If a quantity be recognised to be a directed magnitude, it is, in its mathematical aspect, a vector, and is therefore subject to the same laws as *the* vector, or geometrical vector. So, just as we have the triangle, or parallelogram, or polygon of geometrical vectors, we must have the same property exemplified in the addition of all vectors, as velocity, acceleration, force, &c. This identity of treatment applies not only to the addition property, but to the multiplication and other properties to be later considered. We may symbolise a physical vector by a straight line of given length and direction.

It used to be thought necessary to give demonstrations of the parallelogram of forces, perhaps even before the student knew what force meant. I have some dim recollection of days spent in trying to make out Duchayla's proof, which was certainly elaborate and painstaking, though benumbing. Maxwell, in his treatise, elaborated a demonstration that electric currents compounded according to the vector law. But surely there is something of the vicious circle in such demonstrations. Is it not sufficient to recognise that a quantity is a vector, to know that it follows the laws of the geometrical vector, the addition property of which does not want demonstrating, but only needs pointing out, as in the polygon of vectors?

L 2

Although this is not the place for exercises and examples, yet
it is worth while to point out that by means of the addition
property of vectors a good deal of geometry can be simply
done—better than by Euclid, a considerable part of whose 12
books consists of examples of how not to do it (especially
Book V.). There is a Society for the Improvement of Geometrical
Teaching. I have no knowledge of its work ; but as to the
need of improvement there can be no question whilst the reign
of Euclid continues. My own idea of a useful course is to begin
with arithmetic, and then, not Euclid, but algebra. Next, not
Euclid, but practical geometry, solid as well as plane ; not
demonstrations, but to make acquaintance. Then, not Euclid,
but elementary vectors, conjoined with algebra, and applied to
geometry. Addition first ; then the scalar product. This
covers a large ground. When more advanced, bring in the
vector product. Elementary calculus should go on simultaneously, and come into the vector algebraic geometry after a bit.
Euclid might be an extra course for learned men, like Homer.
But Euclid for children is barbarous.*

The Scalar Product of Two Vectors. Notation and Illustrations.

§ 107. Coming next to the products of vectors, it is to be
noted at the beginning that the ordinary idea of a product in

---

* From *The Electrician*, December 4, 1891, p. 106, I learn that the
correct title of the society above alluded to is the Association for the
Improvement of Geometrical Teaching. "It was founded, we believe,
about ten years ago by a few teachers, who realised that Euclid for
children is, as Mr. Heaviside puts it, simply barbarous. Three pamphlets
have been published by Messrs. Macmillan and Co.—a syllabus of plane
geometry, corresponding to Euclid, Books I. to VI. ; another of modern
plane geometry ; and another of linear dynamics. Messrs. Swan, Sonnen-
schein and Co. have published an elementary geometrical conics, and there
the labours of the Association appear to have stopped. The Association
has had to struggle against the stubborn conventionalism of the modern
schoolmaster, who pleads that he cannot make any change, because of the
Universities. After a considerable fight, Mr. Hamblin Smith's common-
sense proofs of Euclid's problems were accepted by Cambridge examiners.
Small as the visible results of the Association have been, there is a distinct
change of feeling taking place with regard to geometry, both as an educational subject and as an implement of scientific work. At present geometry
is taught as badly as Greek, even in the best public schools ; and the
educational value of Greek is in many respects higher than that of Euclid."

arithmetic and in scalar algebra does not apply to vectors, because they are not scalars. We cannot, therefore, say beforehand what the product of two vectors ought to be, or determine this question by any prior reasoning of a legitimate nature. We must examine in what way vectors enter into combination as products, and then introduce definitions and conventions of notation to give expression to the facts in the simplest and most convenient way. After this, the work is deductive.

By this examination we are led to recognise two distinct kinds of products of a pair of vectors, the scalar product and the vector product. It is with the former of these that we are now immediately concerned.

We *define* the scalar product of a pair of vectors **A** and **B**, whose tensors are A and B, and whose included angle is θ, to be the scalar AB cos θ, and we denote it by **AB**. Thus

$$\mathbf{AB} = \mathrm{AB}\cos\theta \quad . \quad . \quad . \quad . \quad . \quad (12)$$

defines the scalar product and its notation.

To see its full significance and how it works out, let us first apply (12) to the unit reference vectors, **i**, **j**, **k**, which are coperpendicular, by taking **A** and **B** to be one or other of them in turn. First multiplying each of **i**, **j**, **k** by itself, we get

$$\mathbf{i}^2 = 1, \qquad \mathbf{j}^2 = 1, \qquad \mathbf{k}^2 = 1, \quad . \quad . \quad . \quad . \quad (13)$$

with the convention borrowed from scalar algebra that **ii** is equivalently denoted by $\mathbf{i}^2$. This may be called the square of **i**. In each of the three cases (13) the vectors multiplied together are of unit length, and are parallel and concurrent, so that the scalar product, according to (12), is unity.

Similarly, the square of any unit vector is unity. And since $\mathbf{A} = \mathrm{A}\mathbf{A}_1$, the square of any vector is the same as the square of its tensor; or

$$\mathbf{A}^2 = \mathrm{A}^2. \quad . \quad . \quad . \quad . \quad . \quad . \quad (14)$$

Also, the scalar product of any two parallel vectors is the product of their tensors, for $\mathbf{AB} = \mathrm{AB}\,\mathbf{A}_1\mathbf{B}_1$, and $\mathbf{A}_1\mathbf{B}_1$ is now unity.

Next make scalar products of **i**, **j**, **k** in pairs. We get

$$\mathbf{ij} = 0, \qquad \mathbf{jk} = 0, \qquad \mathbf{ki} = 0. \quad . \quad . \quad . \quad . \quad . \quad (15)$$

That is, the scalar product of two perpendicular unit vectors is zero. The same is, of course, true of any two perpendicular vectors. Thus the equation $AB = 0$ means that A is perpendicular to B; unless, indeed, one or other of them is zero. We have also

$$A_1 B_1 = \cos \theta, \quad \cdots \cdots \quad (16)$$

by dividing (12) by AB; or, the scalar product of any two unit vectors is the cosine of the included angle. This is reckoned positively from concurrent coincidence, so that as θ goes from 0 to 2π, $A_1 B_1$ goes from 1 through 0 to -1, and then through 0 again to $+1$.

There is, strictly, no occasion to introduce trigonometry. Or we might make the trigonometry be a simultaneously developed subject. It is, in fact, a branch of vectorial algebra, being scalar developments of parts thereof. We may employ the idea of perpendicular projection simply. Thus, we may say that the scalar product of a pair of unit vectors is the length of the projection of either upon the other; and that the scalar product of any vector A and a unit vector i is the projection of A upon the axis of i; and, comprehensively, that the scalar product AB of any two vectors A and B is the product of the tensor of either into the projection (perpendicularly) upon it of the other. It is the effective product, so to speak. In physical mathematics scalar products frequently have reference to energy, or activity, or connected quantities. Thus, if F be a force and v the velocity of its point of application, their scalar product Fv is the activity of the force; it is the product of the speed and the effective force. When the force and the velocity are perpendicular, the activity is *nil*, although the velocity may be changing—a fact which familiarity does not render less striking. When F and v are parallel (whether concurrent or not), Fv becomes the same as Fv, in the common meaning of a product. A notation that harmonises in this way is obviously a convenient one.

Notice, also, that $AB = BA$, as in scalar algebra.

A frequently occurring operation is the surface integral of the normal component of a vector; for example, to express the amount of induction through a surface. Here the idea of a unit normal vector N is useful. The normal component of B

is then **NB**, and the integral is Σ **NB**, the summation extending over the surface.

Similarly, to express the line-integral of the effective component of a vector along a line, we may let **T** be the unit element of curve ; that is, the unit tangent ; then **TE** is the tangential component of the vector **E**, and Σ**TE** is the integral ; for example, the voltage between two points along the path to which the summation refers, if **E** be the electric force.

Since **ab** is a scalar, it behaves as a scalar, when considered as a whole. Thus, when multiplied by a scalar, x, the result, x**ab** or **ab**x, is scalar, being simply x times **ab**. When multiplied by a vector the result is a vector ; thus, **c.ab** or **ab.c** means **ab** times the vector **c**. The dot here acts rather as a separator than as a sign of multiplication. Thus, to illustrate, **ca.b** means **ca** times the vector **b** ; and, similarly, **a.bc** is **bc** times the vector **a**. But, instead of the dot, we may use brackets to indicate the same thing, thus **c.ab** may be written **c(ab)**. This is, perhaps, preferable in initiatory work, but I think the dot plan is more generally useful.

We have an example of this combination of three vectors in the stress formulæ. Thus, the electric stress vector on the plane whose unit normal is **N**, is expressed by

$$\mathbf{E}.\mathbf{DN} - \mathbf{N}.\tfrac{1}{2}\mathbf{ED},$$

(equation (31), § 72) ; that is, the sum of two vectors, parallel to **E** the electric force and to **N** respectively, whose tensors are **EDN** and $-\tfrac{1}{2}$**ED** respectively, where **D** is the displacement. The interpretation as a tension along **E** combined with an equal lateral pressure, obtained by taking **N** parallel to and then perpendicular to **E** or **D**, has been already discussed.

Fundamental Property of Scalar Products, and Examples.

§ 108. As all the preceding is involved in the definition of a scalar product, and is obvious enough, it may be regarded merely as illustrative. The reader might, in fact, say that he knew it all before in one form or another, trigonometrical or geometrical, and that he did not see any particular advantage in the way of stating it in the notation of scalar products. But we now come to a very striking and remarkable property

of scalar products, which will go far to justify them as working
utilities.

We know that in scalar algebra, when we have a product
xy, we may express x by the sum of any number of other
quantities, and similarly as regards y, and then obtain the
complete product xy by adding together all the component
products obtained by multiplying every element of x into every
element of y; and that this process may be carried out in any
order.

Now, in vector algebra, we know already that there is a
partial similarity, viz., that we can decompose a vector $\mathbf{A}$ into
the sum of any number of others, and similarly as regards $\mathbf{B}$.
The question now is whether the scalar product $\mathbf{AB}$ is the sum
of all the scalar products made up out of the components of $\mathbf{A}$
paired with the components of $\mathbf{B}$, taken in any order. For
example, if

$$\mathbf{A} = \mathbf{a} + \mathbf{b},$$

and

$$\mathbf{B} = \mathbf{c} + \mathbf{d},$$

is

$$\left. \begin{array}{l} \mathbf{AB} = (\mathbf{a} + \mathbf{b})\,(\mathbf{c} + \mathbf{d}) \\ \quad = \mathbf{ac} + \mathbf{ad} + \mathbf{bc} + \mathbf{bd} \, ? \end{array} \right\} \quad \cdots \quad (17)$$

The answer is Yes, and the process of manipulation is the
same as in scalar algebra, that is, as if all the vectors were
scalars. Moreover, this property admits of demonstration in
a sufficiently simple manner as to enable one to see its truth.
Start with the vector $\mathbf{A}$, and first split it up into

$$\mathbf{A} = \mathbf{a} + \mathbf{b} + \mathbf{c}, \quad \cdots \quad (18)$$

two or three vector components being sufficient for illustration.
Now project the vector $\mathbf{A}$ perpendicularly upon any axis, say
that of $\mathbf{i}$. According to our definition of a scalar product, the
projection is $\mathbf{Ai}$. Now it requires no formal demonstration,
but becomes evident as soon as the meaning of the proposition
is correctly conceived, that the projection of $\mathbf{A}$ upon any axis
is the same as the sum of the projections of its component
vectors, $\mathbf{a}$, $\mathbf{b}$, $\mathbf{c}$, on that axis. That is, the latter projections
are either all positive or all negative, and fit together to make
up the projection of $\mathbf{A}$; or else some may be negative and
others positive, when there is overlapping and cancelling, but

still with the same result algebraically.   This is expressed by

$$\mathbf{Ai} = (\mathbf{a} + \mathbf{b} + \mathbf{c})\mathbf{i} = \mathbf{ai} + \mathbf{bi} + \mathbf{ci}, \quad . \quad . \quad . \quad (19)$$

got by multiplying (18) by i.

If we now multiply (19) by any scalar B, so that $\mathbf{Bi} = \mathbf{B}$, which is any vector, since i may have any direction, we obtain

$$\mathbf{AB} = (\mathbf{a} + \mathbf{b} + \mathbf{c})\mathbf{B} = \mathbf{aB} + \mathbf{bB} + \mathbf{cB}, \quad . \quad . \quad (20)$$

the same as if we multiply (18) by B direct.

Similarly, we may split up B into the sum of any number of other vectors, say,

$$\mathbf{B} = \mathbf{d} + \mathbf{e} + \mathbf{f} . \quad . \quad . \quad . \quad . \quad (21)$$

If we substitute this in (20) we have

$$\mathbf{AB} = \mathbf{a}(\mathbf{d} + \mathbf{e} + \mathbf{f}) + \mathbf{b}(\mathbf{d} + \mathbf{e} + \mathbf{f}) + \mathbf{c}(\mathbf{d} + \mathbf{e} + \mathbf{f}).$$

Now make use of the same reasoning which established (19) and (20), applied to the bracketed vectors, and we establish the property fully, with the result

$$\mathbf{AB} = (\mathbf{a} + \mathbf{b} + \mathbf{c})(\mathbf{d} + \mathbf{e} + \mathbf{f})$$
$$= \mathbf{ad} + \mathbf{ae} + \mathbf{af} + \mathbf{bd} + \mathbf{be} + \mathbf{bf} + \mathbf{cd} + \mathbf{ce} + \mathbf{cf} \; ;$$

the expansion being done formally as in scalar algebra in every respect ; since the various terms may be written in any order, and each may be reversed, thus, $\mathbf{ad} = \mathbf{da}$.

I have already remarked that a good deal of geometry may be done by the addition property alone.   The range of application is greatly extended by the use of the scalar product and the fundamental property (17).

To obtain the Cartesian form of **AB**, put the vectors in terms of **i, j, k**; thus

$$\mathbf{AB} = (\mathrm{A}_1\mathbf{i} + \mathrm{A}_2\mathbf{j} + \mathrm{A}_3\mathbf{k})(\mathrm{B}_1\mathbf{i} + \mathrm{B}_2\mathbf{j} + \mathrm{B}_3\mathbf{k}). \quad . \quad (22)$$

Now effect the multiplications, remembering (13) and (15). The result is

$$\mathbf{AB} = \mathrm{A}_1\mathrm{B}_1 + \mathrm{A}_2\mathrm{B}_2 + \mathrm{A}_3\mathrm{B}_3 . \quad . \quad . \quad . \quad (23)$$

For example, the activity of a force is the sum of the activities of its component forces in any three coperpendicular directions.   Or, if **A** and **B** are unit vectors, we express the cosine of the angle between them in terms of the products of the direction cosines of the vectors, each product referring to one axis. In (23) we may, if we please, vectorise the six scalars on the right side.

Since the square of a vector is the square of its tensor, we may express the tensor at once in terms of the tensors and the cosines of the angles between a series of vectors into which the original vector is resolved. Thus, for two,

$$(A + B)^2 = A^2 + 2AB + B^2, \quad \cdot \quad \cdot \quad \cdot \quad (24)$$

$$(A - B)^2 = A^2 - 2AB + B^2 ; \quad \cdot \quad \cdot \quad \cdot \quad (25)$$

and similarly for three,

$$(A + B + C)^2 = A^2 + B^2 + C^2 + 2AB + 2BC + 2CA. \quad (26)$$

Here (24) and (25) apply to a parallelogram, and (26) to a parallelepiped. In (24), (25), if A and B be the vector sides of a parallelogram, then $(A + B)$ and $(A - B)$ are the vector diagonals ; so (24) gives the length of one diagonal and (25) that of the other.

Adding (24) and (25) we obtain

$$(A + B)^2 + (A - B)^2 = 2(A^2 + B^2), \quad \cdot \quad \cdot \quad (27)$$

expressing that the sum of the squares of the diagonals equals the sum of the squares of the four sides. Similarly, by subtraction

$$(A + B)^2 - (A - B)^2 = 4AB, \quad \cdot \quad \cdot \quad \cdot \quad (28)$$

expressing the difference of the squares of the diagonals as four times the scalar product of two vector sides.

Equation (27) also shows that the sum of the squares of the distances of the ends of any diameter of a sphere from a fixed point is constant. For if A is the vector from the fixed point to the sphere's centre, and B the vector from the centre to one end of the diameter, then $A + B$ and $A - B$ are the vectors from the fixed point to the ends of the diameter. Whence, by (27), the proposition.

There is an application of this in the kinetic theory of gases. For when two elastic spheres collide they keep the sum of their kinetic energies constant. If, then, their velocities before collision be $A + B$ and $A - B$, their velocities after collision are indicated by vectors from the origin (in the velocity diagram)

to the ends of some other diameter of the sphere described
upon the line joining their original positions (in the velocity
diagram) as diameter; which is actually the new diameter
depending upon the circumstances of impact.

Reciprocal of a Vector.

§ 109. It is occasionally useful to employ the reciprocal of a
vector in elementary vector algebra. We define the reciprocal
of a vector a to be a vector having the same direction as a,
and whose tensor is the reciprocal of that of a. We may denote
the reciprocal of a by a^{-1} or $1/a$. Thus as $a = aa_1$, we have

$$a^{-1} = \frac{1}{a} = \frac{1}{aa_1} = \frac{a_1}{a}. \quad \ldots \quad (29)$$

Any unit vector is, therefore, its own reciprocal.

The reciprocal of a vector, being a vector, makes scalar products with other vectors. Thus ab^{-1} or a/b means the scalar
product of a and b^{-1}, and we therefore have

$$\frac{a}{b} = ab^{-1} = \frac{aa_1}{bb_1} = \frac{a}{b}a_1b_1 = \frac{ab}{b^2}. \quad \ldots \quad (30)$$

The tensor of a/b is $(a/b)\cos\theta$, where θ is the angle between
a and b, or between their reciprocals, or between either and the
reciprocal of the other.

So a/a or aa^{-1} or $a^{-1}a$ equals unity.

In using reciprocals the defined meaning should be attended to, especially when put in the fractional form. Thus we
easily see that $ab^{-1} = ab/b^2$, because the tensor of b is b^2 times
the tensor of b^{-1}. But we cannot equivalently write a^2/ab,
because this is $(a/b)/\cos\theta$, which is quite a different thing.

Notice that $a^{-1}b^{-1}$ is not the same as $(ab)^{-1}$. The first is
$a^{-1}b^{-1}\cos\theta$, whilst the latter is $a^{-1}b^{-1}/\cos\theta$.

Expression of any Vector as the Sum of Three Independent Vectors.

§ 110. We know that in the equation

$$r = xi + yj + zk,$$

where r is the vector distance of a point from the origin, the
scalars are the lengths of the projections of r upon the axes.

How should we, however, find them in terms of r algebraically?
To find x we must operate on the equation in such a manner as
to cause the j and k terms to disappear. Now this we can do
by multiplying by i. For i is perpendicular to j and k, so that
multiplication by i gives

$$ri = xi^2 = x.$$

Similarly, $rj = y$ and $rk = z$.

From this obvious case we can conclude what to do when the
reference vectors are not perpendicular; for instance, in

$$r = fa + gb + hc, \quad \cdots \cdots \quad (31)$$

where a, b, c are any independent vectors. To find f we must
multiply by a vector perpendicular to b and c, say l, so that
$lb = 0$, and $lc = 0$. Then

$$rl = fal, \quad \text{therefore} \quad f = rl/al.$$

Similarly to isolate g and h, so that we get

$$r = \frac{rl}{al}a + \frac{rm}{bm}b + \frac{rn}{cn}c, \quad \cdots \cdots \quad (32)$$

where l, m, n are vectors normal to the three planes of b,c, c,a
and a,b. This exhibits explicitly the expansion of any vector
in terms of any three independent vectors a, b, c, as the three
edges of a skew parallelepiped. This case, of course, reduces
to the preceding Cartesian case by taking a, b, c to be i, j, k,
when l, m, n will also be i, j, k, or any scalar multiples of the
same.

Observe the peculiarity that auxiliary vectors are used, each
of which is perpendicular to two others, that is, to the plane
containing them. These auxiliary vectors bring us to the
study of the vector product.

The Vector Product of Two Vectors.   Illustrations.

§ 111. The auxiliary vectors just employed in the expansion of
a vector into the sum of three vectors having any independent
directions, are examples of vector products. Two vectors being
given, their vector product is perpendicular to both of them.
Of course, disregarding magnitude, there is but one such

vector, viz., the normal to the plane containing the given
vectors. Nor is the tensor of any consequence in the example
in question, for it will be observed that one of the vectors,
l, m, n, appears both in the numerator and in the denominator of one of the three fractions in equation (32), so that
the values of the fractions are independent of the tensors of
the auxiliary vectors.

But *the* vector product of two vectors has a strictly-fixed
tensor, depending upon those of the component vectors and
their inclination. There is particular advantage in taking the
tensor of the vector product of **a** and **b** to be $ab \sin \theta$. Thus
we *define* the vector product of two vectors **a** and **b** whose
tensors are a and b, and whose included angle is θ, to be a

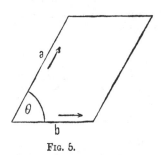

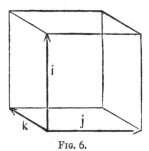

FIG. 5.   FIG. 6.

third vector **c** whose tensor c equals $ab \sin \theta$, and whose direction is perpendicular to the plane of **a** and **b**; the positive
direction of **c** being such that positive or right-handed rotation
about **c** carries the vector **a** to **b**. This vector product is
denoted by

$$c = V\mathbf{ab}, \quad \cdots \cdots \quad (33)$$

and its tensor may be denoted by $V_0\mathbf{ab}$, so that we have

$$c = V_0\mathbf{ab} = ab \sin \theta. \quad \cdots \cdots \quad (34)$$

Similarly, $V_1\mathbf{ab}$ may be used to denote the unit vector
parallel to $V\mathbf{ab}$.

The only troublesome part is to correctly fix which way along
the perpendicular to the plane of **a** and **b** is to be considered
positive. Examples will serve to clinch the matter. Thus, let

a be towards the north, and b towards the east on the earth's surface; then c is straight downwards. Again, in Fig. 5, the direction of c is downwards through the paper, and its tensor is the area of the parallelogram upon a and b.

Let b be fixed, whilst a moves round so as to vary the angle θ. Starting from coincidence, with $\theta = 0$, the tensor of c is zero. It reaches a maximum (downwards) when θ is a right angle, and falls to zero again when θ reaches two quadrants, and a is in the same line with b, though discurrent. After this, in the next two quadrants, the same numerical changes are gone through; but now, the sine being negative, c must be upwards from the paper.

The unit reference vectors i, j, k are so arranged that when, in (33), a is i and b is j, then c is k; noting in Fig. 6, that k is supposed to go downwards, or away from the reader. Thus, in accordance with our definition we have

$$Vij = k, \qquad Vjk = i, \qquad Vki = j, \quad . \quad . \quad (35)$$

because the mutual angles are quadrants and the tensors unity. Observe the preservation of the cyclical order i, j, k in (35). Also, we have

$$Vii = 0, \qquad Vjj = 0, \qquad Vkk = 0, \quad . \quad . \quad (36)$$

because the vectors paired are coincident.

By the definition, the reversal of the order of the letters in a vector product negatives it. Thus

$$Vab = - Vba. \quad . \quad . \quad . \quad . \quad (37)$$

Since the tensor of Vab is $ab \sin \theta$ and the scalar product ab is $ab \cos \theta$, we have

$$(ab)^2 + (Vab)^2 = (ab)^2. \quad . \quad . \quad . \quad (38)$$

Combinations of Three Vectors. The Parallelepipedal Property.

§ 112. A vector product, being a vector, of course combines with other vectors to make scalar and vector products again. Thus cVab, where c is any new vector, means the scalar product of c and Vab; and VcVab means the vector product of c and Vab. These are both important combinations, which occur frequently, and their interpretations and expansions will be given presently.

As illustrative of notation, it may be mentioned that cV_0ab, where, as before explained, V_0ab is the tensor, is obviously a vector parallel to c, but V_0ab times as long. On the other hand, cV_1ab, where V_1ab is the unit vector, is the same as $cVab/V_0ab$; since by dividing by its tensor we unitise a vector. Similarly as regards VcV_1ab. We may also have V_1cVab and V_1cV_1ab and V_0cVab, and various other modifications, whose meanings follow from the definition of a vector product and its notation.

We do not often go further in practical vector algebra than combinations in threes; for instance, on to $dVaVbc$, the scalar product of d and the previously explained $VaVbc$.

The scalar product $cVab$ has an important geometrical illustration. Its value is given by

$$cVab = V_0ab \times c \cos \phi, \quad . \quad . \quad . \quad (39)$$

where ϕ is the angle between c and Vab. This is by the definition of the scalar product, and of the tensor of a vector.

Now refer to Fig. 5 again. We know that V_0ab is the area of the parallelogram. We also know that the volume of a parallelepiped is the product of its base and altitude. Construct, then, a parallelepiped whose three edges meeting at a corner are a, b, and c. The area of one of its bases is V_0ab, and the corresponding altitude is $c \cos \phi$. Therefore, by (39), $cVab$ is the volume of the parallelepiped.

But there are two other bases and two other altitudes to correspond, so there are two other ways of expressing the volume, giving the equalities

$$aVbc = bVca = cVab, \quad . \quad . \quad . \quad (40)$$

in which observe the preservation of cyclical order, done to keep the sign the same throughout, as will be verified a little later.

Semi-Cartesian Expansion of a Vector Product, and Proof of the Fundamental Distributive Principle.

§ 113. The semi-Cartesian expansion of $c = Vab$ is

$$c = i\,(a_2b_3 - a_3b_2) + j\,(a_3b_1 - a_1b_3) + k\,(a_1b_2 - a_2b_1), \quad . \quad (41)$$

in terms of i, j, k and the scalar components of a and b.

To prove this, multiply (41) by **a** and **b** in turns to form the scalar products **ac** and **bc**. Do this by the rule embodied in equation (23). We get

$$\mathbf{ac} = a_1\,(a_2 b_3 - a_3 b_2) + a_2\,(a_3 b_1 - a_1 b_3) + a_3\,(a_1 b_2 - a_2 b_1),$$

$$\mathbf{bc} = b_1\,(a_2 b_3 - a_3 b_2) + b_2\,(a_3 b_1 - a_1 b_3) + b_3\,(a_1 b_2 - a_2 b_1).$$

But, by cancelling, all the terms on the right disappear. That is,

$$\mathbf{ac} = 0, \qquad \mathbf{bc} = 0.$$

From these we know that **c** is perpendicular to **a** and to **b**. It is, therefore, V**ab** itself, or a multiple of the same. To find its tensor, square (41). We get

$$c^2 = (a_2 b_3 - a_3 b_2)^2 + (a_3 b_1 - a_1 b_3)^2 + (a_1 b_2 - a_2 b_1)^2. \quad (42)$$

But this may, by common algebra, be transformed to

$$c^2 = (a_1{}^2 + a_2{}^2 + a_3{}^2)\,(b_1{}^2 + b_2{}^2 + b_3{}^2) - (a_1 b_1 + a_2 b_2 + a_3 b_3)^2. \quad (43)$$

That is, $$\mathbf{c}^2 = \mathbf{a}^2 \mathbf{b}^2 - (\mathbf{ab})^2, \quad \cdots \quad (44)$$

or, $$c^2 = a^2 b^2 - a^2 b^2 \cos^2 \theta = (ab \sin \theta)^2.$$

The tensor c is, therefore, that of V**ab** itself, or else its negative. Equation (44) is the same as (38), in a slightly different form. Equation (41) is proved, except that it remains to be seen whether the right member represents V**ab** or its negative.

To do this, we may remark that the angle between two planes is the same as the angle between the normals to the planes, if the normals coincide when the planes do, so that the three quantities in ( )'s in (41), which are the scalar projections of **c** on the axes, are also the areas of the projections of the area $V_0\mathbf{ab}$ on the planes perpendicular to them. We have therefore only to verify that the projection of any plane on the plane normal to one of the axes is given correctly in sign by (41). Take a rectangle, parallel to the **j**, **k** plane, sides parallel to **j** and **k**, that is, $a_2\mathbf{j}$ and $b_3\mathbf{k}$; so that

$$\mathbf{c} = \mathbf{i} a_2 b_3.$$

Here $a_2 b_3$ is the projection of the rectangle on a parallel plane, and is correctly positive.

Now go back to (41). We have

$$\mathbf{V}ab = \mathbf{i}\,(a_2 b_3 - a_3 b_2) + \dots \quad \bullet \quad \bullet \quad \bullet \quad (45)$$

Similarly
$$\mathbf{V}ac = \mathbf{i}\,(a_2 c_3 - a_3 c_2) + \dots \quad \bullet \quad \bullet \quad \bullet \quad (46)$$

Add these together. We get

$$\mathbf{V}ab + \mathbf{V}ac = \mathbf{i}\,[a_2\,(b_3 + c_3) - a_3\,(b_2 + c_2)] + \dots \quad \bullet \quad (47)$$

But the right member, by definition of a vector product, or by
(41), means $\mathbf{V}a\,(b+c)$. That is,

$$\mathbf{V}a\,(b+c) = \mathbf{V}ab + \mathbf{V}ac. \quad \bullet \quad \bullet \quad \bullet \quad \bullet \quad (48)$$

Similarly
$$\mathbf{V}d\,(b+c) = \mathbf{V}db + \mathbf{V}dc. \quad \bullet \quad \bullet \quad \bullet \quad \bullet \quad (49)$$

So, by adding these again,

$$\mathbf{V}\,(a+d)\,(b+c) = \mathbf{V}ab + \mathbf{V}ac + \mathbf{V}db + \mathbf{V}dc, \quad \bullet \quad (50)$$

a very remarkable and important formula. It shows that the
vector product of two vectors, **A** and **B**, equals the sum of all
the vector products which can be made up out of the component vectors [(a + d) and (b + c) in (50)] into which we may
divide **A** and **B**, provided we keep the components of **A** always
before those of **B**. The necessity of this proviso of course
follows from the reversal of $\mathbf{V}ab$ with the order of a and b.
Although (50) only shows this for two components to each
primary vector, yet the process by which it was obtained evidently applies to any number of components.

Subject to the limitation named, the formula (50) is precisely similar to (17) § 108, relating to the scalar products.
Now the truth of (17) could be seen without much trouble.
A similar proof of (50), on the other hand, would not be at all
easy to follow, owing to the many changes of direction involved in the vector products. This is why I have done it
through **i**, **j**, **k**, which are auxiliaries of the greatest value.
When in doubt and difficulty, fly to **i**, **j**, **k**.

On the other hand, although the establishment of (50) by
geometry without algebra is difficult, and there is preliminary
trouble in fixing the direction of a vector product, yet we see
from (50) that vector products are nearly as easily to be
manipulated algebraically as scalar products.

M

Examples relating to Vector Products.

§ **114.** By means of the formula (41) we can at once obtain the cartesian expansion of the parallelepipedal $c\text{Vab}$. Use equation (23), applied to (41), remembering that c is now any vector, and we get

$$c\text{Vab} = c_1(a_2b_3 - a_3b_2) + c_2(a_3b_1 - a_1b_3) + c_3(a_1b_2 - a_2b_1). \quad (51)$$

By rearrangement of terms on the right side of this, putting the a's outside the brackets, we obtain $a\text{Vbc}$, and putting the b s outside, we obtain $b\text{Vca}$, and thus verify (40) again.

The vector $V c\text{Vab}$ is, perhaps, most simply expanded through i, j, k. First write d for Vab, then by (41) we have

$$Vcd = i(c_2d_3 - c_3d_2) + \ldots .$$

Next put for d_1, d_2, d_3 their values in terms of a's and b's, given in (41), and we have

$$Vc\text{Vab} = i[c_2(a_1b_2 - a_2b_1) - c_3(a_3b_1 - a_1b_3)] + \ldots$$
$$= i[a_1(b_2c_2 + b_3c_3) - b_1(a_2c_2 + a_3c_3)] + \ldots .$$

Next, add and subtract $a_1b_1c_1$, and we get

$$Vc\text{Vab} = i[a_1bc - b_1ac] + \ldots .$$

Lastly, reconvert fully to vectors, and we have

$$Vc\text{Vab} = a.bc - b.ca. \quad \cdot \quad \cdot \quad \cdot \quad (52)$$

This important formula should be remembered. That $Vc\text{Vab}$ is in the plane of a and b is evident beforehand, because Vab is perpendicular to this plane, and multiplying by Vc sends it back into the plane. It is, therefore, expressible in terms of a and b, as in (52), which shows the magnitude of the components.

In some more complex formulæ it is sufficient to remember the principle upon which they are founded, as by its aid they can be recovered at any time. Thus in the case of

$$r = f a + g b + h c, \quad \cdot \quad \cdot \quad \cdot \quad \cdot \quad (53)$$

already treated, equation (31), if we use now Vbc for l, which was, before, any multiple of it, and similarly for m and n, we

have
$$r = \frac{r\text{Vbc}}{a\text{Vbc}} a + \frac{r\text{Vca}}{b\text{Vca}} b + \frac{r\text{Vab}}{c\text{Vab}} c. \quad \cdot \quad \cdot \quad (54)$$

There is no occasion to put a formula like this in the memory, because it is so simply obtained at any time from (53) by multiplying by the auxiliary vectors so as to isolate f, g, h in turn and give their values.

Notice that the three denominators in (54) are equal. Also that by exchanging a and Vbc, b and Vca, c and Vab, we obtain

$$\mathbf{r} = \frac{\mathbf{ra}}{\mathbf{a}\text{Vbc}}\,\text{Vbc} + \frac{\mathbf{rb}}{\mathbf{b}\text{Vca}}\,\text{Vca} + \frac{\mathbf{rc}}{\mathbf{c}\text{Vab}}\text{Vab}, \qquad . \quad (55)$$

and this is also true, as we may at once prove by multiplying it in turns by a, b, and c, each of which operations nullifies two terms on the right. (55) is the expression of r in terms of three vectors, which are normal to the three planes of any three independent vectors, a, b, c, taken in pairs. Here, again, such a formula can be immediately recovered if wanted by attending to the principle concerned.

To obtain the cartesian expansion of any formula containing vectors is usually a quite mechanical operation. The cartesian, or semi-cartesian, representatives of a few fundamental functions being remembered, their substitution in the vector formula is all that is required. If the formula be scalar (although involving as many vectors as we please) the result is a single scalar formula in cartesians. But if it be a vector formula, it reduces to a semi-cartesian vector formula involving i, j, k, giving three scalar equations, one for each component.

The converse process, to put a scalar cartesian investigation into vectorial form, is less easy, though dependent upon the same principles. Here the three component equations have to be reduced to one vector equation. It is very good practice to take a symmetrically written-out cartesian investigation and go through it, boiling it down to a vector investigation, using the unit reference vectors i, j, k, whenever found to be convenient.

The Differentiation of Scalars and Vectors.

§ 115. In the preceding account of vector addition, and of the scalar product and the vector product, the reader has nearly all that is needed for general purposes in geometry and in the usual physical mathematics involving vectors, so far as the algebra itself is concerned. For the addition of differentiations does not usually introduce anything new into the algebra. Thus, in the analysis of varying vectors, we have the same vector algebra with new vectors introduced, these being derived from others by the process of differentiation.

The ideas concerned in the differentiation of a vector with respect to a scalar are essentially the same as in the differentiation of a scalar. Thus, u being a scalar function of x, *i.e.*, a quantity whose value depends on that of x, we know that if Δu is the increment in u corresponding to the increment Δx in x, then the ratio $\Delta u/\Delta x$ usually tends to a definite limiting value when Δx is infinitely reduced, which limit, denoted by du/dx, is called the differential coefficient of u with respect to x. This I should be strongly tempted to call the "differentiant" of u to x, were I not informed on the highest authority that the expression "differentiant" for "differential coefficient" is objectionable. But the differentiating operator d/dx which acts on the operand u may be termed the "differentiator," as has perhaps been done by Sylvester and others. This way of regarding a differential coefficient, splitting du/dx into (d/dx) and u, sometimes leads to great saving of labour, though we are not concerned with it immediately.

The differential coefficient is thus strictly the *rate* of increase of the function with the variable, or the increase of the function per *unit* increase of the variable, on the tacit assumption that the rate of increase of the function keeps the same value throughout the whole unit increment in the variable as it has for the particular value of the variable concerned. This plan, referring to unit increment, is often very useful. The reservation involved, though it must be understood, need not be mentioned.

There is also the method of differentials, which is of some importance in vector analysis, as it can be employed when differential coefficients do not exist. Thus, u being a function of x, whose differential is dx, the corresponding differential of u is

$$du = f(x+dx) - f(x); \quad . \quad . \quad . \quad (56)$$

which, by expanding $f(x+dx)$, reduces to

$$du = \frac{du}{dx}dx, \quad . \quad . \quad . \quad . \quad (57)$$

provided we neglect the squares and higher powers of dx, that is, regard dx as infinitesimal, and therefore du also. And if there be two variables, as in $u = f(x, y)$, then we have

$$du = f(x + dx, y + dy) - f(x, y), \quad \cdot \quad \cdot \quad (58)$$

leading to
$$du = \frac{du}{dx} dx + \frac{du}{dy} dy, \quad \cdot \quad \cdot \quad \cdot \quad \cdot \quad (59)$$

expressing the differential of u in terms of those of x and y.

Now let the operand be a vector function of x, say, **E**. For instance, x may mean distance measured along a straight line or axis in an electrostatic field, where **E**, the electric force, will usually change as we pass along the line. It may change continuously in direction as well as in magnitude ; in any case, the change itself in **E** between two points on the line is a vector, being the vector Δ**E** which must be added to the **E** at the first point to produce that at the second point. Dividing by Δx, the increment in x, we get the vector Δ**E**$/\Delta x$; and this, when the increments are taken smaller and smaller, approximates towards a definite limiting vector denoted by d**E**$/dx$, which is, by the manner of its construction, the rate of increase of **E** with x, or the increase in **E** per unit increase in x.

We have also, in differentials,

$$d\mathbf{E} = f(x + dx) - f(x), \quad \cdot \quad \cdot \quad \cdot \quad (60)$$

leading to
$$d\mathbf{E} = \frac{d\mathbf{E}}{dx} dx, \quad \cdot \quad \cdot \quad \cdot \quad \cdot \quad \cdot \quad (61)$$

understanding that the differentials dx and d**E** are infinitesimal.

Similarly, the rate of increase of the vector d**E**$/dx$ with x is the second differential coefficient $d^2\mathbf{E}/dx^2$; and so on.

Semi-Cartesian Differentiation. Examples of Differentiating Functions of Vectors.

§ 116. In semi-Cartesian form we have

$$\mathbf{E} = \mathbf{i}E_1 + \mathbf{j}E_2 + \mathbf{k}E_3. \quad \cdot \quad \cdot \quad \cdot \quad \cdot \quad (62)$$

Here the scalars E_1, E_2, E_3 are functions of x, whilst the reference vectors **i**, **j**, **k** are constants ; so we have

$$\frac{d\mathbf{E}}{dx} = \mathbf{i}\frac{dE_1}{dx} + \mathbf{j}\frac{dE_2}{dx} + \mathbf{k}\frac{dE_3}{dx}, \quad \cdot \quad \cdot \quad (63)$$

showing the components of the vector d**E**$/dx$.

Similarly, $\dfrac{d^2\mathbf{E}}{dx^2} = \mathbf{i}\dfrac{d^2E_1}{dx^2} + \mathbf{j}\dfrac{d^2E_2}{dx^2} + \mathbf{k}\dfrac{d^2E_3}{dx^2}$ . . . (64)

shows the components of the second differential coefficient, obtained by differentiating (63).

Now, this semi-Cartesian process is quite general. Any vector function may be expressed in the form (62), and when the right member is differentiated, the differentiations are performed upon scalar functions. The same remark applies to (63) and (64), &c. ; so we see that the rules for differentiating vectors, and functions of vectors, with respect to scalar variables, are the same as those for differentiating similar functions of scalars, so far as the independent action of a differentiator on the separate members of a product goes.

Thus, in differentiating a scalar product, $\mathbf{AB}$, with respect to a variable scalar, t, we have, by (23),

$$\frac{d}{dt}\,\mathbf{AB} = \frac{d}{dt}\,(A_1\,B_1 + A_2\,B_2 + A_3\,B_3)$$

$$= (\dot{A}_1\,B_1 + \dot{A}_2\,B_2 + \dot{A}_3\,B_3) + (A_1\,\dot{B}_1 + A_2\,\dot{B}_2 + A_3\,\dot{B}_3) \;;$$

or, re-transforming to vectors,

$$\frac{d}{dt}\mathbf{AB} = \dot{\mathbf{A}}\mathbf{B} + \mathbf{A}\dot{\mathbf{B}}, \quad . \quad . \quad . \quad . \quad (65)$$

just as if $\mathbf{A}$ and $\mathbf{B}$ were scalars. Similarly, we have

$$\frac{d}{dt}\,V\mathbf{AB} = V\dot{\mathbf{A}}\mathbf{B} + V\mathbf{A}\dot{\mathbf{B}}, \quad . \quad . \quad . \quad (66)$$

as we may see immediately by differentiating the semi-Cartesian expansion of $V\mathbf{AB}$, equation (41). We have also

$$\frac{d}{dt}\,\mathbf{A}V\mathbf{BC} = \dot{\mathbf{A}}V\mathbf{BC} + \mathbf{A}V\dot{\mathbf{B}}\mathbf{C} + \mathbf{A}V\mathbf{B}\dot{\mathbf{C}}, \quad . \quad . \quad . \quad (67)$$

and $\dfrac{d}{dt}\,V\mathbf{A}V\mathbf{BC} = V\dot{\mathbf{A}}V\mathbf{BC} + V\mathbf{A}V\dot{\mathbf{B}}\mathbf{C} + V\mathbf{A}V\mathbf{B}\dot{\mathbf{C}}$ . . . (68)

But, although we proceed formally, as in the ordinary differentiation of scalars, as regards the properties peculiar to differentiation, yet we must always remember those which are peculiar to the vector algebra. For example, we must not only on the left sides, but also on the right sides of (67) and

(68), remember the reversal of the sign of a vector product with the order of the letters.

Independently of **i**, **j**, **k**, we may proceed thus, using differentials :—

$$d\,(\mathbf{AB}) = (\mathbf{A} + d\mathbf{A})\,(\mathbf{B} + d\mathbf{B}) - \mathbf{AB} = \mathbf{A}d\mathbf{B} + \mathbf{B}d\mathbf{A}, \quad . \quad (69)$$

omitting the infinitesimal $d\mathbf{A}d\mathbf{B}$ of the second order. Dividing by dt, and proceeding to the limit, we obtain (65).

Similarly we have

$$d\,(\mathbf{VAB}) = \mathbf{V}(\mathbf{A} + d\mathbf{A})\,(\mathbf{B} + d\mathbf{B}) - \mathbf{VAB} = \mathbf{V}(d\mathbf{A})\mathbf{B} + \mathbf{VA}d\mathbf{B}, \quad . \quad (70)$$

omitting the second differential $\mathbf{V}d\mathbf{A}d\mathbf{B}$. So, dividing by dt, we obtain (66).

In getting (69) we have, of course, used the distributive law of scalar products, equation (17) ; and, in getting (70), the similar law of vector products, equation (50).

Motion along a Curve in Space. Tangency and Curvature ; Velocity and Acceleration.

§ 117. Some spacial and motional examples may be here usefully inserted to illustrate the differentiation of vectors. Let

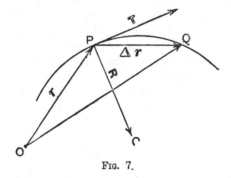

FIG. 7.

there be any curve in space, and let s be length measured along it from some point on the curve. Also let **r** be the vector from any fixed origin O to a point P on the curve. We have then $\mathbf{r} = f(s)$. With the form of the function we have no concern.

Now, consider the meaning of the first differential coefficient $d\mathbf{r}/ds$. First, if the increment Δs is finite, say the *arc* PQ, the

corresponding increment in $\Delta \mathbf{r}$ is the vector *chord* PQ, the new
vector of the curve being OQ, that is $\mathbf{r} + \Delta \mathbf{r}$, instead of OP or $\mathbf{r}$.
We see that $\Delta \mathbf{r}/\Delta s$ is a vector parallel to the chord, whose tensor
is the ratio of the chord to the arc. Now reduce the incre-
ments to nothing. In the limit the direction of $d\mathbf{r}/ds$ is that
of the tangent to the curve at P, and its tensor is unity, since
the chord and the arc tend to equality. That is,

$$\frac{d\mathbf{r}}{ds} = \mathbf{T} = \mathbf{i}\frac{dx}{ds} + \mathbf{j}\frac{dy}{ds} + \mathbf{k}\frac{dz}{ds}, \quad \ldots \quad (71)$$

where $\mathbf{T}$ is the unit vector tangent, and the semi-Cartesian
expression (got by expanding $\mathbf{r}$) is also given. In the figure the
size of $\mathbf{T}$ is quite arbitrary, since the unit of length is arbitrary.

Next, differentiate again with respect to s to obtain the
second differential coefficient $d^2\mathbf{r}/ds^2$ or $d\mathbf{T}/ds$. It is the change
in the unit tangent per unit step along the curve. Now, as the
unit tangent cannot change its length, it can only change its
direction ; and, moreover, this change is a vector perpendicular
to the tangent. It and the two tangents, initial and final, are
in some plane, usually termed the osculating plane. It is the
plane of the curve for the time being, unless it be a plane
curve, when it is the plane of the whole curve. Thus $d\mathbf{T}/ds$,
being at right angles to $\mathbf{T}$, points from the curve towards the
centre of curvature for the time being. Moreover, its tensor
measures the curvature. For the usual measure is $d\theta/ds$,
where $d\theta$ is the angle between the tangents at the ends of ds,
and it may be readily seen that this is the tensor of $d\mathbf{T}/ds$.

Now the reciprocal of the curvature is the radius of curvature. If then $\mathbf{R}$ be the vector from the curve to the centre of
curvature C (Fig. 7), we have

$$\frac{d^2\mathbf{r}}{ds^2} = \frac{d\mathbf{T}}{ds} = \frac{1}{\mathbf{R}}, \quad \ldots \quad (72)$$

and the vector from the origin to the centre of curvature is
$\mathbf{r} + \mathbf{R}$.

Now, referring to the same figure, let a point move along the
curved path. Consider its velocity and acceleration, taking t
the time for independent variable. We have

$$\frac{d\mathbf{r}}{dt} = \frac{d\mathbf{r}}{ds}\frac{ds}{dt} = v\mathbf{T} = \mathbf{v}, \quad \ldots \quad (73)$$

by using (71), and denoting the velocity by v.   Its tensor is v or ds/dt.   The interpretation of (73) is sufficiently obvious, since it says that the motion is (momentarily) along the tangent, at speed v.

Differentiate again to t.   We get

$$\frac{d^2\mathbf{r}}{dt^2} = \frac{dv}{dt}\,\mathbf{T} + v\frac{d\mathbf{T}}{dt}. \quad \cdot \quad \cdot \quad \cdot \quad \cdot \quad \cdot \quad (74)$$

That is, the vector rate of acceleration is exhibited as the sum of two vectors, the first being the tangential component, whose tensor equals the rate of acceleration of speed, whilst the second we may expand thus, by introducing the intermediate variable s :—

$$v\frac{d\mathbf{T}}{dt} = v\frac{d\mathbf{T}}{ds}\frac{ds}{dt} = \frac{v^2}{\mathbf{R}}, \quad \cdot \quad \cdot \quad \cdot \quad \cdot \quad (75)$$

where the third form is got by using (72).   This shows the rate of acceleration perpendicular to the motion.   It is towards the centre of curvature, of amount v^2/R, the well known result.

Tortuosity of a Curve, and Various Forms of Expansion.

§ 118. Again referring to Fig. 7, if the curve be not a plane curve, or be a tortuous curve, the osculating plane undergoes change as we pass along the curve.   It turns round the tangent, and the measure of the tortuosity is the amount of turning per unit step along the curve.   It is $d\phi/ds$, if $d\phi$ is the angle between the two osculating planes at the extremities of ds.   It is equivalently denoted by the tensor of $d\mathbf{N}/ds$, if $\mathbf{N}$ be the unit normal to the osculating plane.   This is similar to the equivalence of $d\theta/ds$ and the tensor of $d\mathbf{T}/ds$ in measuring the curvature before noticed, the equivalent plane in that case being the plane normal to $\mathbf{T}$.

The three vectors, $\mathbf{T}$, $\mathbf{N}$, and $\mathbf{R}_1$, form a unit rectangular system, for we have

$$\mathbf{N} = \mathrm{V}\mathbf{T}\mathbf{R}_1, \quad \cdot \quad \cdot \quad \cdot \quad \cdot \quad \cdot \quad (76)$$

since $\mathbf{T}$ and $\mathbf{R}_1$ are perpendicular unit vectors.   As we move along the curve this system of axes moves as a rigid body, since there is no relative change.   They may, in fact, be imagined to be three perpendicular axes fixed in a rigid body moving along the curve in such a way that these axes keep always coincident with the $\mathbf{T}$, $\mathbf{R}_1$, and $\mathbf{N}$ of the curve. As it moves, the body rotates. It rotates about the $\mathbf{N}$ axis only when the curve is plane, but about the $\mathbf{T}$ axis as well when tortuous. The vector change (infinitesimal) in $\mathbf{T}$ is perpendicular to it, in the osculating plane—that is, the plane whose normal is $\mathbf{N}$, and is directed towards the centre of curvature. And the vector change (infinitesimal) in $\mathbf{N}$ is perpendicular to it, also in the osculating plane, and directed towards the centre of curvature, the vector $\mathbf{R}$ from the curve to the centre of curvature being on the intersection of the two planes mentioned.

Various expressions for the tortuosity may be obtained. Let it be Y the tensor of $\mathbf{Y}$, given by

$$\mathbf{Y} = \frac{d\mathbf{N}}{ds} = \frac{d}{ds} \ \mathrm{V}\mathbf{T}\mathbf{R}_1. \quad \ldots \quad .(77)$$

Then, by (76) and the context, (77) gives one expression. A second is got by performing the differentiation, giving

$$\mathbf{Y} = \mathrm{V} \frac{d\mathbf{T}}{ds} \ \mathbf{R}_1 + \mathrm{V}\mathbf{T}\frac{d\mathbf{R}_1}{ds}. \quad \ldots \quad (78)$$

But the first term on the right is zero, by (72), the two vectors after V being parallel. So

$$\mathbf{Y} = \mathrm{V}\mathbf{T} \frac{d\mathbf{R}_1}{ds} \quad \ldots \quad (79)$$

gives another expression. Or

$$\mathrm{Y} = \mathrm{V}_0\mathbf{T} \frac{d\mathbf{R}_1}{ds}, \quad \ldots \quad (80)$$

considering the tensor only. But, since $\mathbf{Y}$ is parallel to $\mathbf{R}_1$, we have $\mathbf{Y}\mathbf{R}_1 = \mathrm{Y}$; therefore, by (79),

$$\mathrm{Y} = \mathbf{R}_1 \mathrm{V}\mathbf{T}\frac{d\mathbf{R}_1}{ds} \quad \ldots \quad (81)$$

is a third and entirely different expression. That $\mathbf{Y}$ is parallel to $\mathbf{R}_1$ we may prove by (77) and (79). Multiply (77) by $\mathbf{N}$.

Then $$\mathbf{Y}\mathbf{N} = \mathbf{N} \frac{d\mathbf{N}}{ds} = \tfrac{1}{2}\frac{d}{ds} \mathbf{N}^2 = 0 , \quad \ldots \quad (82)$$

since $$\mathbf{N}^2 = 1. \quad \ldots \quad \ldots \quad (83)$$

This shows $\mathbf{Y}$ is perpendicular to $\mathbf{N}$. It is also, by (79), perpendicular to $\mathbf{T}$. So its direction is that of $\mathbf{R}_1$.

By the parallelepipedal property (40) we may also write (81) thus,

$$\mathbf{Y} = -\mathbf{T}\mathbf{V}\mathbf{R}_1 \frac{d\mathbf{R}_1}{ds}. \quad \ldots \ldots \quad (84)$$

Here again, we may write

$$\frac{d\mathbf{R}_1}{ds} = \frac{d}{ds}\frac{\mathbf{R}}{\mathbf{R}} = \frac{1}{\mathbf{R}}\frac{d\mathbf{R}}{ds} + \mathbf{R}\frac{d}{ds}\frac{1}{\mathbf{R}} \quad \ldots \quad (85)$$

Substituting this compound vector in (84), we see that the first of the resulting vector products vanishes because $\mathbf{R}_1$ and $\mathbf{R}^{-1}$ are parallel. There is then left

$$\mathbf{Y} = -\mathbf{T}\mathbf{V}\mathbf{R}\frac{d}{ds}\frac{1}{\mathbf{R}} \quad \ldots \ldots \quad (86)$$

$$= -\mathbf{R}^2\mathbf{T}\mathbf{V}\frac{d\mathbf{T}}{ds}\frac{d^2\mathbf{T}}{ds^2}, \quad \ldots \quad (87)$$

by (72), or, in terms of $\mathbf{r}$ and derivatives,

$$\mathbf{Y} = -\mathbf{R}^2\frac{d\mathbf{r}}{ds}\mathbf{V}\frac{d^2\mathbf{r}}{ds^2}\frac{d^3\mathbf{r}}{ds^3}, \quad \ldots \ldots \quad (88)$$

which is a known symmetrical form, but whose meaning is less easy to understand than several preceding expressions.

Since, by the definition of a vector product we have $\mathbf{V}_0\mathbf{a}_1\mathbf{b} = b$ when $\mathbf{a}_1$ and $\mathbf{b}$ are perpendicular, $\mathbf{a}_1$ being a unit and $\mathbf{b}$ any vector, we may, from the first equation (77) conclude that

$$\mathbf{Y} = \mathbf{V}_0\mathbf{N}\frac{d\mathbf{N}}{ds}, \quad \ldots \ldots \quad (89)$$

which is equivalent to a form given by Thomson and Tait.

Similarly, $$\mathbf{Y} = \mathbf{V}_0\mathbf{T}\frac{d\mathbf{N}}{ds}, \quad \ldots \ldots \quad (90)$$

$\mathbf{Y}$ being perpendicular to $\mathbf{T}$ as well as to $\mathbf{N}$.

The reader may, perhaps, find the above relating to tortuosity hard to follow, though the previous matter relating to tangency, curvature, velocity, and acceleration may be sufficiently plain. The hardness lies in the intrinsic nature of tortuosity, as to which see works on analytical geometry. But the reader who wishes to get a sound working knowledge of vectors should go through the ordinary Cartesian investigations, and turn into vectors. Also, convert the above to Cartesian form through i, j, k, and verify all the results, or correct them, as the case may be. Space is too short here for much detail.

Hamilton's Finite Differentials Inconvenient and Unnecessary.

§ 119. It is now desirable to say a few words regarding the method of treating vector differentiation employed by Hamilton, and followed by Tait. The latter speaks in his treatise (chapter I., § 33, third edition) of the novel difficulties that arise in quaternion differentiation ; and remarks that it is a striking circumstance, when we consider the way in which Newton's original methods in the differential calculus have been decried, to find that Hamilton was *obliged* to employ them, and not the more modern forms, in order to overcome the characteristic difficulties of quaternion differentiation. (For the word "quaternion," we may read vector or vectorial here, because a vector is considered by Hamilton and Tait to be a quaternion, or is often counted as one. This practice is sometimes confusing. Thus the important physical operator ▽ is called a quaternion operator. It is really a vector. It is as unfair to call a vector a quaternion as to call a man a quadruped ; although, four including two, the quadruped might be held (in the matter of legs) to include the biped, or, indeed, the triped, which would be more analogous to a vector. It is also often inconvenient that the name of the science, viz., Quaternions, should be a mere repetition of the name of the operator. There is some gain in clearness by preserving the name "quaternion" for the real quaternion—the quadruped, that is to say.) The matter is illustrated by the motion of a point along a curve. But it does not appear, from this illustration, where the obligation to depart from common usage comes in.

The subject is, however, returned to more generally in Chapter IV., on Differentiation, where we are informed that we "require to employ a definition of a differential somewhat different from the ordinary one, but coinciding with it when applied to functions of mere scalar variables." Here again, however, a most searching examination fails to show me the necessity of the requirement ; or, at least, that the necessity is demonstrated.

To examine this matter, consider for simplicity a function u of a single scalar variable a and a single vector variable $\mathbf{r}$, say

$$u = f(\mathbf{r}, a).$$

Following the common method of infinitesimal differentials, we have

$$du = f(\mathbf{r} + d\mathbf{r}, a + da) - f(\mathbf{r}, a). \quad . \quad . \quad . \quad (91)$$

That is, on giving the infinitesimal increments $d\mathbf{r}$ and da to $\mathbf{r}$ and a we produce the infinitesimal increment du in the function. Here du, $d\mathbf{r}$ and da are the differentials, and it is important to remark, in connection with the following, that they are infinitesimal, and that du is really the increment corresponding to $d\mathbf{r}$ and da.

Now, Hamiltonian differentials have a different meaning. Though they would be denoted by du, $d\mathbf{r}$ and da, yet, being different, we shall here dignify them with brackets, and denote them by (du), $(d\mathbf{r})$, and (da), to distinguish them from the differentials in (91). The Hamiltonian differential (du) is then defined by

$$(du) = n\left[f\left(\mathbf{r} + \frac{(d\mathbf{r})}{n}, a + \frac{(da)}{n} \right) - f(\mathbf{r}, a) \right], \quad (92)$$

on the understanding that n is infinity, and that (on which stress is laid) the differentials $(d\mathbf{r})$ and (da) are finite and perfectly arbitrary.

That this process is circuitous is obvious, but is it necessary? Let us examine into its meaning, and compare (92) with (91). First, divide (92) by n. We then have

$$\frac{(du)}{n} = f\left(\mathbf{r} + \frac{(d\mathbf{r})}{n}, a + \frac{(da)}{n} \right) - f(\mathbf{r}, a). \quad . \quad . \quad (93)$$

Now, we see at a glance, that (93) and (91) become identical if

$$da = \frac{(da)}{n}, \qquad d\mathbf{r} = \frac{(d\mathbf{r})}{n}, \qquad du = \frac{(du)}{n} \quad . \quad . \quad (94)$$

The division of the finite differentials (da) and $(d\mathbf{r})$ by n produces infinitesimal results, and the finite differential (du) is similarly reduced to an infinitesimal. Why then not employ the infinitesimal differentials at once, and avoid the circuitousness?

When we proceed to form a differential coefficient, as du/dt, if $\mathbf{r}$ and a are functions of t, or du/da, if $\mathbf{r}$ is a function of a, we see that their values are equivalently expressed by $(du)/(dt)$ and $(du)/(da)$, using the finite differentials; because this merely says that the value of the fraction du/dt is unaffected when the numerator and denominator are each multiplied by the same quantity.

In the above, the reasoning is the same, whether u is a scalar or a vector function of $\mathbf{r}$ and a.

But whilst the differential coefficient comes out the same, yet the differential (du) is not the increment in u belonging to the finite increments $(d\mathbf{r})$ and (da), nor is it anything like it, exceptions excepted.

The above is as fair and as clear a statement as I can make of the difference between common and finite differentials as applied to vectors; and we see that the Hamiltonian plan is only a roundabout and rather confusing way of expressing what is done instantly with infinitesimal differentials. But then the Hamiltonian plan is said to be obligatory. There's the rub! I can only give my own personal experience. After muddling my way somehow through the lamentable quaternionic Chapter II., the third chapter was tolerably easy; but at the fourth I stuck again, on this very matter of the obligatory nature of the finite differentials. I am no wiser now about it than I was then. Perhaps there is some very profound and occult mystery involved which cannot be revealed to the vulgar in a treatise; Hamilton is a name to conjure by. Or perhaps I "began the study of quaternions too soon!" Or perhaps it is only a fad, after all. Anyway, I am willing to learn, and if I live long enough, may reach a suitable age. At present let us return to common infinitesimal differentials.

Determination of Possibility of Existence of Differential Coefficients.

§ **120.** Although we may employ the method of differentials in any equation involving vectors, using the symbol d, and although we may turn it into an equation of differential coefficients by changing d to the scalar differentiator d/dt (for example), as in § 116, yet we cannot usually turn differentials into differential coefficients when the variable is a vector.

This is not because we cannot form the expressions $du/d\mathbf{r}$ or $d\mathbf{R}/d\mathbf{r}$, but because they are not differential coefficients when the denominators are vector differentials. This will be easily seen from typical examples. (The failure has nothing to do with the previous question of finiteness or otherwise of differentials.) Let

$$u = \mathbf{ab}, \quad \cdots \quad \cdots \quad (95)$$

and let the vector $\mathbf{b}$ vary. Then,

$$du = \mathbf{a}d\mathbf{b}, \quad \cdots \quad \cdots \quad (96)$$

and, therefore, $\dfrac{du}{d\mathbf{b}} = (d\mathbf{b})^{-1}(\mathbf{a}d\mathbf{b}) = (d\mathbf{b})_1[\mathbf{a}(d\mathbf{b})_1], \quad \cdots \quad (97)$

where $(d\mathbf{b})_1$ means the unit $d\mathbf{b}$. Here the imitation differential coefficient $du/d\mathbf{b}$ is a fraud, because its expression on the right side of (97) is not *clear of the differential* $d\mathbf{b}$, whose direction it involves. Nor is equation (97) of any particular use, compared with (96).

Again, in the vector equation

$$\mathbf{R} = V\mathbf{ab} \quad \cdots \quad \cdots \quad (98)$$

let $\mathbf{b}$ be variable. Here we have

$$d\mathbf{R} = V\mathbf{a}d\mathbf{b}, \quad \cdots \quad \cdots \quad (99)$$

and from this equation of differentials we may form $d\mathbf{R}/d\mathbf{b}$ and $Vd\mathbf{R}/d\mathbf{b}$; but neither of them, nor any combination of them, is a differential coefficient clear of differentials. Thus,

$$\frac{d\mathbf{R}}{d\mathbf{b}} = (d\mathbf{b})^{-1}V\mathbf{a}d\mathbf{b} = (d\mathbf{b})_1 V\mathbf{a}(d\mathbf{b})_1 \quad \cdots \quad (100)$$

contains the differential on the right side. And

$$V\frac{d\mathbf{R}}{d\mathbf{b}} = Vd\mathbf{R}(d\mathbf{b})^{-1} = -V(d\mathbf{b})^{-1}V\mathbf{a}d\mathbf{b} = -V(d\mathbf{b})_1 V\mathbf{a}(d\mathbf{b})_1 \quad (101)$$

also contains the differential. As a special result the value of (100) is zero (a parallelepiped in a plane): and by equation (52) we may develope (101) to

$$(d\mathbf{b})_1\{\mathbf{a}(d\mathbf{b})_1\} - \mathbf{a}. \quad \cdots \quad \cdots \quad (102)$$

But it is (99) that is the really useful fundamental equation, the developments (100) to (102) being useless for differentiating purposes, although true and interpretable geometrically.

Generally, if

$$\mathbf{R} = \mathbf{i}R_1 + \mathbf{j}R_2 + \mathbf{k}R_3 \quad \cdot \quad \cdot \quad \cdot \quad \cdot \quad (103)$$

is a function of

$$\mathbf{r} = \mathbf{i}x + \mathbf{j}y + \mathbf{k}z, \quad \cdot \quad \cdot \quad \cdot \quad \cdot \quad (104)$$

then R_1, R_2, R_3 are functions, usually different, of x, y and z, and the ratio of $d\mathbf{R}$ to $d\mathbf{r}$ (even in the most general quaternionic sense), cannot possibly be freed of the differentials dx, dy, dz.

Even when we reduce the generality of the problem by making $\mathbf{R}$ and $\mathbf{r}$ coplanar, say in the plane of $\mathbf{i}$, $\mathbf{j}$, with $z = 0$ and $R_3 = 0$, $d\mathbf{R}/d\mathbf{r}$ is not usually a definite differential coefficient. It only becomes one by special relations between the differential coefficients of R_1 and R_2 with respect to x and y, viz :—

$$\frac{dR_1}{dx} = \frac{dR_2}{dy}, \qquad \frac{dR_1}{dy} = -\frac{dR_2}{dx} \quad \cdot \quad \cdot \quad \cdot \quad (105)$$

This is the vectorial basis of the large and important branch of modern mathematics called the Theory of Functions, which appears in electric and magnetic bidimensional problems.

Variation of the Size and Ort of a Vector.

§ 121. Sometimes it is convenient to have the variation of a vector exhibited in terms of the variations of its tensor and unit vector, or its size and ort. Thus, from

$$\mathbf{R} = R\mathbf{R}_1,$$

we produce, by differentiating,

$$d\mathbf{R} = Rd\mathbf{R}_1 + \mathbf{R}_1 dR. \quad \cdot \quad \cdot \quad \cdot \quad \cdot \quad (106)$$

If there be no variation of direction, then

$$d\mathbf{R} = \mathbf{R}_1 dR ; \quad \cdot \quad \cdot \quad \cdot \quad \cdot \quad \cdot \quad (107)$$

and if the tensor does not vary, whilst the direction does, then

$$d\mathbf{R} = Rd\mathbf{R}_1. \quad \cdot \quad \cdot \quad \cdot \quad \cdot \quad \cdot \quad (108)$$

Dividing (107) by the scalar dR, we get a proper differential coefficient

$$\frac{d\mathbf{R}}{dR} = \mathbf{R}_1, \quad \cdot \quad \cdot \quad \cdot \quad , \quad \cdot \quad \cdot \quad (109)$$

provided $d\mathbf{R}_1 = 0$.   From (108),

$$\frac{d\mathbf{R}}{d\mathbf{R}_1} = \mathbf{R}, \qquad V\frac{d\mathbf{R}}{d\mathbf{R}_1} = 0, \quad \cdot \quad \cdot \quad \cdot \quad (110)$$

provided $d\mathbf{R} = 0$, which makes $d\mathbf{R}$ perpendicular to $\mathbf{R}$.   But it is as well to avoid forms of this kind (with a vector variation in the denominator), even when by some limitation, as above, we can get rid of the differential.   Of course, in (106), without any limitation, we can change d to d/dt (or other scalar differentiator) ; thus,

$$\dot{\mathbf{R}} = \mathbf{R}\dot{\mathbf{R}}_1 + \mathbf{R}_1\dot{\mathbf{R}}. \quad \cdot \quad \cdot \quad \cdot \quad \cdot \quad (111)$$

It is this (scalar) kind of differentiation that nearly always occurs in physical applications, and there is no abstrusity about it.   See § 116.   Nor is there about differentials, provided we do not attempt to make differential coefficients with respect to vectors, or make the differentials themselves finite.

We have also $\qquad \mathbf{R}^2 = \mathbf{R}^2$,

which, on differentiation, gives

$$\mathbf{R}d\mathbf{R} = \mathbf{R}d\mathbf{R} ; \quad \cdot \quad \cdot \quad \cdot \quad \cdot \quad (112)$$

or, dividing by R,

$$d\mathbf{R} = \mathbf{R}_1 d\mathbf{R}, \quad \cdot \quad \cdot \quad \cdot \quad \cdot \quad (113)$$

which says that the increment in the tensor is the component along the vector of its vector change $d\mathbf{R}$, whose full expression is given by (106).   Using (106) in (113) we get

$$\mathbf{R}d\mathbf{R}_1 = 0. \quad \cdot \quad \cdot \quad \cdot \quad \cdot \quad \cdot \quad (114)$$

Using (113) in (106) we get

$$d\mathbf{R} = \mathbf{R}d\mathbf{R}_1 + \mathbf{R}_1(\mathbf{R}_1 d\mathbf{R}) \quad \cdot \quad \cdot \quad \cdot \quad (115)$$

or, dividing by R,

$$\frac{d\mathbf{R}}{\mathbf{R}} = d\mathbf{R}_1 + \frac{\mathbf{R}_1 d\mathbf{R}}{\mathbf{R}}, \quad \cdot \quad \cdot \quad \cdot \quad \cdot \quad (116)$$

all of which results may be geometrically interpreted.

In differentiating the reciprocal of a vector, proceed thus :—

$$d\frac{1}{\mathbf{R}} = d\frac{\mathbf{R}_1}{\mathbf{R}} = \mathbf{R}_1 d\frac{1}{\mathbf{R}} + \frac{d\mathbf{R}_1}{\mathbf{R}}, \quad \cdot \quad \cdot \quad \cdot \quad (117)$$

where, for d may be written d/dt, &c.

N

Another way is

$$d\frac{1}{\mathbf{R}} = d\frac{\mathbf{R}}{\mathrm{R}^2} = \frac{d\mathbf{R}}{\mathrm{R}^2} + \mathbf{R}d\frac{1}{\mathrm{R}^2}. \quad \cdots \quad (118)$$

It may easily be shown, by (106), that these results are equivalent.

Enough has now been said about the meaning and effect of differentiating vectors. The important vector differentiator ∇ deserves and demands a separate treatment.

Preliminary on ∇. Axial Differentiation. Differentiation referred to Moving Matter.

§ 122. The Hamiltonian vector ∇ occurs in all physical mathematics involving three dimensions, when treated vectorially. It is defined by

$$\nabla = \mathbf{i}\frac{d}{dx} + \mathbf{j}\frac{d}{dy} + \mathbf{k}\frac{d}{dz}, \quad \cdots \quad (119)$$

$$= \mathbf{i}\nabla_1 + \mathbf{j}\nabla_2 + \mathbf{k}\nabla_3,$$

the second form being merely an equivalent one, often more convenient, and easier to set up. We see that the Hamiltonian is a fictitious vector, inasmuch as the tensors of its components are not magnitudes, but are differentiators. As, however, these differentiators are scalar—not scalar magnitudes, but scalar operators, having nothing vectorial about them—the Hamiltonian, in virtue of $\mathbf{i}$, $\mathbf{j}$, $\mathbf{k}$, behaves just like any other vector, provided its differentiating functions are simultaneously attended to. Of course, an operand is always implied, which may be either scalar or vector.

The Hamiltonian has been called Nabla, from its alleged resemblance to an Assyrian harp—presumably, only the frame thereof is meant, without any means of evoking melody. On the other hand, that better known musical instrument, the Triangle, perhaps equally well resembles ∇, and it does not want strings to play upon.

First notice that

$$\mathbf{i}\nabla = \nabla_1, \qquad \mathbf{j}\nabla = \nabla_2, \qquad \mathbf{k}\nabla = \nabla_3, \quad \cdot \cdot \quad (120)$$

and that these are all special cases of the scalar product

$$\mathbf{N}\nabla = N_1\nabla_1 + N_2\nabla_2 + N_3\nabla_3, \quad \cdots \quad (121)$$

where $\mathbf{N}$ is any unit vector. Here $\mathbf{N}\nabla$, or its equivalent Cartesian expansion, means differentiation with respect to length, say s, measured along the axis of $\mathbf{N}$, usually denoted by

$$\frac{d}{ds} = l\frac{d}{dx} + m\frac{d}{dy} + n\frac{d}{dz}, \quad \ldots \quad (122)$$

or by similar forms. This operation was termed by Maxwell differentiation with respect to an axis, which we may more conveniently call axial differentiation. We see, therefore, that $\mathbf{N}\nabla$, or the $\mathbf{N}$ component (scalar) of ∇, is an axial differentiator, the axis being that of $\mathbf{N}$. Thus,

$$\mathbf{N}\nabla.\mathrm{P} = \frac{d\mathrm{P}}{ds}, \qquad \mathbf{N}\nabla.\mathbf{A} = \frac{d\mathbf{A}}{ds}, \quad \ldots \quad (123)$$

P being scalar and $\mathbf{A}$ vector. Also

$$\mathbf{N}\nabla.\mathbf{AB} = \mathbf{A}.\mathbf{N}\nabla.\mathbf{B} + \mathbf{B}.\mathbf{N}\nabla.\mathbf{A}, \quad \ldots \quad (124)$$

$$\mathbf{N}\nabla.\mathrm{V}\mathbf{AB} = \mathrm{V}(\mathbf{N}\nabla.\mathbf{A})\mathbf{B} + \mathrm{V}\mathbf{A}.\mathbf{N}\nabla.\mathbf{B}, \quad \ldots \quad (125)$$

and so on, simply because $\mathbf{N}\nabla$ is a differentiator. Expanded in Cartesians, (124) contains the eighteen differential coefficients of the components of $\mathbf{A}$ and $\mathbf{B}$ with respect to x, y, and z. The last two equations will also serve to illustrate notation. The dots are merely put in to act as separators, and keep the proper symbols connected. Mere blank spacing would do, but would be troublesome to work. We may, however, use brackets equivalently.

Thus, $\mathbf{A}.\mathbf{N}\nabla.\mathbf{B}$ may be written $\mathbf{A}(\mathbf{N}\nabla)\mathbf{B}$, and this form may be used with advantage if found easier to read. Of course, $\mathbf{N}\nabla$ being scalar, on its insertion between the two members of $\mathbf{AB}$, the result remains a scalar product. Similarly, $\mathrm{V}\mathbf{A}.\mathbf{N}\nabla.\mathbf{B}$, arising from $\mathrm{V}\mathbf{AB}$, is still a vector product.

A somewhat more general operator is $\mathbf{v}\nabla$, where $\mathbf{v}$ is any vector. It is plainly v times the corresponding axial differentiator, or $v.\mathbf{v}_1\nabla$. Thus, expanded,

$$\mathbf{v}\nabla.\mathbf{D} = (v_1\nabla_1 + v_2\nabla_2 + v_3\nabla_3)(\mathrm{i}\mathrm{D}_1 + \mathrm{j}\mathrm{D}_2 + \mathrm{k}\mathrm{D}_3), \quad . \quad (126)$$

which we may separate into $\mathbf{i}$, $\mathbf{j}$, $\mathbf{k}$ terms if we like.

When $\mathbf{v}$ is the velocity of moving matter, $\mathbf{v}\nabla$ comes frequently into use. Let, for example, w denote the measure of some property of the moving matter (here a scalar, but it may equally well be a vector), and, therefore, a function of position and of time. Its rate of change with the time at a *fixed* point in space is dw/dt, and the matter to which this refers is changing. But if we wish to know the rate of time-change of w for the *same* portion of matter, we must go thus. Let $\delta w/\delta t$ denote the result, then

$$\frac{\delta w}{\delta t} = \frac{dw}{dt} + \frac{dw}{dx}\frac{dx}{dt} + \frac{dw}{dy}\frac{dy}{dt} + \frac{dw}{dz}\frac{dz}{dt},$$

by elementary calculus. Or

$$\frac{\delta w}{\delta t} = \frac{dw}{dt} + \mathbf{v}\nabla.w, \quad \cdots \quad (127)$$

where $\mathbf{v}$ is the velocity of the matter.

This is the equation extensively employed in hydrokinetics. In elastic solid theory the term $\mathbf{v}\nabla.w$ is generally omitted, being often small compared with the first term on the right. But, in special applications, it may happen that any one of the three terms in (127) vanishes. Thus the term on the left side obviously vanishes when w keeps the same in the same matter. If, however, for w we substitute a vector function, say $\mathbf{A}$, even though it may not suffer change at first sight in the same matter, yet the rotational part of the motion will alter it by turning it round, so that $\delta\mathbf{A}/\delta t$ will not vanish.

The operator $\delta/\delta t$ is distributive, like d/dt and $\mathbf{v}\nabla$, as exemplified in (124), (125).

Motion of a Rigid Body. Resolution of a Spin into other Spins.

§ 123. If the moving matter is so connected that it moves as a rigid body, we have a further development. If we set a rigid body spinning about a fixed axis whose direction is defined by the unit vector $\mathbf{a}_1$, with angular speed a, the speed of any point P in the body equals the product of its perpendicular distance from the axis into the angular speed, and the direction of P's motion is perpendicular to the plane through the axis containing P. That is, a given particle describes a circle round the axis. This is obviously necessitated by the rigid con-

nection of the parts of the body (or detached bodies).   The
velocity of P is therefore expressed by

$$\mathbf{v} = V\mathbf{ar}, \quad . \quad . \quad . \quad . \quad . \quad (128)$$

if $\mathbf{a} = a\mathbf{a}_1$, and $\mathbf{r}$ is the vector to P from any point O on the
axis.   This is by the definition of a vector product.

Here we may observe a striking advantage possessed by the
vectorial method.   For (128), obtained by elementary considerations, proves without any more ado that angular velocities
about different axes compound like displacements, translational
velocities, or forces, or, in short, like *vectors*.   Thus, in (128)
we see that the velocity of P is the vector product of $\mathbf{a}$ and $\mathbf{r}$.
The latter fixes the position of P.   The former depends on the
spin, its tensor being the angular speed, and its direction that
of the axis of spin.   Now every vector, as before seen, may be
expressed as the sum of others, according to the rule of vector
addition.   But every vector has its species.   Thus, $\mathbf{r}$ being a
vector distance, its components are vector distances.   Similarly,
$\mathbf{a}$ being a vector axis of spin with tensor equal to the angular
speed, the components of $\mathbf{a}$, for instance, the three terms on
the right of

$$\mathbf{a} = \mathbf{i}a_1 + \mathbf{j}a_2 + \mathbf{k}a_3,$$

are vectors of the same nature.   Therefore any spin may be
replaced by other spins about other axes, according to the
vector law.   Of course the body cannot spin about more than
one axis at once, but its motion is the same as if it could and
did.   This property is notoriously difficult to understand by
Cartesian mathematics.

The most general kind of motion the rigid body can possess
is obtained by imposing upon the rotational any translational
velocity common to all parts.   Let $\mathbf{q}$ be this translational
velocity, then, instead of (128),

$$\mathbf{v} = \mathbf{q} + V\mathbf{ar}. \quad . \quad . \quad . \quad . \quad (129)$$

Here observe that $\mathbf{q}$ is the velocity of the point O, or of any
other point on the axis of rotation passing through the point O.
We may infer from this that if we shift the origin O to Q, a
point on a parallel axis, equation (129) will still be true, provided $\mathbf{r}$ means the vector from the new origin Q to P, and $\mathbf{q}$
means the translational velocity of Q itself.   Of course, the

axis of rotation is now through Q. We may formally prove it thus. Take

$$\mathbf{r} = \mathbf{h} + \mathbf{R}, \quad \ldots \quad \ldots \quad (130)$$

where $\mathbf{h}$ is the vector from O to Q and $\mathbf{R}$ the vector from Q to P. Put (130) in (129), then

$$\mathbf{v} = \mathbf{q} + \mathrm{Vah} + \mathrm{VaR}. \quad \ldots \quad \ldots \quad (131)$$

But here $\mathbf{q} + \mathrm{Vah}$ is, by (129), the velocity of Q, say $\mathbf{u}$, so we have

$$\mathbf{v} = \mathbf{u} + \mathrm{VaR}, \quad \ldots \quad \ldots \quad (132)$$

showing the velocity of P to be the sum of the translational velocity $\mathbf{u}$ at *any* point Q, plus the velocity at P due to rotation about the axis through Q.

All motion is relative. But observe the absolute character of the spin $\mathbf{a}$. It is absolute just because it involves and depends entirely upon relative motions.

The translational velocity $\mathbf{u}$ of the point Q consists of a motion along the axis of rotation, combined with a transverse motion. The latter may be got rid of by shifting the axis. For if $\mathbf{w}$ is the velocity of Q transverse to the axis, if we go to the distance w/a from the axis through Q, keeping in the plane through the axis perpendicular to $\mathbf{w}$, we shall reach a place where the velocity due to the rotation about the Q axis is either $+\mathbf{w}$ or $-\mathbf{w}$, according to which side of the Q axis we go. Choosing the latter, where the transverse motion is cancelled, and transferring the axis to it we see that the general motion of a rigid body consists of a spin about a certain axis (which is termed the central axis), combined with a translation along this axis. That is, it is a screw motion. It may, however, be more convenient not to employ the screw motion with a shifting axis, but to use (132), and let $\mathbf{u}$ and $\mathbf{a}$ be the translational and rotational velocities of a point Q fixed in the body.

We may also have simultaneous spins about axes which do not meet. Thus, let $\mathbf{R}$ be the vector from any origin O to a point Q about which there is a spin $\mathbf{a}$, and $\mathbf{r}$ the vector from O to any other point P. Then the velocity of P due to the spin is, by the preceding,

$$\mathbf{v} = \mathrm{Va}(\mathbf{r} - \mathbf{R}). \quad \ldots \quad \ldots \quad (133)$$

Next let there be any number of points of the type Q, each with its spin vector **a**. The velocity at P is then the sum of terms of the type (133), in which **r** is the same for all. The result is therefore

$$\mathbf{v} = \mathbf{VAr} - \Sigma\,\mathbf{VaR}, \quad . \quad . \quad . \quad (134)$$

where **A** is the sum of the **a**'s, or *the* spin of the body, whilst the *minus* summation expresses the velocity at O. For example, two equal spins, **a**, in opposite senses about parallel axes at distance h combine to make a mere translational velocity perpendicular to the plane of the axes; speed ah. I give these examples, in passing, to illustrate the working of vectors. These transformations are effected with a facility and a simplicity of ideas which put to shame the Cartesian processes. It may, indeed, be regarded as indicative of a mental deficiency to be unable to readily work the Cartesian processes—which is in fact my own case. But that does not alter the fact that if a man is not skilful in Cartesians, he may get along very well in vectors, and that the skilful mathematician who can play with Cartesians of great complexity, could as easily do far more difficult work in vectors, if he would only get over the elements, and accustom himself to the vectors as he had to do to the other method.

Motion of Systems of Displacement, &c.

§ 124. Going back to (127), we should, when the property whose time-variation is followed belongs to matter moving as a rigid body, employ in it the special reckoning of **v** of equation (132), giving the velocity of any point P in terms of **u** and **a** at an invariable point Q in the body. Thus, we have

$$\frac{dw}{dt} = \frac{\delta w}{\delta t} - (\mathbf{u} + \mathbf{VaR})\nabla.w. \quad . \quad . \quad (135)$$

But equation (127) applies not merely to the case of matter moving through space, carrying some property with it, but also when the matter itself is fixed, whilst some measurable property or quality moves through the matter, or is transferred from one part of the matter to another. And here we may leave out the consideration of matter altogether, and think only of some stationary medium which can support and through which the phenomenon can be transferred.

Thus, we may have a system of electric displacement $\mathbf{D}$ in a dielectric, which may be moving through it independently of any motion it may possess, and apply (127) to calculate $\dot{\mathbf{D}}$. It is, however, only in relatively simple cases that this can be followed up. For it is implied that we know the motion of the displacement and how it changes in itself, and these may be things to be found out.

But should the displacement system move as a rigid body, the matter is greatly simplified. This will occur when the sources of the displacement (electrification, for instance) move as a rigid body, and the motion is of a steady type, so long continued that the displacement itself (in the moving system) is steady. Writing $\mathbf{D}$ for w in (135), we shall have

$$\frac{d\mathbf{D}}{dt} = \mathbf{Va}\mathbf{D} - (\mathbf{u} + \mathbf{Va}\mathbf{R})\nabla.\mathbf{D} \quad . \quad . \quad . \quad (136)$$

Here $\mathbf{Va}\mathbf{D}$ is the rate of time-variation of the displacement in a moving element of the displacement system, caused by the rotation. It is zero when the displacement is parallel to the axis of spin, and a maximum when it points straight away from the axis, like the spoke of a wheel.

The displacement system is, however, not necessarily or usually the same when in steady motion as when at rest. In the latter case only can it be regarded as known initially. Set it steadily moving and it will (by reason of the self-induction) be changed to another displacement system. Nevertheless (136), along with the electromagnetic equations, enable us to find the new displacement system.

Should, however, the velocities of the connected sources be only a very small fraction of the speed of propagation of disturbances through the medium, the re-adjustment of the displacement as the sources move takes place practically instantaneously (as if the speed of propagation were infinite), so that the displacement system remains unchanged, preserving its stationary type, whilst it moves through the medium as a rigid body, in rigid connection with the sources. (This is obviously quite incorrect at a great distance from the sources, but there the effects themselves are insensible.) In this case, too, it is clearly unnecessary that the motion of the sources should be steady. Thus, in (136), $\mathbf{D}$ becomes the known displacement system (of equilibrium), moving as defined by u and a, which may vary anyhow.

Similar considerations apply to systems of induction, moving with their sources. A general caution, however, is necessary when there are conductors or other bodies in the field, not containing sources. To keep them from having currents induced in them, or in other ways upsetting the regularity of the moving system, they should also partake in the motion, by rigid connection.

Motion of a Strain-Figure.

§ 125. Equation (136) also applies to the motion through a medium of a " strain-figure," treated of by Dr. C. V. Burton in the current number of the *Phil. Mag.* (February, 1892). It is similar to that of slow motion of an electrical displacement system. Imagine a stationary infinite elastic medium with inertia, somehow set into a strained state, and let the speed of propagation of disturbances of strain be practically infinite. If then the strain-figure can move about through the medium, it will do so as a rigid body, provided the sources of the strain do so. In the above, D may signify the displacement (ordinary), whose variation in space constitutes the strain. Then $\dot{D}$ is the velocity of displacement, and is expressed in terms of the velocity of the strain-figure. From the expression for the kinetic energy of the complete strain-figure, the mechanical forces concerned can be deduced. As this is to be done on dynamical principles involving Newton's laws, we may expect beforehand that a strain-figure symmetrical with respect to a centre will behave as a Newtonian point-mass ; as does a similar electrical displacement figure. But, in general, without symmetry, the calculation of the forces concerned would be very troublesome indeed.

But the real difficulty appears to me to be rather of a physical than a mathematical nature. We have first to get some idea of how the strain-figure is kept up. Let it be stationary first. Then the strain-figure is referred to a forced or unnatural state of the medium in certain places. At any rate, we require intrinsic " sources " somewhere, and perhaps it might for this reason be convenient to consider

these portions of the medium (with the sources) to constitute
the atom or molecule, rather than the whole strain-figure.

Now, having got a stationary strain-figure, how is it to be
set moving? Three ways suggest themselves. First, a bodily
motion of the medium carrying the strain-figure with it. This
is plainly inadmissible. Next a motion of the atomic portions
only of the medium (with the sources) through the rest of the
medium, either disturbing it to some extent near by, or not
disturbing it at all—slipping through, so to speak. This
would carry on the strain-figure. But it is inadmissible, if it
be our object not to move the medium at all in any part.
Thirdly, we may keep the whole medium at rest, and cause
the sources themselves to move through it, so that the atomic
portions of the medium change. But there is no means for
doing this presented to our consideration. It lies beyond
the dynamical question of the forces on an atom brought
into play if the strain-figure can move in the manner supposed. The second course above, on the other hand, is more
intelligible, although it implies something akin to liquidity.

These objections are, however, only made suggestively.
The matter is sufficiently important to be deserving of a
thorough threshing out.

Space-Variation or Slope ∇P of a Scalar Function.

§ 126. The simplest and most easily understood effect of ∇
upon a function is when it acts upon a scalar, say P. This
implies that P is a scalar function of position, or of x, y, z. In
the most important applications P is single-valued, as when it
represents density, or pressure, or temperature, or electric
potential, or the corresponding magnetic potential of magnets.
But P may also be multiplex, as when it is the magnetic
potential outside a linear electric current. At present let P be
simplex.

By (119) we have

$$\nabla P = \mathbf{i}.\nabla_1 P + \mathbf{j}.\nabla_2 P + \mathbf{k}.\nabla_3 P. \quad . \quad . \quad . \quad (137)$$

Of course, since ∇ is vector and P scalar, the result is a vector.
Its meaning is easily found. From the fact indicated in (137)
that the rectangular scalar components of ∇P are the rates of
increase of P along the axes of $\mathbf{i}$, $\mathbf{j}$, $\mathbf{k}$, we may conclude that

the component of ∇P in any direction is the rate of increase of P in that direction. Thus, **N** being any unit vector, we have, by (137),

$$\mathbf{N}\nabla P = N_1\nabla_1 P + N_2\nabla_2 P + N_3\nabla_3 P. \qquad . \quad (138)$$

Comparing with (121), we see that

$$\mathbf{N}\nabla P = (\mathbf{N}\nabla)P = d\mathrm{P}/ds, \qquad . \quad . \quad . \quad (139)$$

by (123), s being length measured along **N**.

The identity of $\mathbf{N}\nabla P$ and $\mathbf{N}\nabla.P$ is tolerably obvious algebraically. It is not, however, true when for P is substituted a vector, as we shall see later.

If we take P = constant, we obtain the equation to a surface. If, then, **T** is any tangent to the surface at a chosen point, that is, any line in the tangent plane perpendicular to the normal, we shall have **T**∇P = 0. The direction of ∇P itself is therefore that of the normal to the surface; or the lines of ∇P cut the equipotential surfaces perpendicularly, and pass from one to the next (infinitely close) by the shortest paths. The distance between any two consecutive equipotential surfaces of a series having a common difference of potential varies inversely as the magnitude of ∇P, and ∇P itself is the vector showing at once the direction and the rate of the fastest increase of P. No perfectly satisfactory name has been found for this slope, or space-variation of a scalar function. Comparing P with height above the level on a hillside, ∇P shows the greatest slope upwards. But the illustration is inadequate, since on the hillside we are confined to a surface.

The tensor of ∇P is given by

$$(\nabla P)^2 = (\nabla_1 P)^2 + (\nabla_2 P)^2 + (\nabla_3 P)^2, \qquad . \quad . \quad (140)$$

as with any other vector.

We may also here notice the vector product $\mathbf{VN}\nabla$ in its effect on a scalar. We have, by the semi-Cartesian formula for a vector product,

$$\mathbf{VN}\nabla = \mathbf{i}(N_2\nabla_3 - N_3\nabla_2) + \mathbf{j}(N_3\nabla_1 - N_1\nabla_3) + \mathbf{k}(N_1\nabla_2 - N_2\nabla_1). (141)$$

Thus, when the operand is scalar, as P, we shall have

$$(\mathbf{VN}\nabla)P = \mathbf{VN}\nabla P. \qquad . \quad . \quad . \quad . \quad (142)$$

The result is a vector perpendicular to the plane of $\mathbf{N}$ and ∇P, as in any other vector product, vanishing when they are parallel, and a maximum when they are perpendicular.

Scalar Product ∇D.   The Theorem of Divergence.

§ 127.   When the operand of ∇ is a vector, say $\mathbf{D}$, we have both the scalar product and the vector product to consider. Taking the former alone first, we have

$$\text{div } \mathbf{D} = \nabla \mathbf{D} = \nabla_1 \mathbf{D}_1 + \nabla_2 \mathbf{D}_2 + \nabla_3 \mathbf{D}_3. \quad . \quad . \quad (143)$$

This function of $\mathbf{D}$ is called its divergence, and is a very important function in physical mathematics.   Its general signification will be best appreciated by a consideration of the Theorem of Divergence.

Let liquid be in motion.   The continuity of existence of the matter imposes certain restrictions on the motion.   The current is $m\mathbf{q}$ per unit area, where m is the density and $\mathbf{q}$ the velocity.   But, to further simplify, let $m = 1$, making the liquid incompressible.   Then the current is simply $\mathbf{q}$, which measures the amount of liquid crossing unit area of any surface perpendicular to $\mathbf{q}$, per second.   But if the surface be not perpendicular to $\mathbf{q}$, the effective flux is only $\mathbf{Nq}$, the normal component of $\mathbf{q}$, if $\mathbf{N}$ denotes the unit normal to the surface.   Therefore, $\Sigma \mathbf{Nq}$, or the summation of $\mathbf{Nq}$ over any surface, expresses the total flux of liquid through the surface.

Now suppose the surface (fixed in space) is closed.   Then the summation represents the amount of liquid leaving the enclosed region per second through its boundary, if $\mathbf{N}$ be the outward normal.   This amount is evidently zero, because of the assumed incompressibility.   If then we observe, or state, that $\Sigma \mathbf{Nq}$ is not zero, either the fluid is compressible, or else there must be sources of liquid within the region.   Adopting the latter idea, because it simplifies the reasoning, we see that the summation $\Sigma \mathbf{Nq}$ is the appropriate measure of the total strength of the sources in the region, since it is the rate at which liquid is being generated therein, or the rate of supply.   The position of the internal sources is quite immaterial, and so is the shape of the region, and its size, or the manner in which a source sends out its liquid (i.e., equably or not in all directions).   So we have

$$\Sigma \mathbf{Nq} = \Sigma \rho, \quad . \quad . \quad . \quad . \quad . \quad (144)$$

if ρ represents the strength of a source, and the Σ includes them all.

When the sources are distributed continuously, so that ρ is a continuous function of position, its appropriate reckoning is per unit volume. That this is the divergence of $\mathbf{q}$ is clear enough, according to the explanations relating to divergence already given, (§ 51). That it is the same as $\nabla\mathbf{q}$, as in (143), we may prove at once by the Cartesian form there exhibited.

We have to reckon the flux outward through the sides of a unit cube. Take its edges parallel to $\mathbf{i}$, $\mathbf{j}$, $\mathbf{k}$ respectively. Then there are two sides whose outward normals are $-\mathbf{i}$ and $\mathbf{i}$, and their distance apart is unity. The outward fluxes due to them are therefore $-\mathbf{i}\mathbf{q}$, or $-q_1$, and $q_1 + \nabla_1 q_1$, whose sum is $\nabla_1 q_1$. Similarly the two sides whose normals are $\pm\mathbf{j}$ contribute $\nabla_2 q_2$, and the remaining sides contribute $\nabla_3 q_3$. Comparing with (143), we verify the Cartesian form of the divergence of a vector.

But we may have any number of other special forms, according to the co-ordinates we may choose to employ for calculating purposes, such as spherical, columnar, &c., and the most ready way to find the corresponding form is by the immediate application of the idea of divergence to the volume-element concerned. For purposes of reasoning, however, it is best to entirely eliminate the idea of co-ordinates. Divergence is independent of co-ordinates.

The sources need not be so distributed as to give rise to a finite volume-density. We may have, within the region concerned, surface-, line-, or point-sources, and the principle concerned is the same throughout. Thus the density of a surface-source is measured by the sum of the normal fluxes on its two sides per unit area, that is, by $\Sigma \mathbf{N}\mathbf{q}$ applied to the two sides of the unit area, if the flux wholly proceeds outwards. This is, however, not fully general, as there may be a flux in the surface itself, so that the full measure of the surface-density has to include the divergence of the surface-flux, to be found by calculating the flux leaving the unit area across its bounding line. Similar considerations apply to linear sources. In the case of a point-source the measure of the strength is the flux outward through a closed surface enclosing the point infinitely near it—the surface of the point, so to speak. Any other surface enclosing the point will do, provided there are no other sources brought in by the change.   There may also be multiplex sources.   Thus, a pair of equal point-sources, one a source, the other a sink, would be equivalent to no source at all if brought infinitely near one another; but if the reduction in distance be accompanied by a corresponding increase in strength of the sources, the final result is not zero.   This is the case of a magnetised particle on the theory of magnetic matter; but it is not necessary or desirable to complicate matters by entering upon special peculiarities of discontinuity in considering the Divergence Theorem.   Its general form is

$$\Sigma \, \mathbf{ND} = \Sigma \, \mathrm{div} \, \mathbf{D}, \quad . \quad . \quad . \quad . \quad (145)$$

when the vector $\mathbf{D}$, to which it is applied, admits of finite differentiation ; and the special meanings to be attached to the divergence of $\mathbf{D}$, in order to satisfy the principle concerned, may be understood.

Although a material analogy, as above, is very useful, it is not necessary.   Any distributed vector magnitude will have the same peculiarities of divergence as the flux of a liquid.   If it be the motion of a real expansible fluid that is in question, then the divergence of its velocity represents the rate of expansion.   It would, however, be very inconvenient to have to carry out this analogy in electric or magnetic applications; an incompressible liquid, with sources and sinks to take the place of expansions and contractions, is far more manageable.

Extension of the Theorem of Divergence.

§ 128. The following way of viewing the Divergence Theorem, apart from material analogies, is important.   Consider the summation $\Sigma \, \mathbf{ND}$ of the normal component of a vector $\mathbf{D}$ over any closed surface.   Divide the region enclosed into two regions.   Their bounding surfaces have a portion in common. If, then, we sum up the quantity $\mathbf{ND}$ for both regions (over their boundaries, of course), the result will be the original $\Sigma \, \mathbf{ND}$ for the complete region.   The normal is always to be reckoned positive outwards from a region, so that on the surface common to the two smaller regions $\mathbf{N}$ is + for one and – for the other region, and $\Sigma \, \mathbf{ND}$ for one is the negative of that for the other, so far as the common surface goes.

Since this process of division may be carried on indefinitely, we see that the summation ΣND for the boundary of any region equals the sum of the similar summations applied to the surfaces of all the elementary regions into which we may divide the original. That is,

$$\Sigma \, \mathbf{ND} = \Sigma \phi(\mathbf{D}), . \quad . \quad . \quad . \quad . \quad (146)$$

where, on the right side, we have a volume-summation whose elementary part $\phi(\mathbf{D})$ is the same quantity ΣND as before, belonging now, however, to the elementary volume in question. We have already identified $\phi(\mathbf{D})$ with the divergence of $\mathbf{D}$.

But if this demonstration be examined, it will be seen that the validity of the process whereby we pass from a surface- to a volume-summation, depends solely upon the quantity summed up, viz., **ND**, changing its sign with **N**. We may therefore at once give to the Divergence Theorem a wide extension, making it, instead of (146), take this form :—

$$\Sigma \, \mathrm{F}(\mathbf{N}) = \Sigma f(\mathbf{N}). \quad . \quad . \quad . \quad . \quad (147)$$

Here, on the left side, we have a surface-, and on the right side a volume-summation. The function $\mathrm{F}(\mathbf{N})$, where **N** is the out ward normal, is any function which changes sign with **N**. The other function $f(\mathbf{N})$, the element of the volume-summation, is the value of $\Sigma \mathrm{F}(\mathbf{N})$ for the surface of the element of volume. Thus, by considering a cubical element,

$$f(\mathbf{N}) = \nabla_1 \mathrm{F}(\mathbf{i}) + \nabla_2 \mathrm{F}(\mathbf{j}) + \nabla_3 \mathrm{F}(\mathbf{k}) \quad . \quad . \quad (148)$$

is the Cartesian form, should $\mathrm{F}(\mathbf{N})$ be a scalar function, or the semi-Cartesian form should it be a vector function. A few examples of the general theorem (147) will be given later.   In the meantime the other effect of ∇ should be considered.

Vector Product $\nabla\nabla$E, or the Curl of a Vector.   The Theorem of Version, and its Extension.

§ 129. The vector product of ∇ and a real vector, say **E**, is given in semi-Cartesian form by

$$\nabla\nabla\mathbf{E} = \mathbf{i}(\nabla_2 \mathrm{E}_3 - \nabla_3 \mathrm{E}_2) + \mathbf{j}(\nabla_3 \mathrm{E}_1 - \nabla_1 \mathrm{E}_3) + \mathbf{k}(\nabla_1 \mathrm{E}_2 - \nabla_2 \mathrm{E}_1)$$
$$= \mathrm{curl} \ \mathbf{E}. \quad . \quad . \quad . \quad . \quad (149)$$

As before, with respect to the divergence of a vector, we can best appreciate the significance of this formula by the general property involved, expressed by the Theorem of Version.

On any surface draw a closed curve or circuit. Let, for distinctness, **E** be electric force. Calculate the voltage in the circuit due to **E**. The effective force per unit length of circuit is the tangential component of **E**, or **TE**, if **T** is the unit tangent. The voltage in the circuit is, therefore, Σ **TE**, the summation being circuital, or a line-integration extended once round (along) the closed curve.

Now draw on the surface a line joining any two points of the circuit. Two circuits are thus made, having a portion in common. Reckon up the voltage in each of the smaller circuits and add them together. The result is the voltage in the first circuit, if we rotate the same way in both the smaller circuits as in the original, because the common portion contributes voltage equally and oppositely to the two smaller circuits.

This process may be carried on to any extent by drawing fresh lines on the surface. We therefore have the result that the voltage in the circuit bounding a surface equals the sum of the voltages in the elementary circuits bounding the elements of surface. Or

$$\Sigma \, \textbf{TE} = \Sigma \, \theta(\textbf{E}), \quad . \quad . \quad . \quad . \quad (150)$$

where on the left side we have a circuital summation, and on the right side an equivalent surface-summation, in which $\theta(\textbf{E})$, the quantity summed, is the value of Σ **TE**, that is, the voltage, in the circuit bounding the particular element of surface concerned.

To find the form of θ in terms of **i**, **j**, **k**, &c., we need only calculate the voltages in unit square elements of surface taken successively with edges parallel to **j**, **k**, to **k**, **i**, and to **i**, **j**. In the first case, when the normal to the square circuit, or the axis of the circuit, is **i**, the voltage is

$$\nabla_2 E_3 - \nabla_3 E_2,$$

that is, by (149), the **i** component of V∇**E**, and therefore the normal component, since here **N** = **i**. Similarly when **N** = **j**, and the axis of the circuit is **j**, the voltage in it is expressed by the coefficient of j in (149). And when $N = k$, the voltage is the coefficient of k in (149). Thus, in any case, the voltage in an elementary circuit of unit area is the normal component of $V\triangledown E$, that is, $NV\triangledown E$. So (150) becomes

$$\Sigma\, TE = \Sigma\, NV\triangledown E = \Sigma\, N \text{ curl } E, \quad . \quad . \quad (151)$$

expressing the Theorem of Version, sometimes termed Stoke's Theorem.

When E is not electric force, $\Sigma\, TE$ is not circuital voltage, but circuital something else; but this does not affect the general application of the theorem.

This theorem is particularly important in electromagnetism because it is involved in the two fundamental laws thereof, what were termed the First and Second Circuital Laws, or Laws of Circuitation, connecting together the electric and magnetic forces and their time-variations. (*See* §§ 33 to 36, Chap. II., and later.)

If the vector whose curl is taken be velocity q in a moving fluid, then curl q represents twice the spin or vector angular velocity of the fluid immediately surrounding the point in question; its direction being that of the axis of rotation, and magnitude twice the rate of rotation. But I have not made use of the fluid analogy in describing and proving the Version Theorem, because it is not of material assistance.

Since the validity of the process whereby we pass from a circuital summation of the tangential component of a vector to an equivalent surface-summation depends upon the fact that TE changes sign with T, we may generalise the theorem thus :—

$$\Sigma\, F(T) = \Sigma f(T), \quad . \quad . \quad . \quad . \quad (152)$$

where on the left we have a circuital and on the right a surface-summation, and $F(T)$ is such a function of T as to change sign with T; whilst $f(T)$, the element of the surface-summation, is the value of the former $\Sigma\, F(T)$ for the particular element of surface in question. Of this general theorem a few examples will be given later.

A few words regarding $\triangledown$, div and curl, terminologically considered, may be useful. Since divergence and curl are expressible in terms of vex (a provisional name for $\triangledown$, which has been suggested to me), why not use the vex operator only, like

o

the quaternionists? The reasons, which are weighty, should be obvious.

In the first place, we require a convenient language for describing or referring to processes and results, expressing approximately their essential meaning without being too mathematical. Now ∇ alone is not convenient for this purpose. The scalar product of ∇ and $\mathbf{D}$ conveys no such distinct idea as does divergence; nor does the vector product of ∇ and $\mathbf{E}$ speak so plainly as the curl or rotation of $\mathbf{E}$.

Besides, the three results of ∇, exemplified in ∇P, and $\nabla \mathbf{D}$, and $V\nabla \mathbf{E}$, are so remarkably different in their algebraical development and in their meaning, that it is desirable, even in the algebra, to very distinctly separate them in representation. Therefore, in the preceding part of the present work (as in all former papers), the symbol ∇ is only prefixed to a scalar, as in ∇P, the space variation of P, whilst for the scalar and vector product are employed div and curl, in the formulæ as well as in descriptive matter.

There are, however, cases when it may be desirable to use ∇ and $V\nabla$ applied to vectors in formulæ, namely, when the combinations of symbols are not so simple that their meaning and effect can be readily seen, and when it is required to perform transformations also not readily recognisable. The utility of ∇ in its vectorial significance then becomes apparent, for one may use it alone, temporarily if desired, and work it as a vector, remembering, however, its other functions. Of this, too, some examples should be given.

Five Examples of the Operation of ∇ in Transforming from Surface to Volume Summations.

§ 130. Returning to the theorem (147), or

$$\Sigma F(\mathbf{N}) = \Sigma f, \quad \cdots \quad (153)$$

let us take a few of the simplest cases that present themselves. Given any odd function F of $\mathbf{N}$, the normal outwards from a closed surface over which the summation on the left side extends, we convert it to an equivalent summation throughout the enclosed region by making f, the quantity summed, be the value of $\Sigma F(N)$ for the surface of the element of volume. This last is most conveniently a unit cube, so that

$$f = \nabla_1 F(i) + \nabla_2 F(j) + \nabla_3 F(k), \quad . \quad . \quad . \quad (154)$$

as in (148).

(a). The simplest case of all is $F(N) = N$ itself. Then, by (154), or by considering that the vector normals to the six faces of a cube balance one another in pairs, we have

$$\Sigma N = 0, \quad . \quad . \quad . \quad . \quad . \quad (155)$$

expressing that a closed surface has no resultant orientation ; or, that a normal pull applied to every part of a closed surface, of uniform amount per unit area, has no resultant.

(b). If we multiply by p, any scalar function of position, we have the same case again if p be constant. When negative, it makes a well-known elementary hydrostatic result. But when p is not constant, then, by (154),

$$f = \nabla_1 i p + \nabla_2 j p + \nabla_3 k p = \nabla p, \quad . \quad . \quad . \quad (156)$$

so that we have

$$\Sigma N p = \Sigma \nabla p. \quad . \quad . \quad . \quad . \quad (157)$$

These are, of course, vector summations. The sum of the surface tractions equals the sum of the bodily forces ∇p arising from the space-variation of the internal tension. Take p negative to indicate pressure.

(c). Take $F(N) = ND$, where D is a vector function of position. This gives the most valuable theorem of divergence,

$$\Sigma ND = \Sigma \nabla D = \Sigma \operatorname{div} D, \quad . \quad . \quad . \quad (158)$$

already discussed.

(d). Take $F(N) = NDP$, where P is a scalar. Here, by (154), or by (158), we shall find

$$\Sigma NDP = \Sigma \nabla (DP). \quad . \quad . \quad . \quad . \quad (159)$$

Here ∇ has to differentiate both D and P, thus,

$$\nabla . DP = \nabla D . P + D \nabla P, \quad . \quad . \quad . \quad (160)$$

o 2

so that the previous equation may be written

$$\Sigma\, \mathrm{NDP} = \Sigma\, \mathrm{P\ div\ D} + \Sigma\, \mathrm{D}\nabla\mathrm{P}, \quad . \quad . \quad . \quad (161)$$

which is a form of Green's Theorem relating to electrostatic energy. **D** may be the displacement in one system of electrification and P the potential in another. The quantity $-\Sigma\,\mathrm{D}\nabla\mathrm{P}$ is their mutual energy; and this is, by (161), equivalently expressed by the sum of products of every charge in one system into the potential due to the other.

(e). Take $\mathrm{F(N)} = \mathrm{VNH}$.   Then, by (154),

$$f = \nabla_1\mathrm{ViH} + \nabla_2\mathrm{VjH} + \nabla_3\mathrm{VkH}.$$

But here we may shift the ∇'s to the other side of the V's, because they are scalars; this produces

$$f = \mathrm{V}\nabla\mathrm{H},$$

so that we have

$$\Sigma\,\mathrm{VNH} = \Sigma\,\mathrm{V}\nabla\mathrm{H} = \Sigma\ \text{curl }\mathrm{H}. \quad . \quad . \quad . \quad (162)$$

The interpretation may be more readily perceived by reversing the direction of the normal.   Take $\mathrm{N} = -\mathrm{n}$, so that n is the normal drawn inward from the boundary.   Then

$$\Sigma\ \text{curl }\mathrm{H} + \Sigma\,\mathrm{VnH} = 0. \quad . \quad . \quad . \quad . \quad (163)$$

If **H** be magnetic force, curl **H** is the electric current-density in the region.   Now VnH is the surface equivalent of the bodily curl **H**.   Ignore altogether the magnetic force outside the region, if there be any.   Then the circuitation of **H** gives the current through a circuit.   Applied to elementary circuits wholly within the region, the result is curl **H**.   But at the boundary, where **H** suddenly ceases, there is a surface-current as well.   To find its expression, apply the process of circuitation to a circuit consisting of two parallel lines of unit length, infinitely close together, but on opposite sides of the boundary, joined by infinitely short cross-pieces.   Only the unit line inside the region contributes anything to the circuitation; and by taking it to coincide with **H**, so as to make the circuitation a maximum, we find that VnH represents the surface-density of current.   So, if **J** be current, we have, by (163), $\Sigma\,\mathrm{J} = 0$ for any region, by itself.   The surface-distribution and the volume-distribution of current are complementary ; that is, they are properly joined together to make up a circuital distribution. Thus,

$$n\nabla\nabla H = -\nabla VnH, \quad \ldots \quad (164)$$

where on the left side we have the divergence (at the surface) of the internal current, and on the right the equal convergence of the surface-current.

Five Examples of the Operation of ∇ in Transforming from Circuital to Surface Summations.

§ 131. Next, take a few examples of the extended Theorem of Version (152), viz. :—

$$\Sigma F(T) = \Sigma f, \quad \ldots \quad (165)$$

where now, on the left side, we have the circuital summation of an odd function of **T**, and on the right an equivalent surface-summation, whose elementary part f is the value of $F(T)$ for the circuit bounding the element of surface.

Taking three elements of surface to be unit squares, whose normals are **i**, **j**, **k**, we readily see that the corresponding f's are

$$\nabla_2 F(k) - \nabla_3 F(j), \quad \nabla_3 F(i) - \nabla_1 F(k), \quad \nabla_1 F(j) - \nabla_2 F(i). \quad (166)$$

By means of these we can see the special form assumed by f in any case.

(*a*). Thus, take $F(T) = T$ itself, the unit tangent. Then we have

$$\Sigma T = 0, \quad \ldots \quad (167)$$

merely expressing the fundamental property of adding vectors, that the sum of any vectors forming, when put end to end, a circuit, is zero.

(*b*). Take $F(T) = TP$, where P is a scalar function of position. Then, with normal **i**, we have, by the first of (166),

$$f = \nabla_2 kP - \nabla_3 jP = Vi\nabla.P ; \quad \ldots \quad (168)$$

so, writing **N** for **i**, we obtain

$$\Sigma TP = \Sigma VN\nabla.P = \Sigma VN\nabla.P \quad \ldots \quad (169)$$
$$= \Sigma V(N.\nabla P)$$

The quantity summed over the surface is, therefore, the surface representative of the curl of ∇P. This has no volume representative, its value being then zero.

(c). Take $F(\mathbf{T}) = \mathbf{T}H$.   Here, with normal $\mathbf{i}$ to the element of surface, we have

$$f = \nabla_2 \mathbf{k}H - \nabla_3 \mathbf{j}H = \nabla_2 H_3 - \nabla_3 H_2 = \mathbf{i}V\nabla H,$$

by (149).   Therefore, putting $\mathbf{N}$ for $\mathbf{i}$,

$$\Sigma\, \mathbf{T}H = \Sigma\, \mathbf{N}V\nabla H = \Sigma\, \mathbf{N}\, \text{curl } H \quad . \quad . \quad (170)$$

the Version Theorem again.   But observe that, by the transformation (164), we may also write it

$$\Sigma\, \mathbf{T}H = \Sigma\, V\mathbf{N}\nabla.H, \quad . \quad . \quad . \quad (171)$$

similarly to (169), in which the operand is a scalar.   This is mnemonically useful, but (170) is more practically useful.

(d). Take $F(\mathbf{T}) = V\mathbf{T}H$.   Here, with normal $\mathbf{i}$, the first of (166) gives

$$f = \nabla_2 V\mathbf{k}H - \nabla_3 V\mathbf{j}H = V(\mathbf{k}\nabla_2 - \mathbf{j}\nabla_3)H.$$

But here we have $\mathbf{k}\nabla_2 - \mathbf{j}\nabla_3 = V\mathbf{i}\nabla,$

so that $f = V(V\mathbf{i}\nabla)H$;

and therefore, generally, putting $\mathbf{N}$ for $\mathbf{i}$,

$$\Sigma\, V\mathbf{T}H = \Sigma\, V(V\mathbf{N}\nabla)H. \quad . \quad . \quad (172)$$

(e). Let the quantity in the circuital summation be a vector of length P (a scalar function of position) drawn perpendicularly to the plane of $\mathbf{T}$ and $\mathbf{N}$.   That is,

$$F(\mathbf{T}) = (V\mathbf{T}\mathbf{N})P.$$

We then find, taking $\mathbf{N} = \mathbf{i}$, and using the first of (166),

$$f = \nabla_2 V\mathbf{k}\mathbf{i}.P - \nabla_3 V\mathbf{j}\mathbf{i}.P = (\mathbf{j}\nabla_2 + \mathbf{k}\nabla_3)P = \nabla P - \mathbf{i}(\mathbf{i}\nabla)P.$$

In general, therefore,

$$\Sigma\, V\mathbf{T}\mathbf{N}.P = \Sigma(\nabla P - \mathbf{N}.\mathbf{N}\nabla P) = \Sigma\, \nabla_s P = \Sigma\, V(V\mathbf{N}\nabla)\mathbf{N}.P \quad . \quad (173)$$

The element of the surface-summation is $\nabla_s P$, meaning the slope of P on the surface itself, disregarding any variation it may have out of the surface. The last form of (173) involves the transformation formula (52).

Observe that in all the above examples,

$$\Sigma \, F(\mathbf{N}) = \Sigma \, F(\nabla), \quad \cdots \quad (174)$$

when we pass from a closed surface to the enclosed region; and that

$$\Sigma \, F(\mathbf{T}) = \Sigma \, F(V\mathbf{N}\nabla), \quad \cdots \quad (175)$$

when we pass from a circuit to the surface it bounds. Thus, **N** becomes ∇, and **T** becomes $V\mathbf{N}\nabla$. But I cannot recommend anyone to be satisfied with such condensed symbolism alone. It is much more instructive to go more into detail, as in the above examples, and see how the transformations occur, bearing in mind the elementary reasoning upon which the passage from one kind of summation to another is based (§§ 128, 129).

Nine Examples of the Differentiating Effects of ∇.

§ 132. The following examples relate principally to the modifications introduced by the differentiating functions of ∇.

(*a*). We have, by the parallelepipedal property,

$$\mathbf{N}V\nabla\mathbf{E} = \nabla V\mathbf{E}\mathbf{N} = \mathbf{E}V\mathbf{N}\nabla, \quad \cdots \quad (176)$$

when ∇ is a common vector. The equalities remain true when ∇ is vex, provided we consistently employ the differentiating power in the three forms. Thus, the first form, expressing the **N** component of curl **E**, is not open to misconception. But in the second form, expressing the divergence of $V\mathbf{E}\mathbf{N}$, since **N** follows ∇, we must understand that **N** is supposed to remain constant. In the third form, again, the operand **E** precedes the differentiator. We must either, then, assume that ∇ acts backwards, or else, which is preferable, change the third form to $V\mathbf{N}\nabla.\mathbf{E}$, the scalar product of $V\mathbf{N}\nabla$ and **E**; or $(V\mathbf{N}\nabla)\,\mathbf{E}$, if that be plainer.

(*b*). Suppose, however, that both vectors in the vector product are variable. Thus, required the divergence of $V\mathbf{E}\mathbf{H}$, expanded vectorially. We have

$$\nabla V\mathbf{E}\mathbf{H} = \mathbf{E}V\mathbf{H}\nabla = \mathbf{H}V\nabla\mathbf{E}, \quad \cdots \quad (177)$$

where the first form alone is entirely unambiguous. But we may use either of the others, provided the differentiating power of ∇ is made to act on both $\mathbf{E}$ and $\mathbf{H}$. But if we keep to the plainer and more usual convention that the operand is to follow the operator, then the third form, in which $\mathbf{E}$ alone is differentiated, gives one part of the result, whilst the second form, or rather, its equivalent $-\mathbf{E}V\nabla\mathbf{H}$, wherein $\mathbf{H}$ alone is differentiated, gives the rest. So we have, complete, and without ambiguity,

$$\operatorname{div} \mathbf{VEH} = \mathbf{H} \operatorname{curl} \mathbf{E} - \mathbf{E} \operatorname{curl} \mathbf{H}, \quad . \quad . \quad . \quad (178)$$

a very important transformation. It is concerned in the deduction of the equation of activity from the two circuital laws of electromagnetism.

(c). In these circuital laws we have also to consider the curl of a vector product, viz., the curl of the motional electric force in one law, and the curl of the motional magnetic force in the other. Taking the former, we have

$$\operatorname{curl} \mathbf{VqB} = \mathbf{V}\nabla\mathbf{VqB}, \quad . \quad . \quad . \quad . \quad (179)$$

where $\mathbf{B}$ is the induction and $\mathbf{q}$ the velocity of the medium supporting it. Apply the elementary transformation (52) to (179). It gives

$$\mathbf{V}\nabla\mathbf{VqB} = \mathbf{q}.\nabla\mathbf{B} - \mathbf{B}.\nabla\mathbf{q}, \quad . \quad . \quad (180)$$

when ∇ is a mere vector. But on the left side both $\mathbf{q}$ and $\mathbf{B}$ have to be differentiated; therefore the same is true in both terms on the right side. This gives

$$\mathbf{V}\nabla\mathbf{VqB} = \mathbf{q} \operatorname{div} \mathbf{B} + \mathbf{B}\nabla.\mathbf{q}$$
$$- \mathbf{B} \operatorname{div} \mathbf{q} - \mathbf{q}\nabla.\mathbf{B}, \quad . \quad . \quad (181)$$

without ambiguity or need of reservation. That is to say, as in the $\mathbf{q}.\nabla\mathbf{B}$ of (180) both $\mathbf{q}$ and $\mathbf{B}$ have to be differentiated, we get $\mathbf{q} \operatorname{div} \mathbf{B}$ when $\mathbf{B}$ alone, and $\mathbf{B}\nabla.\mathbf{q}$ when $\mathbf{q}$ alone is differentiated. Similarly for the other term in (180).

Or we might write ∇_q when $\mathbf{q}$ alone, and ∇_B when $\mathbf{B}$ alone suffers differentiation. Then, fully,

$$\mathbf{V}\nabla\mathbf{VqB} = \mathbf{V}\nabla_q\mathbf{VqB} + \mathbf{V}\nabla_B\mathbf{VqB}, \quad . \quad . \quad (182)$$

$$\mathbf{V}\nabla_q\mathbf{VqB} = \mathbf{B}\nabla.\mathbf{q} - \mathbf{B} \operatorname{div} \mathbf{q}, \quad . \quad . \quad . \quad (183)$$

$$\mathbf{V}\nabla_B\mathbf{VqB} = \mathbf{q} \operatorname{div} \mathbf{B} - \mathbf{q}\nabla.\mathbf{B}. \quad . \quad . \quad . \quad (184)$$

Here the sum of (183) and (184) gives (181). The meaning of $B\nabla$ and $q\nabla$ has been already explained.

(*d*). Equation (181) may be applied to the circuital laws. Take the second, for example, in the form (4), § 66,

$$-\operatorname{curl}(\mathbf{E} - \mathbf{e}_0) = \mathbf{K} + \mathbf{B} + \mathbf{w}\sigma - \operatorname{curl}V\mathbf{q}\mathbf{B}, \quad . \quad (185)$$

and suppose that $\mathbf{w} = \mathbf{q}$, or that sources move with the medium. Then, by (181), we cancel the convective term $\mathbf{w}\sigma$. Further, we have $\dot{\mathbf{B}} + \mathbf{q}\nabla.\mathbf{B} = \delta\mathbf{B}/\delta t$, by (127), § 122, so that (185) becomes

$$-\operatorname{curl}(\mathbf{E} - \mathbf{e}_0) = \mathbf{K} + \delta\mathbf{B}/\delta t + \mathbf{B}\operatorname{div}\mathbf{q} - \mathbf{B}\nabla.\mathbf{q}, \quad . \quad (186)$$

and the corresponding form of the first law (equation (3), § 66), is

$$\operatorname{curl}(\mathbf{H} - \mathbf{h}_0) = \mathbf{C} + \delta\mathbf{D}/\delta t + \mathbf{D}\operatorname{div}\mathbf{q} - \mathbf{D}\nabla.\mathbf{q}. \quad . \quad (187)$$

The time-variations refer to the same (moving) portion of the medium now. But if we wish to indicate the movement of electrification, &c., through the medium, that is, have relative motion $\mathbf{u} - \mathbf{q}$ of ρ (and $\mathbf{w} - \mathbf{q}$ of σ) with respect to the ether then to the right side of (187) add the term $(\mathbf{u} - \mathbf{q})\rho$, and to the right ride of (186) add $(\mathbf{w} - \mathbf{q})\sigma$.

It is desirable to preserve the velocities $\mathbf{u}$ and $\mathbf{w}$, or else the relative velocities, as well as the velocity $\mathbf{q}$ of the medium, in order to facilitate the construction and comprehension of problems relating to electromagnetic waves, which, although abstract and far removed from practice, are of a sufficiently simple nature to enable one to follow the course of events.

(*e*). We have already had the divergence of the product of a scalar and vector under consideration. Now examine its curl. Thus,

$$\operatorname{curl}\mathbf{D}P = V\nabla(\mathbf{D}P)$$
$$= V\nabla\mathbf{D}.P - V\mathbf{D}\nabla.P$$
$$= P\operatorname{curl}\mathbf{D} - V\mathbf{D}\nabla P. \quad . \quad . \quad (188)$$

Here ∇ can only make a vector product with $\mathbf{D}$, because P is scalar. On the other hand, both P and $\mathbf{D}$ suffer differentiation. So in the second line we have ∇ both before and after $\mathbf{D}$.

(*f*). The divergence of the curl of any vector is zero. That is, $\operatorname{div}\operatorname{curl}\mathbf{H} = 0$, or $\nabla V\nabla\mathbf{H} = 0$. . . (189)

If ∇ here were a real vector (189) would mean that the volume of a parallelepiped vanishes when two edges coincide.

(g). A somewhat similar case is presented by the vanishing of the curl of a polar force. Thus,

$$\text{curl } \nabla P = 0, \quad \text{or} \quad V \nabla \cdot \nabla P = 0. \quad . \quad . \quad . \quad (190)$$

Of course $V \nabla \nabla$ is zero. But the scalar product $\nabla \nabla$, or ∇^2, is the Laplacean operator,

$$\nabla^2 = \nabla_1{}^2 + \nabla_2{}^2 + \nabla_3{}^2, \quad . \quad . \quad . \quad . \quad (191)$$

which occurs frequently.

(h). Let the operation curl be done twice on a vector. Thus,

$$(\text{curl})^2 \mathbf{A} = V \nabla V \nabla \mathbf{A}$$
$$= \nabla \cdot \nabla \mathbf{A} - \nabla^2 \mathbf{A}, \quad . \quad . \quad . \quad (192)$$

by the transforming formula (52). Or

$$\nabla^2 \mathbf{A} = \nabla \text{ div } \mathbf{A} - \text{curl}^2 \mathbf{A}. \quad . \quad . \quad . \quad (193)$$

Thus there are two principal forms. If the vector $\mathbf{A}$ has no curl, then $\nabla^2 \mathbf{A}$ is the slope of its divergence. If, on the other hand, it has no divergence, then $-\nabla^2$ has the same effect exactly as taking the curl twice.

(i). In the case of the operand being a scalar, then we have

$$\nabla^2 P = \text{div } \nabla P, \quad . \quad . \quad . \quad . \quad (194)$$

the divergence of the slope of the scalar.

The Potential of a Scalar or Vector. The Characteristic Equation of a Potential, and its Solution.

§ 133. The last equation brings us to the theory of potentials. There are several senses in which the word potential has been employed, to enumerate which would be valueless here. For our present purpose we may conveniently fix its meaning by defining the potential at A of a quantity ρ at B to be the quantity $\rho/4\pi r$, where r is the distance from B to A. This is the rational potential, of course.

When ρ is distributed throughout space, whether at points, or over surfaces, or throughout volumes, the potential at any point is the sum of the potentials of all the elements of ρ. That is,

$$P = \text{pot } \rho = \Sigma \rho/4\pi r, \quad . \quad . \quad . \quad . \quad (195)$$

if P is the potential of ρ.

We may use the same definition when it is a vector that has to be potted, or potentialised. Thus, if **A** is the potential of **C**, then

$$\mathbf{A} = \text{pot } \mathbf{C} = \Sigma \mathbf{C}/4\pi r. \quad . \quad . \quad . \quad (196)$$

The summation is now a vector summation. Also, pot means " potential," or " the potential of," and has no more to do with kettle than the trigonometrical sin has to do with the un-mentionable one. It seems unnecessary to say so, but one cannot be too particular.

We may connect these potentials with $\triangledown$ as follows :—Given that

$$\text{div } \mathbf{F} = \rho, \quad . \quad . \quad . \quad . \quad (197)$$

that is to say, that the divergence of a vector **F** is ρ. The meaning of divergence has been explained more than once; both its intrinsic and its vectorial meaning. Now, if the vector **F** be explicitly given, it is clear that ρ is known definitely, since it is derived from **F** by differentiation, which should, perhaps, be regarded as a direct process, rather than inverse. But if it be ρ that is given, **F** is not immediately determinable, unless we subject **F** to limitations. For we may construct any number of different **F**'s to satisfy (197). Let every elementary source ρ of **F** send out the quantity ρ of **F**, according to my rational theory of sources already explained ; that is, the unitarian system of *one* "line of force" to the *unit* "pole." Then, by the manner of construction, the resultant **F** will satisfy (197), and it will do so independently of the way we choose to let a source send out the flux it generates, whether equably or not.

But if it be done equally in all directions, so that $\rho/4\pi r^2$ is the intensity of the "force" at distance r from the point-source ρ, and $\mathbf{r}_1 \rho/4\pi r^2$ the vector force to correspond, where $\mathbf{r}_1$ is a unit vector drawn from ρ towards the point under considera-tion, making the resultant **F** be

$$\mathbf{F} = \Sigma(\rho/4\pi r^2)\mathbf{r}_1, \quad . \quad . \quad . \quad . \quad (198)$$

we obtain a special solution of (197) which has a remarkable property, viz.,

$$\text{curl } \mathbf{F} = 0, \quad \ldots \ldots \quad (199)$$

so that if $\mathbf{F}$ be electric force, the voltage in any circuit is zero. The meaning of curl, I may observe, has been explained more than once ; both its intrinsic and its vectorial meaning. Those who seek can find. If they will not take the trouble to seek or to remember it is of no consequence to them. There are plenty of other things they may concern themselves about; perhaps more profitably.

The property (199) is visibly true in the case of a single source. It is therefore separately true for the fields of all the sources, and therefore, by summation, is true for the complete $\mathbf{F}$.

But (198) does not give the only vector which has no curl and a given divergence. For a constant vector (that is, constant throughout all space), has no curl and no divergence, unless we go to the very end of space to find the sources. Of course this constant solution has no relation to the sources ρ, and may be wholly ignored. If allowed, $\mathbf{F}$ would not vanish at an infinite distance from the sources. Remembering this, and excluding the constant solution, we may say that (198) is *the* solution of (197) and (199).

That there is no other solution may be proved analytically by Green's Theorem. But we do not really need any appeal to analysis of that kind, if the intrinsic meanings of divergence and curl are understood. For the admission that there could be a second solution, say, $\mathbf{F} + \mathbf{f}$, where $\mathbf{F}$ is the solution (198), would, by (197) and (199), imply that the vector $\mathbf{f}$ had no divergence, and also no curl anywhere. But the first of these conditions means that $\mathbf{f}$ is entirely circuital, if existent at all. The second denies that it is circuital. So $\mathbf{f}$ is non-existent.

Now observe that

$$- \nabla \left(1/4\pi r \right) = \mathbf{r}_1 / 4\pi r^2,$$

or the slope of the scalar $1/4\pi r$ is the vector with tensor $1/4\pi r^2$ and direction $\mathbf{r}_1$. It follows from this that

$$- \nabla \operatorname{pot} \rho = \mathbf{F}. \quad \ldots \ldots \quad (200)$$

when a single point-source is in question. Therefore, by summation, the same is true for any distribution of sources, or

$$\mathbf{F} = -\nabla P = -\nabla \operatorname{pot} \rho, \quad . \quad . \quad . \quad (201)$$

where P is the potential of ρ, as defined by (195), and $\mathbf{F}$ is as in (198). The slope of P, if by this we understand slope downwards, or vector rate of fastest decrease, is therefore the same vector as was constructed to solve (197) subject to (199).

Taking the divergence of (201), we have, by (197) and (194),

$$\rho = \operatorname{div} \mathbf{F} = -\nabla^2 P = -\nabla^2 \operatorname{pot} \rho. \quad . \quad . \quad (202)$$

A solution of the characteristic equation of P, or

$$\nabla^2 P = -\rho, \quad . \quad . \quad . \quad . \quad (203)$$

is therefore (195), and it is *the* solution vanishing at an infinite distance from the sources.

If we start from (203), we should first use (194), and make it

$$\operatorname{div}(-\nabla P) = \rho. \quad . \quad . \quad . \quad . \quad (204)$$

Then, by (190), we see that $-\nabla P$ has no curl, so that we have again the two equations (197), (199) to consider, as above.

The consideration of $\mathbf{F}$ rather than of P has many advantages for purposes of reasoning, as distinguished from calculation. This is true even in statical problems; for instance, when $\mathbf{F}$ is electrostatic force, and P the corresponding potential. When we proceed further, to kinetic problems, when $\mathbf{F}$ can no longer be wholly expressed as the slope of a potential, the utility of considering P at all, even for calculating purposes, becomes sometimes very questionable, and the consideration is sometimes certainly useless and misleading.

From (202) we see that $-\nabla^2$ and pot are reciprocal. In another form, $-\nabla^{-2}$ and pot are equivalent; or $(\operatorname{pot})^{-1}$ and $-\nabla^2$ are equivalent. The property has only been proved for a scalar function, having a scalar potential. But since any vector $\mathbf{C}$ may be written $\mathbf{i}C_1 + \mathbf{j}C_2 + \mathbf{k}C_3$, and the property is true for the three scalars C_1, &c., it is also true for the vector $\mathbf{C}$. Thus, explicitly,

$$-\nabla^2 \operatorname{pot} \mathbf{C} = -\nabla^2 \operatorname{pot} (\mathbf{i}C_1 + \mathbf{j}C_2 + \mathbf{k}C_3)$$

$$= \mathbf{i}(-\nabla^2 \operatorname{pot} C_1) + \mathbf{j}(-\nabla^2 \operatorname{pot} C_2) + \mathbf{k}(-\nabla^2 \operatorname{pot} C_3),$$

because the reference vectors **i**, &c., are constant vectors. So, if A_1 is the potential of C_1, A_2 of C_2, and A_3 of C_3, which makes **A** be the potential of **C**, according to (196), we shall have

$$A = \text{pot } C, \quad \cdots \cdots \cdots \quad (205)$$

$$-\nabla^2 A = C, \quad \cdots \cdots \cdots \quad (206)$$

$$-\nabla^2 \text{pot } C = C. \quad \cdots \cdots \cdots \quad (207)$$

In short, the characteristic equation of the vector **A** merely unites, from the above point of view, the characteristics of the components, so that pot and $-\nabla^2$ are reciprocal when the operand is a vector, as well as when it is a scalar.

Connections of Potential, Curl, Divergence, and Slope. Separation of a Vector into Circuital and Divergent Parts. A Series of Circuital Vectors.

§ 134. But the above gives a very partial and imperfect view of the general theory of potentials. There are numerous other relations between a vector and its associated functions. For instance, if in (206), **A** be circuital, then the Laplacean ∇^2 may, by (193), be replaced by $-(\text{curl})^2$. That is, if A_1 be circuital, and be the potential of C_1, then,

$$\text{curl}^2 A_1 = C_1, \quad \cdots \cdots \quad (208)$$

or
$$\text{curl}^2 \text{pot } C_1 = C_1. \quad \cdots \cdots \quad (209)$$

Here, then, we have replaced the scalar operation $-\nabla^2$ by the vector operation curl done twice. Of course, C_1 is also circuital, as is proved by (189).

Again, let the **A** of (206) be polar, or wholly divergent, and be now called A_2, the potential of C_2; then, by (193), we shall have

$$-\nabla \operatorname{div} A_2 = C_2, \quad \cdots \cdots \quad (210)$$

or
$$-\nabla \operatorname{div} \text{pot } C_2 = C_2. \quad \cdots \cdots \quad (211)$$

Here again we have replaced ∇^2 by a double operation, first div and then ∇. This is similar to the passage from (203) to (204), only done in the reverse manner. In (210), by (190), C_2 is polar, because A_2 is.

Conversely, we see that the potential of a circuital vector is also circuital, and that the potential of a polar vector is also polar.

Now A_1 has no divergence, so it may be added on to the divergent A_2 in (210) without affecting its truth. Thus, if $A = A_1 + A_2$, we have

$$-\nabla \operatorname{div} A = C_2. \quad \ldots \quad (212)$$

Similarly, A_2 has no curl, so may be added to the A_1 in (208), giving

$$\operatorname{curl}^2 A = C_1. \quad \ldots \quad (213)$$

Here remember that A is the potential of C or $C_1 + C_2$.

These equations supply one way of effecting the division of a vector A of general type (having both curl and divergence) into two vectors, one of which, A_1, is circuital, whilst the other A_2 is polar. For A_2 is the potential of C_2, so, by (212),

$$-\operatorname{pot} \nabla \operatorname{div} A = A_2 \quad \ldots \quad (214)$$

separates A_2 from A. Similarly, A_1 is the potential of C_1, so by (213),

$$\operatorname{pot} \operatorname{curl}^2 A = A_1 \quad \ldots \quad (215)$$

separates A_1 from A, and therefore A_2 from A by a different method. There are many other ways of splitting A into circuital and divergent parts. The one most easily understood, apart from the mathematics, is the following. Go over the whole field of A and measure its divergence. If we find that there is no divergence, then we do not need to go further, for we know that A is circuital already; that is, $A = A_1$, and $A_2 = 0$. But should there be divergence, say B_2, so that

$$\operatorname{div} A = \operatorname{div} A_2 = B_2, \quad \ldots \quad (216)$$

then construct the flux A_2 corresponding to the divergence B_2 according to the method already explained with respect to (197); thus,

$$A_2 = \Sigma \left(B_2 / 4\pi r^2 \right) r_1$$
$$= -\nabla \operatorname{pot} B_2, \quad \ldots \quad (217)$$

by (198) and (200). Knowing A_2, we know A_1, or $A - A_2$.

Or we might vary the process thus. First measure the curl of A. This is the same as the curl of A_1 because the curl of A_2 is zero. Let, then,

$$\operatorname{curl} A = \operatorname{curl} A_1 = B_1, \quad \ldots \quad (218)$$

and construct the circuital solution of this equation; that is to say, regarding $\mathbf{B}_1$ as given, find $\mathbf{A}_1$. It is given by

$$\mathbf{A}_1 = \text{curl pot } \mathbf{B}_1. \quad \ldots \quad (219)$$

For $\mathbf{A}_1$ as thus defined is evidently circuital, in the first place; and next, by taking the curl, we produce

$$\text{curl } \mathbf{A}_1 = \text{curl}^2 \text{ pot } \mathbf{B}_1 = \mathbf{B}_1, \quad \ldots \quad (220)$$

which is the given datum. Here we use $\text{curl}^2 \text{ pot} = 1$, because the operand is circuital, as in (209).

But instead of (219) we may write

$$\mathbf{A}_1 = \text{pot curl } \mathbf{B}_1, \quad \ldots \quad (221)$$

showing an entirely different way of going from $\mathbf{B}_1$ to $\mathbf{A}_1$. For $\mathbf{A}_1$ as thus constructed is circuital; and, since $\text{curl } \mathbf{B}_1 = \mathbf{C}_1$, (221) is the same as

$$\mathbf{A}_1 = \text{pot } \mathbf{C}_1,$$

which was our definition of $\mathbf{A}_1$ in terms of $\mathbf{C}_1$.

Thus pot curl and curl pot are equivalent when the operand is circuital, as above. They are, however, also equivalent when the operand is general, or both circuital and divergent, because if any divergent vector be added to the $\mathbf{B}_1$ in the right members of (219) or (221), the operation of curl to which it is subjected renders its introduction inoperative. We therefore have

$$\text{pot curl } \mathbf{C} = \text{curl pot } \mathbf{C}, \quad \ldots \quad (222)$$

where $\mathbf{C}$ is any vector. We have also the similar exchangeability,

$$\text{pot curl}^2 \mathbf{C} = \text{curl}^2 \text{ pot } \mathbf{C}, \quad \ldots \quad (223)$$

where $\mathbf{C}$ is any vector. For, either way, the result is the circuital part of $\mathbf{C}$, or $\mathbf{C}_1$.

These results, though puzzling at first from their variety, are yet capable of being brought under rapid mental control by bringing them together in a compact form. Thus, start with any circuital vector $\mathbf{A}_1$. Let $\mathbf{B}_1 = \text{curl } \mathbf{A}_1$, $\mathbf{C}_1 = \text{curl } \mathbf{B}_1$, $\mathbf{D}_1 = \text{curl } \mathbf{C}_1$, &c. We have a series of vectors

$$\mathbf{A}_1, \quad \mathbf{B}_1, \quad \mathbf{C}_1, \quad \mathbf{D}_1, \quad \mathbf{E}_1, \quad \ldots$$

which are all circuital, and any one of which is the curl of the preceding. We thus pass down the series one step at a time by means of the operation of curling ; for example,

$$D_1 = \operatorname{curl} C_1. \quad \cdots \quad \cdots \quad (224)$$

If, however, we wish to go down two steps, we do not need to go first one step, as above, and then another, also as above ; but can make a double step in one operation by means of the Laplacean ∇^2. Thus,

$$-\nabla^2 B_1 = D_1. \quad \cdots \quad \cdots \quad (225)$$

Now go the other way. If we wish to rise up two steps, we can do it in one operation by potting ; thus,

$$B_1 = \operatorname{pot} D_1. \quad \cdots \quad \cdots \quad (226)$$

If we wish to go up only one step, we may do it by (224), (225), (226) combined ; that is, either go down one step first, and then up two, as in

$$B_1 = \operatorname{pot} D_1 = \operatorname{pot} \operatorname{curl} C_1 ; \quad \cdots \quad (227)$$

or else, first go up two steps and down one ; thus,

$$B_1 = \operatorname{curl} A_1 = \operatorname{curl} \operatorname{pot} C_1. \quad \cdots \quad (228)$$

There are other less important combinations. But if we wish to make one step up directly, without making use of the double step, we must do it by the Ampèrean formula, already used, whereby we pass direct from electric current to its magnetic force, which, in rational units, is (when applied to any pair of neighbours C_1 and D_1 in the above series),

$$C_1 = \Sigma (V D_1 r_1)/4\pi r^2, \quad \cdots \quad (229)$$

where r_1 is a unit vector from the element D_1 to the point at distance r therefrom, where C_1 is reckoned. We have now a complete scheme, so far as the circuital vectors are concerned.

A Series of Divergent Vectors.

§ 135. Deferring temporarily a vectorial proof of the last formula (229), which is the only unproved formula in the con- nections of the series of circuital vectors, it will now be convenient to bring together the connections of the divergent vectors and associated quantities. We saw the advantage of the systematic arrangement of the connected circuital vectors

P

to be like producing a harmonious chord out of apparently
disconnected tones. The advantage is much greater in the
divergent series, on account of the less uniform relations involved and the greater need of a system to bring them under
rapid mental control. In the circuital series, four kinds of
operation were involved; but in the divergent series there are
six. The chord will be found to be perfect, though of greater
complexity.

Thus, let

$$\mathbf{A}_2, \qquad B_2, \qquad \mathbf{C}_2, \qquad D_2, \qquad \mathbf{E}_2, \quad . \ . \ . \ .$$

be a series of vectors and scalars connected as follows :—Start
with $\mathbf{A}_2$, which is to be any divergent vector; that is, having
no curl. Let B_2 be its divergence; $\mathbf{C}_2$ the slope of B_2; D_2
the divergence of $\mathbf{C}_2$; $\mathbf{E}_2$ the slope of D_2, &c. Then $\mathbf{A}_2, \mathbf{C}_2,$
$\mathbf{E}_2, \ . \ . \ .$ are all divergent vectors. But they are separated
from one another by two steps instead of one, as was the case
in the circuital series last treated. The intermediate quantities
are scalars. Instead, also, of the single operation of curl
which suffices, in the circuital series, to carry us from any
vector to the following one, we now have two distinct operations; viz., that of slope, when we pass from a scalar to the
next vector, as in

$$\mathbf{C}_2 = - \nabla B_2 ; \quad . \ . \ . \ . \ . \quad (230)$$

and that of divergence, when we pass from a vector to the next
scalar, as in

$$D_2 = \operatorname{div} \mathbf{C}_2. \quad . \ . \ . \ . \quad (231)$$

But if we wish to go down two steps at once, we can do so
by means of the Laplacean operator, whether the operand be a
scalar, as in

$$- \nabla^2 B_2 = D_2, \quad . \ . \ . \ . \ . \quad (232)$$

or else a vector, as in

$$- \nabla^2 \mathbf{C}_2 = \mathbf{E}_2. \quad . \ . \ . \ . \quad (233)$$

In this respect, then, we have the same property as in the
circuital series.

We have also identity of operation in going up two steps at
once, whether from a scalar to the next higher scalar, as in

$$B_2 = \operatorname{pot} D_2, \quad . \ . \ . \ . \quad (234)$$

or from a vector to the next higher vector, as in

$$\mathbf{C}_2 = \operatorname{pot} \mathbf{E}_2 ; \quad . \quad . \quad . \quad . \quad (235)$$

any member of the series being the potential of the second after, as in the circuital series.

Next, to go up one step only, we may utilise the preceding in two ways. First go up two steps and then down one, as in

$$\mathbf{B}_2 = \operatorname{div} \mathbf{A}_2 = \operatorname{div} \operatorname{pot} \mathbf{C}_2, \quad . \quad . \quad . \quad (236)$$

where we pass from the vector $\mathbf{C}_2$ to the scalar $\mathbf{B}_2$ by pot first (up two steps), and then by div (down one step); and also as in

$$\mathbf{C}_2 = - \nabla \mathbf{B}_2 = - \nabla \operatorname{pot} \mathbf{D}_2, \quad . \quad . \quad . \quad (237)$$

where we pass from the scalar $\mathbf{D}_2$ to the vector $\mathbf{C}_2$ by pot first (up two steps), and then by $- \nabla$ (down one).

Or, secondly, we may first go down one step and then up two, as in

$$\mathbf{B}_2 = \operatorname{pot} \mathbf{D}_2 = \operatorname{pot} \operatorname{div} \mathbf{C}_2, \quad . \quad . \quad . \quad (238)$$

when rising from the vector $\mathbf{C}_2$ to the scalar $\mathbf{B}_2$; or as in

$$\mathbf{C}_2 = \operatorname{pot} \mathbf{E}_2 = - \operatorname{pot} \nabla \mathbf{D}_2, \quad . \quad . \quad . \quad (239)$$

when rising from the scalar $\mathbf{D}_2$ to the vector $\mathbf{C}_2$.

Finally, if we wish to rise up one step at once, without using the double step either up or down, we can do it by means of

$$\mathbf{C}_2 = (\operatorname{div})^{-1} \mathbf{D}_2 = \Sigma \, \mathbf{r}_1 \mathbf{D}_2 / 4 \pi r^2, \quad . \quad . \quad (240)$$

when we rise from a scalar $\mathbf{D}_2$ to a vector $\mathbf{C}_2$, which is, in fact, the fundamental formula of the inverse-square law upon which our potential investigations are based. But in rising from a vector $\mathbf{C}_2$ to the scalar $\mathbf{B}_2$ just above it in the series, we require to use a different process, namely,

$$\mathbf{B}_2 = (- \nabla)^{-1} \mathbf{C}_2 = - \Sigma \, \mathbf{r}_1 \mathbf{C}_2 / 4 \pi r^2. \quad . \quad . \quad (241)$$

In these formulæ (240), (241), $\mathbf{r}_1$ is a unit vector from the element in the summation towards the place of the resultant ; that is, from $\mathbf{D}_2$ to $\mathbf{C}_2$ in (240), and from $\mathbf{C}_2$ to $\mathbf{B}_2$ in (241).

We now have a complete scheme for the divergent vectors as we had before for the circuital series. On comparing them we see that they are alike in the double steps, either up or down, but differ in the single steps. There is but one kind of

step up and but one kind down in the circuital series, whereas
there are two kinds up and two kinds down in the divergent
series. The down step in the circuital series is always done by
curl; the up step, shown in (229), may be denoted by (curl)$^{-1}$.
In the divergent series the down steps are done by $-\nabla$ and by
div; their inverses may be denoted by $(-\nabla)^{-1}$ and (div)$^{-1}$.
It is now the nature of these inverse operations (229), (240), and
(241) that remains to be elucidated vectorially. The first is
the Ampèrean formula rationalized, whereby we rise from
electric current to its magnetic force; by the second we rise
from (for instance) electrostatic force to the electrostatic
potential; or, with a slight change (of sign), from intensity of
magnetisation to magnetic potential; in the third we rise
from (for instance), electrification to electrostatic displacement.

The Operation inverse to Divergence.

§ 136. Let p and q be scalar functions, and consider the
space-variation of their product. We have

$$\nabla(pq) = p\nabla q + q\nabla p, \quad . \quad . \quad . \quad . \quad (242)$$

a formula not previously used, but which is seen to be true
by observing that it is true for each of the three components
of ∇. Now integrate through any region. We know that

$$\Sigma \mathbf{N} pq = \Sigma \nabla(pq), \quad . \quad . \quad . \quad . \quad (243)$$

if $\mathbf{N}$ is the unit normal outwards from the boundary of the
region, so that the left member is a surface-summation, whilst
the right member is a volume-summation throughout the
region bounded by the surface. Equation (243) is, in fact, a
case of (157), with pq substituted for p. If, then, the surface-
summation vanishes, we shall have a simultaneous evanescence
of the right member of (243), and therefore, by (242),

$$\Sigma p\nabla q = -\Sigma q\nabla p. \quad . \quad . \quad . \quad . \quad (244)$$

All the work done by a vector-analyst is exhibited in (244)
itself, viz., the transfer of the symbol ∇ from one operand
to the other with change of sign, converts the integral of
$p\nabla q$ into that of $-q\nabla p$. The previous remarks contain the
justification of the process.

Now $-\nabla q$ is a polar or divergent vector, so may be any one of our divergent series, say $\mathbf{C}_2$, when q itself becomes $\mathbf{B}_2$. Then

$$\Sigma p\mathbf{C}_2 = \Sigma \mathbf{B}_2\nabla p. \quad . \quad . \quad . \quad . \quad (245)$$

Lastly, let p have the special value $1/4\pi r$; then (245) is the same as

$$\text{pot } \mathbf{C}_2 = \mathbf{A}_2 = \Sigma \mathbf{B}_2\nabla p, \quad . \quad . \quad . \quad (246)$$

which exhibits the divergent vector $\mathbf{A}_2$ in terms of its divergence $\mathbf{B}_2$. It is the same as (240), since $\nabla p = \mathbf{r}_1/4\pi r^2$, if $\mathbf{r}_1$ is the unit vector from $\mathbf{B}_2$ to $\mathbf{A}_2$.

The Operation inverse to Slope.

§ 137. Next, substitute for q in pq a vector, say $\mathbf{g}$. The new quantity $p\mathbf{g}$ has, being a vector, both curl and divergence, in general. Considering the latter first, we have

$$\text{div } p\mathbf{g} = p \text{ div } \mathbf{g} + \mathbf{g}\nabla p, \quad . \quad . \quad . \quad (247)$$

which is an example of (160). Integrating throughout any volume, we have

$$\Sigma \mathbf{N} p\mathbf{g} = \Sigma \text{div } p\mathbf{g}, \quad . \quad . \quad . \quad (248)$$

as in (159), where $\mathbf{N}$ is as before. So, if the surface-integral vanishes, we obtain, by (247),

$$\Sigma p \text{ div } \mathbf{g} = -\Sigma \mathbf{g}\nabla p, \quad . \quad . \quad . \quad (249)$$

and, in this transformation, all the vector-analyst has to do is to shift the operator ∇ from one operand to the other, and change the sign.

The vector $\mathbf{g}$ here has no restriction imposed upon it. It may therefore be of the general type $\mathbf{C} = \mathbf{C}_1 + \mathbf{C}_2$, giving

$$\Sigma p\mathbf{D}_2 = -\Sigma \mathbf{C}\nabla p = -\Sigma \mathbf{C}_2\nabla p . \quad . \quad . \quad (250)$$

Here the portion $\Sigma \mathbf{C}_1 \nabla p$ vanishes because $\mathbf{C}_1$ is circuital and ∇p is polar, which is one of the important theorems in analysis that become visibly true by following the tubes of the circuital flux in performing the summation, when the summation is seen to vanish separately for every tube. (*See* § 87.)

If in (250) we give p the special value $1/4\pi r$, viz., the potential due to a unit source at distance r, we obtain

$$\text{pot } \mathbf{D}_2 = \mathbf{B}_2 = -\Sigma \mathbf{C}\nabla p = -\Sigma \mathbf{C}_2\nabla p, \quad . \quad . \quad (251)$$

showing how to pass up one step in the divergent series from a vector to the preceding scalar. It is the same as (241), remembering the value of ∇p.

The Operation inverse to Curl.

§ 138. Thirdly, we have the curl of $p\mathbf{g}$ to consider. Here, by (188),

$$\operatorname{curl} p\mathbf{g} = p \operatorname{curl} \mathbf{g} - \mathrm{V}\mathbf{g}\nabla p, \quad . \quad . \quad . \quad (252)$$

Integrating throughout any region, we obtain

$$\Sigma \, \mathrm{V}\mathbf{N}p\mathbf{g} = \Sigma \operatorname{curl} p\mathbf{g}, \quad . \quad . \quad . \quad (253)$$

which is a case of (162), with $p\mathbf{g}$ put for $\mathbf{H}$. So, if the surface-summation vanishes, we obtain, by (252),

$$\Sigma p \operatorname{curl} \mathbf{g} = \Sigma \, \mathrm{V}\mathbf{g}\nabla p, \quad . \quad . \quad . \quad (254)$$

where the symbol $\mathrm{V}\nabla$ is moved from $\mathbf{g}$ to p, with a change of sign, as before. In this, take $p = 1/4\pi r$; then, since there is no restriction upon $\mathbf{g}$, we get, taking $\mathbf{g} = \mathbf{C}$,

$$\operatorname{pot} \mathbf{D}_1 = \mathbf{B}_1 = \Sigma \, \mathrm{V}\mathbf{C}\nabla p = \Sigma \, \mathrm{V}\mathbf{C}_1\nabla p, \quad . \quad . \quad (255)$$

which is the companion to (251), showing how to pass up one step in the circuital series, from $\mathbf{C}_1$ to $\mathbf{B}_1$. This is equivalent to (229). The divergent part of $\mathbf{C}$ contributes nothing. That is to say, for example, the magnetic force due to a completely divergent distribution of electric current, according to Ampère's formula for the magnetic force of a current *element*, is zero. We might, indeed, argue from this, that there could not be such a kind of electric current ; that is to say, that the current must be circuital, since the mathematical machinery itself, constructed on old ideas, automatically rejects the want of circuitality, and refuses to admit the purely divergent part as contributory to magnetic force. This is a perfectly valid argument, provided the test of the existence of electric current be the existence of magnetic force, which is tantamount to what Maxwell insisted upon, in another form.

For instance, if we calculate by (255) the magnetic force due to a supposititious current element at a point, simply by removing the sign of summation, we obtain the magnetic force of a rational current element, a system of circuital current resembling the induction due to a magnetised particle. (*See* § 62.)

Remarks on the inverse Operations.

§ 139. Returning to the three single up-step formulæ, it will be observed that in every case we base the proof upon the vanishing of a surface-integral, viz., $\Sigma \mathbf{N}pq$, or $\Sigma \mathbf{N}pg$, or $\Sigma \mathbf{V N}pg$. Now it is obvious that these are true if the surface be taken wholly outside the region occupied by the quantity q or g. The value pq or pg is made zero all over the surface of summation. This is what occurs in the applications made. In (246) the space-summation has to include all the B_2, which represents q; it is sufficient, therefore, for the surface to completely enclose all B_2. Again, in (255), the enclosure by the surface of all C, which represents g, will ensure the vanishing of the surface-integral; and in (251) the same. Now, it is usual to imagine the surface to be at infinity, so that the space-summations extend over all space. This is also most convenient, in general. But caution is sometimes necessary when the quantity to be summed extends to infinity. The surface-summation then may, or may not, vanish, according to circumstances. Even if the quantity summed becomes infinitely small at infinity, the summation may still not vanish. To illustrate this, it is sufficient to mention the case of a single point-source. The surface-integral of the flux it produces is finite always, being the measure of the strength of the source. At infinity the flux may be infinitely attenuated, but the surface is simultaneously infinitely magnified. But in the application of the above processes to practical cases in electromagnetism it is usually quite easy to see that the integral over the surface at infinity vanishes, owing either to the actual absence of anything to be summed, or else, when there is an infinitely attenuated quantity to be summed, of its being, per unit area, of lower dimensions than $1/r^2$. The quantity p itself attenuates to nothing, as well as q or g, if they are existent at all at infinity.

The three up-step formulæ may, by inspection, be transformed so as to involve entirely different operations. Thus, in (241), put p for $1/4\pi r$. Then we have

$$B_2 = - \operatorname{pot} (\mathbf{C}_2/r). \quad \cdots \quad (256)$$

Do the same in (240), and we obtain

$$\mathbf{C}_2 = \operatorname{pot} (\mathbf{D}_2/r). \quad \cdots \quad (257)$$

Finally, (229) gives

$$C_1 = \text{pot} \ (V D_1 / r). \quad . \quad . \quad . \quad . \quad (258)$$

We may sometimes utilise these formulæ for purposes of calculation, should the integrations to be performed be amenable
to practical treatment. But there is a caution to be mentioned. The quantities whose potentials are calculated by the
last three formulæ are not functions of position, which are definitely distributed in space, and are independent of the position
of the point where the potential is reckoned, but are definite only
when this point is fixed. As you pass to another point, the
potential there is that of a different distribution from that
belonging to the first point. These quantities, C_2/r, &c., are
therefore not subject to the various properties of the circuital
and divergent vectors already considered. For example, C_2 is
purely divergent, but D_2/r has curl; or, C_1 is circuital, whilst
$V D_1/r$ has divergence, and so on.

Integration "by parts."   Energy Equivalences in the Circuital Series.

§ 140. The transformations (244), (249), (254), or the more
general ones containing the omitted surface-summations, expressed by putting the sign of summation before every term in
(242), (247), (252), and converting the summation on the left
to a surface-summation by the introduction of **N**, as in (243),
(248), and (253), are examples of what is, in the Cartesian
mathematics, called integration "by parts." It is usually a
very tedious and uninforming process, that is, in Cartesians,
with its triple and double $\int$'s, its dS and $dx \ dy \ dz$, and its
l, m, n. The vectorial methods go straight to the mark at
once, avoid a large amount of quite useless work, and enable
you to keep your attention fixed upon the actual magnitudes
concerned, and their essential relations, instead of being dis-
tracted by a crowd of coordinates and components.

There are, of course, many other cases which arise of this
"by parts" integration. One of the most important in connection with the circuital series of vectors, is the following :—
Substitute for p in the preceding, a vector, say **f**, then we shall
have

$$\Sigma \mathbf{f} \ \text{curl} \ \mathbf{g} = \Sigma \mathbf{g} \ \text{curl} \ \mathbf{f}, \quad . \quad . \quad . \quad (259)$$

the summations being throughout all space, or, at any rate

through enough of it to include all the quantities summed. The proof is that since

$$\operatorname{div} Vgf = f \operatorname{curl} g - g \operatorname{curl} f, \quad . \quad . \quad . \quad (260)$$

which is (178) again, we have, by space-summation,

$$\Sigma f \operatorname{curl} g = \Sigma g \operatorname{curl} f + \Sigma \operatorname{div} Vgf, \quad . \quad . \quad (261)$$

and the third term may be at once turned into the surface-summation $\Sigma NVgf$. If this vanish, as usual, then (259) follows.

Applying this result to the series of circuital vectors A_1, A_2, A_3, &c., each of which is the curl of the preceding, we obtain various equivalences. Thus,

$$\Sigma D_1{}^2 = \Sigma C_1 E_1 = \Sigma B_1 F_1 = \Sigma A_1 G_1, \quad . \quad . \quad (262)$$

through all space. Operate on one member by curl and on the other by $(\operatorname{curl})^{-1}$, to pass from one form to the next. Also

$$\Sigma D_1 E_1 = \Sigma C_1 F_1 = \Sigma B_1 G_1 = \Sigma A_1 H_1, \quad . \quad (263)$$

in a similar manner ; and so on.

In any of these summations we may convert either of the circuital vectors involved to a general vector by adding any divergent vector. For we see, by (259) that if f is polar, the right summation vanishes, and therefore so does the left. So $\Sigma D_1{}^2$ is the same as $\Sigma D_1 D$, for $\Sigma D_1 D_2 = 0$. Similarly, $\Sigma C_1 F_1$ is the same as ΣCF_1, or the same as $\Sigma C_1 F$, since $\Sigma C_1 F_2 = 0$, and $\Sigma C_2 F_1 = 0$. It will be understood here that the suffix 1 refers to the circuital vectors entirely, and the suffix 2 to the divergent vectors, and that $C = C_1 + C_2$, &c.

But we cannot generalise *both* vectors at the same time. Take CF for example. We have,

$$\Sigma CF = \Sigma (C_1 + C_2)(F_1 + F_2) = \Sigma C_1 F_1 + \Sigma C_2 F_2, \quad . \quad (264)$$

so that the summation of the scalar product of the two divergent vectors now enters. The series of divergent vectors and their intermediate scalars have properties similar to (262), (263).

Energy and other Equivalences in the Divergent Series.

§ 141. These properties of the space-integrals of scalar products in the divergent series are formally obtainable from the corresponding ones in the circuital series by changing the suffix from

1 to 2 ; at the same time changing the type from clarendon to roman should the quantity typified be a scalar.   Thus, analogous to (262), (263), we have

$$\Sigma \, \mathbf{C}_2{}^2 = \Sigma \, B_2 D_2 = \Sigma \, \mathbf{A}_2 \mathbf{E}_2 = \dots , \quad . \quad . \quad (265)$$

$$\Sigma \, D_2{}^2 = \Sigma \, \mathbf{C}_2 \mathbf{E}_2 = \Sigma \, B_2 F_2 = \dots , \quad . \quad . \quad (266)$$

starting from the square of a vector in the first set, and from the square of a scalar in the second.   These transformations all rest upon (249); that is, we pass from any form to the next by the exchange of $-\nabla$ and div between the factors.   For example,

$$\Sigma \, D_2{}^2 = \quad \Sigma \, D_2 \operatorname{div} \mathbf{C}_2, \quad \text{by definition of } D_2,$$

$$= -\Sigma \, \mathbf{C}_2 \nabla D_2, \qquad \text{by integration, using (249),}$$

$$= \quad \Sigma \, \mathbf{C}_2 \mathbf{E}_2, \qquad \text{by definition of } \mathbf{E}_2.$$

Similarly in all the rest.   The preceding equations conveniently summarise them.

But it will be observed that there is another way of pairing terms in the divergent series, viz., a vector with a scalar.   The corresponding transformations do not work so symmetrically as the previous.   For instance,

$$\Sigma \, \mathbf{C}_2 D_2 = -\Sigma \, \nabla B_2 . D_2, \quad \text{by definition of } \mathbf{C}_2,$$

$$= \quad \Sigma \, B_2 \nabla D_2, \qquad \text{by (244),}$$

$$= -\Sigma \, B_2 \mathbf{E}_2, \qquad \text{by definition of } \mathbf{E}_2,$$

$$= -\Sigma \operatorname{div} \mathbf{A}_2 . \mathbf{E}_2, \quad \text{by definition of } B_2,$$

$$= \quad \Sigma \, \mathbf{A}_2 \nabla . \mathbf{E}_2, \quad \text{by integration.} \quad . \quad (267)$$

Here the symmetry breaks down.   The last transformation depends upon

$$\Sigma \, (\mathbf{Nf}) g = \Sigma \, (\nabla f) g = \Sigma \, (g \operatorname{div} \mathbf{f} + \mathbf{f} \nabla . g), \quad . \quad . \quad (268)$$

which is a case of the theorem (147) or (153).   Note that in the second form ∇ has to differentiate both $\mathbf{f}$ and g, so that the full expression is in the third form.   If the surface-integral vanishes at infinity, we have

$$\Sigma \, g \operatorname{div} \mathbf{f} = -\Sigma \, \mathbf{f} \nabla . g, \quad . \quad . \quad . \quad (269)$$

which is the transformation used in getting (267). We have also, by going the other way,

$$\Sigma\, \mathbf{C}_2 \mathbf{D}_2 = \Sigma\, \mathbf{C}_2 \operatorname{div} \mathbf{C}_2 = - \Sigma\, \mathbf{C}_2 \nabla . \mathbf{C}_2, \quad . \quad . \text{ (270)}$$

by using the same transformation (269). Next take $\mathbf{D}_2 \mathbf{E}_2$. Here the vector is the slope of the scalar it is multiplied by, whereas in the former case, the divergence of the vector was the associated scalar. So it now goes quite differently. Thus,

$$\Sigma\, \mathbf{D}_2 \mathbf{E}_2 = - \Sigma\, \mathbf{D}_2 \nabla \mathbf{D}_2 = - \Sigma\, \nabla \tfrac{1}{2} \mathbf{D}_2{}^2 = - \Sigma\, \mathbf{N} \tfrac{1}{2} \mathbf{D}_2{}^2 \quad . \text{ (271)}$$

The result is therefore zero, if the surface-integral at infinity vanishes.

The Isotropic Elastic Solid. Relation of Displacement to Force through the Potential.

§ 142. It usually happens that potentials present themselves in physical mathematics as auxiliary functions introduced to facilitate calculations relating to other quantities. But in the theory of the elastic solid, the potential function presents itself in a very direct and neat manner; besides that, the elastic theory presents excellent illustrations of the above transformations and the general theory of ∇. If the solid be homogeneous and isotropic, there are but two elastic constants, the rigidity n and the coefficient of resistance k to compression or expansion. Let also $m = k + \tfrac{1}{3}n$, in Thomson and Tait's notation; then the equation of motion is

$$\mathbf{f} + n \nabla^2 \mathbf{G} + m \nabla \operatorname{div} \mathbf{G} = \rho \ddot{\mathbf{G}}, \quad . \quad . \quad . \text{ (272)}$$

where $\mathbf{G}$ is the displacement, and $\mathbf{f}$ the impressed force per unit volume, whilst ρ is the density. Whatever limitations may need to be put upon the values of m and n in treating of solids as we find them, in speculations relating to ethers they may have any values not making the stored energy negative. Thus m may, as we shall see presently, go down to the negative value $- n$, keeping n positive.

Split $\mathbf{f}$ into $\mathbf{f}_1 + \mathbf{f}_2$, and $\mathbf{G}$ into $\mathbf{G}_1 + \mathbf{G}_2$, where $\mathbf{f}_1$ and $\mathbf{G}_1$ are circuital, and the others divergent. Then (272) may be split into two equations, one circuital, the other divergent. Thus, remembering (193),

$$\mathbf{f}_1 + n \nabla^2 \mathbf{G}_1 = \rho\, \ddot{\mathbf{G}}_1, \quad . \quad . \quad . \text{ (273)}$$

$$\mathbf{f}_2 + (n + m) \nabla^2 \mathbf{G}_2 = \rho \ddot{\mathbf{G}}_2. \quad . \quad . \quad . \text{ (274)}$$

In the circuital equation k does not appear, so that the propagation of circuital disturbances depends only upon the rigidity and density ; speed, $(n/\rho)^{\frac{1}{2}}$.   In the divergent equation both n and k appear, and the speed of disturbances of this class is $\{(n+m)/\rho\}^{\frac{1}{2}}$.   In another form, if we take the curl of (272), the divergence goes out, so that the curl of $\mathbf{G}$, which is twice the rotation, has the curl of $\mathbf{f}$ for source, and is propagated independently of compressibility.

And if we take the divergence of (272), we see that the divergence of $\mathbf{G}$, or the expansion, has the divergence of $\mathbf{f}$ for source, and the rate of its propagation depends upon both rigidity and compressibility.   But when $m = 0$, there is but one speed of propagation, and a complete amalgamation of the two kinds of disturbances.

In equilibrium, when equilibrium is possible, the right members of the last three equations vanish, since they represent "rate of acceleration of momentum" (a long-winded expression).   We therefore have

$$\mathbf{f}_1 = -n\nabla^2\mathbf{G}_1 , \quad . \quad . \quad . \quad . \quad (275)$$

$$\mathbf{f}_2 = -(n+m)\nabla^2\mathbf{G}_2. \quad . \quad . \quad (276)$$

The solutions are visible by inspection.   We see that $n\mathbf{G}_1$ is the potential of $\mathbf{f}_1$, and $(n+m)\mathbf{G}_2$ the potential of $\mathbf{f}_2$.   That is, the displacement produced by circuital impressed force is its potential divided by n ; and the displacement produced by divergent impressed force is its potential divided by $n+m$; whilst, in the general case, we must split the impressed force into circuital and divergent parts, and then re-unite them in different proportions to obtain the resultant displacement. Symbolically,

$$n\mathbf{G}_1 = \text{pot } \mathbf{f}_1 \quad . \quad . \quad . \quad . \quad (277)$$

$$(m+n)\mathbf{G}_2 = \text{pot } \mathbf{f}_2, \quad . \quad . \quad . \quad . \quad (278)$$

$$\mathbf{G} = \frac{\text{pot } \mathbf{f}_1}{n} + \frac{\text{pot } \mathbf{f}_2}{m+n}. \quad . \quad . \quad . \quad (279)$$

In the case of incompressibility k, and therefore m, is infinite, so that $\mathbf{G}_2$ is zero, and the displacement is $\text{pot } \mathbf{f}_1/n$, whatever $\mathbf{f}_2$ may be.   For $\mathbf{f}_2$ is balanced by difference of pressure, which is set up instantly.   The corresponding speed is infinite, but there is no displacement.

If, again, $m = 0$, and there is but one speed, the displacement is the potential (divided by n) of the impressed force, whatever be its type.

On the other hand, if $n + m = 0$, we see by (274) that there is no steady state possible due to divergent force, as f_2 is then employed simply in accelerating momentum on the spot. The corresponding speed is zero. The circuital disturbances are propagated as before. This is the case of Sir W. Thomson's contractile ether, in which the wave of normal disturbance is abolished by making the speed zero.

The Stored Energy and the Stress in the Elastic Solid. The Forceless and Torqueless Stress.

§ 143. We may also find expressions for the stored energy from the equation of motion. The work done and stored by the impressed force f is $\frac{1}{2}fG$, though, if f be put on suddenly, an equal amount is dissipated, since, G being the final displacement, fG is the work then done by f, per unit volume. Now

$$\Sigma \tfrac{1}{2}fG = \Sigma \tfrac{1}{2}f_1G_1 + \Sigma \tfrac{1}{2}f_2G_2,$$

by (264). Calling the first part U_1, and the second part U_2, we have, by (273), (274),

$$U_1 = - \Sigma \tfrac{1}{2}nG_1\nabla^2G_1 = \Sigma \tfrac{1}{2}n(\text{curl } G)^2 \quad . \quad . \quad . \quad . \quad (280)$$

$$U_2 = - \Sigma \tfrac{1}{2}(m+n)G_2\nabla^2G_2 = \Sigma \tfrac{1}{2}(m+n)(\text{div } G)^2. \quad (281)$$

The transformations here used are (259) for U_1 and (249) for U_2. We see that the energy stored is expressed in terms of the squares of the rotation and of the expansion. In the contractile ether U_2 is zero, and the stored energy is U_1. We do not correctly localise the energy by the above formulæ, but only express the total amounts. For correct localisation we need to know the stress and the distortion. Now the distortion is a function of the variation of displacement, so is known in terms of G. Can we, however, find the stress itself from the equation of motion? If not, we can come very close to it. Thus, in a state of equilibrium, $F + f = 0$, if F is the force arising from the stress. So, by (275), (276),

$$F_1 = n\nabla^2G_1, \quad . \quad . \quad . \quad . \quad . \quad . \quad (282)$$

$$F_2 = (n+m)\nabla^2G_2, \quad . \quad . \quad . \quad . \quad (283)$$

and $\qquad F = n\nabla^2G + m\nabla^2G_2. \quad . \quad . \quad . \quad (284)$

Now let $\mathbf{P_N}$ be the stress on the plane whose normal is $\mathbf{N}$. We have, by consideration of the equilibrium of a unit cube,

$$\mathbf{FN} = \operatorname{div} \mathbf{P_N}, \quad \cdots \quad \cdots \quad (285)$$

to express the relation between an irrotational stress and the force arising from it. Therefore, applying this to (284),

$$\operatorname{div} \mathbf{P_N} = n\nabla^2 \mathbf{GN} + m\nabla^2 \mathbf{G_2 N}. \quad \cdots \quad (286)$$

Here the divergence of $\mathbf{P_N}$ is given; find $\mathbf{P_N}$ itself. The immediate answer is, by (201), or (214),

$$\mathbf{P_N} = -\nabla \operatorname{pot} \operatorname{div} \mathbf{P_N},$$

or
$$\mathbf{P_N} = n\nabla \mathbf{GN} + m\nabla \mathbf{G_2 N}, \quad \cdots \quad (287)$$

provided $\mathbf{P_N}$ is divergent. That is, we have constructed a stress-vector giving the proper force required. But, without interfering with this essential property, we might add on to the right side of (287) any circuital vector. That we must do so now, we may see by remembering that the stress must be irrotational, or produce no torque, and then by finding that (287) does give a torque. To show this, consider the first part only of the stress (287), say with $m = 0$. Then

$$\mathbf{NP_M} - \mathbf{MP_N} = n(\mathbf{N}\nabla.\mathbf{MG} - \mathbf{M}\nabla.\mathbf{NG}).$$

Here take $\mathbf{M} = \mathbf{j}$ and $\mathbf{N} = \mathbf{k}$; then

$$\mathbf{kP_j} - \mathbf{jP_k} = n(\nabla_3 G_2 - \nabla_2 G_3) = -n\mathbf{i} \operatorname{curl} \mathbf{G},$$

by (149). This gives the $\mathbf{i}$ component of the torque, which is therefore $-n \operatorname{curl} \mathbf{G}$. But $\mathbf{G_2}$ has no curl, therefore the second part of (287) produces no torque. We require, therefore, to add to $\mathbf{P_N}$ in (287) a stress giving no force, but a torque $n \operatorname{curl} \mathbf{G}$. Such a stress is

$$\mathbf{X_N} = n \operatorname{curl} \mathbf{VGN} \quad \cdots \quad \cdots \quad (288)$$

as may be tested in the above manner.

This brings us from (287) to

$$\mathbf{P_N} = n(\nabla \mathbf{GN} + \operatorname{curl} \mathbf{VGN}) + m\nabla \mathbf{G_2 N}, \quad \cdots \quad (289)$$

which is a stress-vector giving the correct force and no torque.

[It should be noted here that since (285) applies to an irrotational stress, the process employed is only justified when we finally get rid of the torque, as in (289), (290). If the stress is rotational, the divergence of $\mathbf{P_N}$ is really the $\mathbf{N}$ component of the force due to the conjugate stress. Thus the force due to $n\nabla.\mathbf{GN}$ is $n\nabla\operatorname{div}\mathbf{G}$, and the torque $-n\operatorname{curl}\mathbf{G}$. The force due to $m\nabla.\mathbf{G_2N}$ is $m\nabla\operatorname{div}\mathbf{G}$, with no torque. The force due to $\operatorname{curl}\mathbf{VGN}$ is $-n\operatorname{curl}^2\mathbf{G}$, and the torque $n\operatorname{curl}\mathbf{G}$. Adding together the three stresses and the three forces we obtain the stress (289), with no torque and the correct force.]

But on comparison with the real stress deduced from the elastic properties of a solid, which is

$$\mathbf{P_N} = n(\nabla.\mathbf{GN} + \mathbf{N}\nabla.\mathbf{G}) + \mathbf{N}(m-n)\operatorname{div}\mathbf{G} \quad . \quad . \quad (290)$$

we see that they do not agree. Yet they can only differ by a stress which produces neither force nor torque. And we know already, by (288), that if $\mathbf{G} = \mathbf{G_2}$, the modified stress will give no force or torque. In fact, on comparing (289), (290), we find that their difference is of this nature, (290) being equivalent to

$$\mathbf{P_N} = n(\nabla\mathbf{GN} + \operatorname{curl}\mathbf{VGN}) + m(\nabla\mathbf{G_2N} - \operatorname{curl}\mathbf{VG_2N}), \quad (291)$$

where the first and third terms on the right are sufficient to give the correct force, but with a torque, which is, however, cancelled by the second term; whilst the fourth term is apparently (so far as force and torque are concerned) like a fifth wheel to the coach, off the ground. If we inquire under what circumstances the real stress can assume this singular form, we shall find that $m + n = 0$ and $\operatorname{curl}^- \mathbf{G} = 0$ will do it. With these conditions, only the circuital part of $\mathbf{P_N}$ is now left, and (291) reduces to

$$\mathbf{P_N} = 2n\operatorname{curl}\mathbf{VG_2N}. \quad . \quad . \quad . \quad (292)$$

It is the case of irrotational displacement in the contractile ether, previously referred to, and is entirely remote from the real elastic solid. The noteworthy thing is that we cannot apparently conclude what the stress is, even when the force and torque to correspond are everywhere given, owing to the forceless and torqueless stress coming into the formulæ. The form (292) is convenient for showing at sight that there is no force. It may, by (181), remembering the constancy of $\mathbf{N}$, be expanded to

$$\mathbf{P_N} = 2n(\mathbf{N}\nabla.\mathbf{G_2} - \mathbf{N}\operatorname{div}\mathbf{G_2}), \quad . \quad . \quad (293)$$

where put $\mathbf{N} = \mathbf{i}, \mathbf{j}, \mathbf{k}$ in turns to get the three stresses on the planes having these normals. One-third of the sum of the normal tractions on these planes is $-(4n/3)\,\mathrm{div}\,\mathbf{G}_2$, or, which is the same, $+k\,\mathrm{div}\,\mathbf{G}$. It is the negative of the pressure. But in the real stress (291) itself, the fourth part, from its having the negative sign prefixed, correctly associates pressure and compression. But this forceless and torqueless stress does not contribute anything to the total energy. The amount of stored energy is

$$\Sigma \tfrac{1}{2}\mathbf{f}\mathbf{G} = -\Sigma \tfrac{1}{2}\mathbf{F}\mathbf{G} = -\tfrac{1}{2}\Sigma\,(F_1G_1 + F_2G_2 + F_3G_3)$$
$$= -\tfrac{1}{2}\Sigma\,(G_1\,\mathrm{div}\,\mathbf{P}_1 + G_2\,\mathrm{div}\,\mathbf{P}_2 + G_3\,\mathrm{div}\,\mathbf{P}_3)$$
$$= \tfrac{1}{2}\Sigma\,(\mathbf{P}_1\nabla G_1 + \mathbf{P}_2\nabla G_2 + \mathbf{P}_3\nabla G_3), \quad . \quad . \quad . \quad (294)$$

where the first line needs no remark, the second is got by (285), and the third by the common transformation (249), $\mathbf{P}_1$, etc., being the stresses on the $\mathbf{i}, \mathbf{j}, \mathbf{k}$ planes. The final form in (294) shows the correct distribution of the energy in terms of the stress and the distortion. Now, if in the stress $\mathbf{P}_N$ be included any terms giving rise to no force, the above transformation shows that they contribute nothing on the whole to the energy of distortion. This is the case with the fourth term in (293), and would also be the case with the second term, only that we have no right to ignore it, on account of the torque thereby brought in. If the solid be not unbounded, it is sufficient for its boundary to be at rest for the same principles to apply.

[The practical meaning is that in the contractile ether the energy of distortion due to any irrotational displacement is zero on the whole, the sum of the positive amounts in certain parts being equal to the sum of the negative amounts in the rest.]

Other Forms for the Displacement in terms of the Applied Forcive.

§144. The simple solution (279) may, of course, receive many other forms. If it be desired to find the displacement due to an explicitly given impressed forcive, it is a matter of some importance to select a method of obtaining it which shall not be unnecessarily difficult in execution; for different processes leading to the same result may vary greatly in readiness of application. Now, if the impressed forcive be either circuital or divergent, we do not need to modify. But if of a mixed type, then it may be desirable. Put $f_1 = f - f_2$, then an alternative form is

$$G = \text{pot} \left(\frac{f}{n} - \frac{m f_2}{n(m+n)} \right) ; \quad . \quad . \quad . \quad (295)$$

and now, when f is given, it is only the part pot f_2 that needs development. We have

$$f_2 = \Sigma \, r_1 \, \text{div} \, f / 4\pi r^2 = \Sigma \, \nabla p \, . \, \text{div} \, f, \quad . \quad . \quad (296)$$

as before explained. Now, if we notice that

$$\nabla^2 r = 2/r, \quad . \quad . \quad . \quad . \quad (297)$$

which may be proved by differentiating, we can further conclude that

$$\text{pot} \, (r_1 / 4\pi r^2) = r_1 / 8\pi, \quad . \quad . \quad . \quad (298)$$

a very curious result. The potential of a radial vector following the inverse-square law of intensity is a radial vector with a constant tensor in all space. Using this result, we have

$$\text{pot} \, f_2 = \Sigma \, r_1 \, \text{div} \, f / 8\pi, \quad . \quad . \quad . \quad (299)$$

so that (295) becomes

$$G = \frac{1}{n} \text{pot} \, f - \frac{m}{n(m+n)} \Sigma \frac{r_1 \, \text{div} \, f}{8\pi}, \quad . \quad . \quad (300)$$

wherein f alone appears on the right side.

Another form is got by using

$$f_2 = \nabla \Sigma \, f \nabla p, \qquad \therefore \, \text{pot} \, f_2 = \nabla \text{pot} \, \Sigma \, f \nabla p. \quad . \quad (301)$$

But a better one is

$$\text{pot} \, f_2 = \Sigma f \nabla \frac{dr}{ds} / 8\pi, \quad . \quad . \quad . \quad (302)$$

where s is length measured along f, and f is the tensor of the latter. This makes

$$G = \frac{1}{n} \text{pot} f - \frac{m/8\pi}{n(m+n)} \Sigma f \nabla \frac{dr}{ds}, \quad . \quad . \quad (303)$$

which is generally suitable for practical calculation.

A considerably more complex form is given in Thomson and Tait. It may be obtained from (303) by means of the identity

$$\nabla \frac{dr}{ds} = \frac{2}{3r}\mathbf{s}_1 - \frac{r^2}{3} \nabla \frac{d}{ds} \frac{1}{r}, \quad . \quad . \quad . \quad (304)$$

where $\mathbf{s}_1$ is unit $\mathbf{s}$, and therefore parallel to $\mathbf{f}$, making

$$\mathbf{G} = \text{pot } \mathbf{f} \left(\frac{1}{n} - \frac{m}{3n\,(m+n)}\right) + \frac{m}{24\pi n\,(m+n)} \, \Sigma f r^2 \, \nabla \frac{d}{ds} \frac{1}{r}, (305)$$

the same as Thomson and Tait's formulæ when expanded in cartesians. But this is a gratuitous complication, as (303) is simpler in expression and in application. Of course (300) is simpler still in expression, and the practical choice may lie between it and (303), or (295).

Notice that an impressed forcive of the divergent type with a single point-source produces not merely uniform radial displacement, as per (298), but an infinite discontinuity, in fact, a disruption, at the source itself. It is a very extreme case, of course.

The Elastic Solid generalised to include Elastic, Dissipative, and Inertial Resistance to Translation, Rotation, Expansion, and Distortion.

§ 145. The elastic solid with two elastic constants (k and n) has not been found sufficiently elastic to supply a thoroughly satisfactory analogy with Maxwell's ether, though partial analogies may readily be found. Other kinds of elasticity than resistance to compression and distortion, and other kinds of resistance than elastic resistance, present themselves to the consideration of searchers for analogies between the propagation of disturbances in Maxwell's ether and in the brutally simple elastic solid of theory, which is, however, known to fairly represent real solids within a certain range.

There are four distinct ideas involved in the displacement of a small portion of matter, viz., the translation as a whole, the rotation as a whole, the change of size, and the change of shape. These separate themselves from one another naturally. As regards the mere translational motion, if it be only inertially resisted, we have the equation of motion

$$\mathbf{f} + \mathbf{F} = \rho\frac{\delta\mathbf{q}}{\delta t}, \quad . \quad . \quad . \quad . \quad . \quad (306)$$

where ρ is the density, $\mathbf{q}$ the velocity, $\mathbf{f}$ impressed force (per unit volume), $\mathbf{F}$ the force arising from the stress associated with the strain (including rotation, distortion, and expansion), and $\delta/\delta t$ the time-differentiator for moving matter. This equation is constructed on simple Newtonian principles. We may, however, wish to have elastic and frictional resistance to translation, as well as inertial resistance. Then generalize the above to

$$\mathbf{f} + \mathbf{F} = \left(\rho_0 + \rho_1\frac{d}{dt} + \rho_2\frac{d^2}{dt^2}\right)\mathbf{G}, \quad . \quad . \quad (307)$$

where $\mathbf{G}$ is the displacement, when small departures from equilibrium are concerned. Here ρ_2 is the ρ of (306), ρ_1 is the frictionality (Lord Kelvin's word for coefficient of friction), and ρ_0 is an elastic constant. The displacement from equilibrium calls into action a back force $\rho_0\mathbf{G}$ proportional to the displacement, with storage of potential energy; a force $\rho_1\dot{\mathbf{G}}$ proportional to the velocity, with waste of energy; and a force $\rho_2\ddot{\mathbf{G}}$ proportional to the acceleration, with associated kinetic energy. The potential energy is $\frac{1}{2}\rho_0\mathbf{G}^2$, the rate of waste $\rho_1\dot{\mathbf{G}}^2$, and the kinetic energy $\frac{1}{2}\rho_2\dot{\mathbf{G}}^2$.

As regards the force $\mathbf{F}$, this is given by (284), when the stress is irrotational, and the elastic constants are n and k. But, in general, we may proceed thus. First separate the rotation from the strain vector. We have

$$\mathbf{N}\nabla.\mathbf{G} = \tfrac{1}{2}(\mathbf{N}\nabla.\mathbf{G} + \nabla.\mathbf{N}\mathbf{G}) + \tfrac{1}{2}(\mathbf{N}\nabla.\mathbf{G} - \nabla.\mathbf{N}\mathbf{G})$$

$$= \mathbf{p}_N - \tfrac{1}{2}\mathbf{V}\mathbf{N}\operatorname{curl}\mathbf{G}. \quad . \quad . \quad . \quad . \quad (308)$$

Here $\mathbf{N}\nabla.\mathbf{G}$ means the variation of displacement per unit distance along $\mathbf{N}$, which is any unit vector, so that by giving $\mathbf{N}$ all directions (practically only three) the complete state of strain is known. This strain vector is above analysed into the strain $\mathbf{p}_N$ without rotation, and the second part depending upon rotation. But $\mathbf{p}_N$ includes the expansion, as well as the distortion, or mere change of shape. To exhibit the distortion without expansion, one-third of the expansion (vectorised) must be deducted. Thus

$$\mathbf{N}\nabla.\mathbf{G} = (\mathbf{p}_N - \tfrac{1}{3}\mathbf{N}\operatorname{div}\mathbf{G}) + \tfrac{1}{3}\mathbf{N}\operatorname{div}\mathbf{G} - \tfrac{1}{2}\mathbf{V}\mathbf{N}\operatorname{curl}\mathbf{G} \quad (309)$$

shows the separation of the strain into a distortion (without change of size), an expansion, and a rotation, which are naturally independent.

If distortion, expansion, and rotation are all elastically resisted, three independent elastic constants (in an isotropic medium) intervene between the above strain and the corresponding stress $\mathbf{P_N}$ upon the plane whose normal is $\mathbf{N}$. To obtain $\mathbf{P_N}$, multiply the distortional part by $2n$, the expansional part by $3k$, and the rotational part by 2ν, and add the results. Thus,

$$\mathbf{P_N} = 2n(\mathbf{p_N} - \tfrac{1}{3}\mathbf{N}\operatorname{div}\mathbf{G}) + \mathbf{N}k\operatorname{div}\mathbf{G} - \nu\mathbf{V N}\operatorname{curl}\mathbf{G}. \quad (310)$$

Compare with (290). Here n and k are as before, whilst ν is a new elastic constant connected with the rotation, and which was previously assumed to have the value zero. That is, there was assumed to be no resistance to rotation. The corresponding torque is

$$\mathbf{S} = 2\nu\operatorname{curl}\mathbf{G}. \quad\quad (311)$$

With the rotational part of the stress is also associated translational force, given by

$$-\operatorname{curl}\tfrac{1}{2}\mathbf{S} = -\nu\operatorname{curl}^2\mathbf{G}. \quad (312)$$

Adding this on to the former expression for the force (or deriving the force from (310) directly), we find that

$$\mathbf{F} = n(\nabla^2\mathbf{G} + \tfrac{1}{3}\nabla\operatorname{div}\mathbf{G}) + k\nabla\operatorname{div}\mathbf{G} - \nu\operatorname{curl}^2\mathbf{G} \quad (313)$$

represents the translational force to be used in the equation of motion (307). We should notice here that the quantity $k\operatorname{div}\mathbf{G}$ in (310) represents a uniform tension. It equals one-third of the sum of the normal tractions on any three mutually perpendicular planes. Its negative represents the pressure, or $p = -k\operatorname{div}\mathbf{G}$. Now when there is incompressibility, k is infinite, and $\operatorname{div}\mathbf{G}$ is zero. But their product usually remains finite. So in any case we may replace the k term in (310) by $-\mathbf{N}p$, and the k term in (313) by $-\nabla p$.

The energy of the strain is

$$U = \tfrac{1}{2}(\mathbf{P}_1\nabla_1 + \mathbf{P}_2\nabla_2 + \mathbf{P}_3\nabla_3)\mathbf{G}, \quad (314)$$

or one-half the sum of the scalar products of the stress vector and the strain vector for three perpendicular planes, $\mathbf{P}_1, \mathbf{P}_2, \mathbf{P}_3$

being the stresses on the **i**, **j**, **k** planes respectively. On reckoning up, by (310), (309), we find the energies of expansion, rotation, and distortion are all independent, and that U is their sum, given by

$$U = n[\mathbf{p_1}^2 + \mathbf{p_2}^2 + \mathbf{p_3}^2 - \tfrac{1}{3}(\text{div } \mathbf{G})^2] + \tfrac{1}{2}k(\text{div } \mathbf{G})^2 + \tfrac{1}{2}\nu(\text{curl } \mathbf{G})^2, \quad (315)$$

where the ν term is the energy of rotation ($= \tfrac{1}{2}$ torque × rotation), the k term is the energy of expansion, and the n term the energy of distortion.

Now the stress (310) is derived from the strain (309) by the introduction of elastic resistances only. There is, however, no reason why we should limit the nature of the resistance in this way. Consider the n term only for example, relating to distortion. There may also be dissipative or frictional resistance to distortion. To exhibit it, change n to $n_0 + n_1(d/dt)$ in (310), (313). Then n_0 will be the rigidity and n_1 the viscosity, or coefficient of frictional resistance to distortion. For instance, if we abolish n_0 and ν, and retain n_1 and k, we have the stress in a real viscous fluid according to Stokes's theory. It may be that there is not a complete disappearance of the rigidity in a fluid; if so, then retain both n_0 and n_1. When the rigidity is marked, as in a solid, n_0 is the important part of n; whilst in a fluid it is the other part. It is, however, not at all to be expected that the expression of the viscosity of solids by n_1, if true at all, would extend beyond the small range of approximately perfect elasticity. Lord Kelvin's experiments on the subsidence of the oscillations of wires tended to show that a different law was followed than that corresponding to the viscosity of fluids, and that elastic fatigue was concerned in the matter.

There may also be, conceivably, inertial resistance to distortion, and not only elastic, but also frictional and inertial resistance to rotation, and the same with respect to expansion. These will be brought in by generalising the three elastic constants k, n, and ν of (310) thus :—

$$\left.\begin{aligned} k &= k_0 + k_1(d/dt) + k_2(d^2/dt^2), \\ n &= n_0 + n_1(d/dt) + n_2(d^2/dt^2), \\ \nu &= \nu_0 + \nu_1(d/dt) + \nu_2(d^2/dt^2), \end{aligned}\right\} \quad . \quad . \quad (316)$$

where of the nine coefficients on the right side, the first three are elastic constants, the second three dissipative constants, the last three inertial constants; the elastic constants involving potential energy, the dissipative constants waste of energy, and the inertial constants kinetic energy, similarly to (307) as regards translation. Out of these twelve constants only four are in use, or seven, if v_0 and v_1, and ρ_1 be included, which additionals have been speculatively employed. This does not represent finality by any means; but sufficient has been said on the subject of generalising the elastic constants to emphasize the fact that there is plenty of scope for investigation in the theory of the motion of media of unknown internal constitution, which, externally viewed, involve the four elements translation, rotation, expansion and distortion.

The activity of the stress vector $\mathbf{P_N}$ is $\mathbf{P_N}q$, and the flux of energy is (see §§ 68, 72 also), if q be the speed, or tensor of the velocity,

$$q(\mathrm{U}+\mathrm{T}) - \mathbf{Q}_q q, \quad . \quad . \quad . \quad . \quad (317)$$

where U and T are the complete potential and kinetic energies, per unit volume, and $\mathbf{Q_N}$ is the stress conjugate to $\mathbf{P_N}$, to obtain which change the sign of the last (rotational) term in (310). The convergence of the energy flux represents the work done and stored or wasted in the unit volume. Considering the term $-\mathbf{Q}_q q$ only (the rest being the convective flux of energy), its convergence is

$$\operatorname{div}(\mathbf{Q}_1 q_1 + \mathbf{Q}_2 q_2 + \mathbf{Q}_3 q_3) = \nabla_1(\mathbf{P}_1 q) + \nabla_2(\mathbf{P}_2 q) + \nabla_3(\mathbf{P}_3 q)$$

$$= q_1 \operatorname{div}\mathbf{Q}_1 + q_2 \operatorname{div}\mathbf{Q}_2 + q_3 \operatorname{div}\mathbf{Q}_3 + \mathbf{Q}_1\nabla q_1 + \mathbf{Q}_2\nabla q_2 + \mathbf{Q}_3\nabla q_3$$

$$= \mathbf{F}q + (\mathbf{P}_1\nabla_1 + \mathbf{P}_2\nabla_2 + \mathbf{P}_3\nabla_3)q, \quad . \quad . \quad (318)$$

where $\mathbf{F}$ is the force as in (313). $\mathbf{F}q$ is the translational activity, which may, by (307), be employed in increasing kinetic and potential energy or be wasted. The rest represents the rate of increase of the sum of the three potential energies when there is merely elastic resistance; but if we use (316) we obtain also the sum of the rates of waste of energy, through k_1, n_1, v_1, and also the rate of increase of the sum of the kinetic energies, through k_2, n_2, v_2.

Thus we have, by (310),

$$(\mathbf{P}_1\nabla_1 + \mathbf{P}_2\nabla_2 + \mathbf{P}_3\nabla_3)\mathbf{q}$$

$$= \frac{d\mathbf{q}}{dx}\Big(2n(\mathbf{p}_1 - \tfrac{1}{3}\mathbf{i}\,\mathrm{div}\,\mathbf{G}) + k\mathbf{i}\,\mathrm{div}\,\mathbf{G} - \nu\mathbf{V}\mathbf{i}\,\mathrm{curl}\,\mathbf{G} \Big)$$

$$+ \frac{d\mathbf{q}}{dy}\Big(2n(\mathbf{p}_2 - \tfrac{1}{3}\mathbf{j}\,\mathrm{div}\,\mathbf{G}) + k\mathbf{j}\,\mathrm{div}\,\mathbf{G} - \nu\mathbf{V}\mathbf{j}\,\mathrm{curl}\,\mathbf{G} \Big)$$

$$+ \frac{d\mathbf{q}}{dz}\Big(2n(\mathbf{p}_3 - \tfrac{1}{3}\mathbf{k}\,\mathrm{div}\,\mathbf{G}) + k\mathbf{k}\,\mathrm{div}\,\mathbf{G} - \nu\mathbf{V}\mathbf{k}\,\mathrm{curl}\,\mathbf{G} \Big)$$

$$= 2n(\mathbf{p}_1\nabla_1\mathbf{q} + \mathbf{p}_2\nabla_2\mathbf{q} + \mathbf{p}_3\nabla_3\mathbf{q}) - \frac{2n}{3}\,\mathrm{div}\,\mathbf{G}\,\mathrm{div}\,\mathbf{q}$$

$$+ k\,\mathrm{div}\,\mathbf{G}\,\mathrm{div}\,\mathbf{q} + \nu\,\mathrm{curl}\,\mathbf{G}\,\mathrm{curl}\,\mathbf{q} \quad . \quad . \quad . \quad (319)$$

Take for illustration, the rotational term $\nu\,\mathrm{curl}\,\mathbf{G}\,\mathrm{curl}\,\mathbf{q}$. This is simply $(d/dt)\{\tfrac{1}{2}\nu\,(\mathrm{curl}\,\mathbf{G})^2\}$, or the rate of increase of the rotational energy, when ν is a constant, but when we use the last of (316) it becomes

$$\frac{d}{dt}\Big(\tfrac{1}{2}\nu_0(\mathrm{curl}\,\mathbf{G})^2\Big) + \nu_1(\mathrm{curl}\,\mathbf{q})^2 + \frac{d}{dt}\Big(\tfrac{1}{2}\nu_2(\mathrm{curl}\,\mathbf{q})^2\Big), \quad (320)$$

showing the rates of increase of potential and kinetic energy, and the rate of waste (the middle term) due to rotational friction. Similarly as regards the other terms in (319). The expansion terms give

$$\frac{d}{dt}\Big(\tfrac{1}{2}k_0(\mathrm{div}\,\mathbf{G})^2\Big) + k_1(\mathrm{div}\,\mathbf{q})^2 + \frac{d}{dt}\Big(\tfrac{1}{2}k_2(\mathrm{div}\,\mathbf{q})^2\Big), \quad . \quad (321)$$

where the middle term is the rate of waste, and the others rates of increase of potential and kinetic energies. Finally, the distortional terms of (319) give

$$\frac{d}{dt}n_0\Big(\mathbf{p}_1{}^2 + \mathbf{p}_2{}^2 + \mathbf{p}_3{}^2 - \tfrac{1}{3}(\mathrm{div}\,\mathbf{G})^2\Big)$$

$$+ 2n_1\Big(\dot{\mathbf{p}}_1{}^2 + \dot{\mathbf{p}}_2{}^2 + \dot{\mathbf{p}}_3{}^2 - \tfrac{1}{3}(\mathrm{div}\,\mathbf{q})^2\Big)$$

$$+ \frac{d}{dt}n_2\Big(\dot{\mathbf{p}}_1{}^2 + \dot{\mathbf{p}}_2{}^2 + \dot{\mathbf{p}}_3{}^2 - \tfrac{1}{3}(\mathrm{div}\,\mathbf{q})^2\Big), \quad . \quad . \quad (322)$$

where the first line is the rate of increase of the potential energy, the second line the frictional rate of waste, and the third line the rate of increase of kinetic energy. We thus complete the energy relations of the stress with the generalised elastic constants, and so far as small motions are concerned, the equations are manageable.

Electromagnetic and Elastic Solid Comparisons. First Example : Magnetic Force compared with Velocity in an incompressible Solid with Distortional Elasticity.

§ 146. Comparisons between the propagation of electromagnetic disturbances in Maxwell's ether, or in a homogeneous isotropic dielectric, whether conducting or not, and the propagation of motion in an elastic solid, either simple or generalised, may be made in a variety of ways. They all break down sooner or later, but are nevertheless useful as far as they go. A few cases will be now considered, based on the preceding. First take the case of a non-conducting dielectric at rest, and compare it with the regular elastic solid made incompressible by infinite resistance to compression. The incompressibility includes inexpansibility.

In the dielectric we have, if p stands for d/dt, for convenience,

$$\operatorname{curl}(\mathbf{H} - \mathbf{h}) = cp\mathbf{E}, \qquad \operatorname{curl}(\mathbf{e} - \mathbf{E}) = \mu p\mathbf{H}, \quad . \quad (323)$$

the circuital equations connecting $\mathbf{E}$ and $\mathbf{H}$ (§§ 38, 24, 66), where $\mathbf{e}$ and $\mathbf{h}$ are impressed, and μ, c are the inductivity and permittivity. Now, the equation of motion in an incompressible solid is, by (273),

$$\mathbf{f}_1 = (\rho p^2 - n\nabla^2)\,\mathbf{G} = \left(\rho p - \frac{n}{p}\nabla^2\right)\mathbf{q}, \quad . \quad . \quad (324)$$

where ρ and n are the density and rigidity, $\mathbf{G}$ the displacement, $\mathbf{q}$ the velocity, and $\mathbf{f}_1$ the circuital part of the impressed forcive, the divergent part being inoperative. Or we may take the impressed forcive to be circuital to begin with.

Now, let $\mathbf{h} = 0$, and $\mathbf{e}$ be finite. If it has no curl, it does nothing, as before shown (§ 89). The source of disturbance is curl $\mathbf{e}$. This being the case in the dielectric, and $\mathbf{f}_1$ being the source of motion in the solid, it will be convenient to take $\mathbf{f}_1$ and curl $\mathbf{e}$ as corresponding. Let the latter be (temporarily) $\mathbf{f}$. Then (323) gives (remembering that div $\mathbf{H} = 0$),

$$\mathbf{f} = \mu p\mathbf{H} + \operatorname{curl}\mathbf{E} = \mu p\mathbf{H} + \operatorname{curl}^2 \mathbf{H}/cp,$$

or,
$$\mathbf{f} = \left(\mu p - \frac{\nabla^2}{cp}\right)\mathbf{H} = \left(\mu p^2 - \frac{\nabla^2}{c}\right)\frac{\mathbf{H}}{p}. \quad . \quad (325)$$

Comparing this with the second of (324), and assuming the equality of f and f_1, we see that H is velocity, μ density, c^{-1} rigidity; and so on. H/p means the time-integral of H. Or, more explicitly, let Z be the time-integral of H, so that $H = pZ$; then

$$f = (\mu p^2 - c^{-1}\nabla^2)Z, \quad \ldots \quad (326)$$

which compares with the first of (324).   Thus

	$Z = H/p$ corresponds to G,		(spacial displ.)	
(mag. force)	$H = pZ$	,,	q,	(velocity),
(el. current)	curl H	,,	curl q,	($2 \times$ spin),
(el. displ.)	$D = cE$	,,	curl G,	($2 \times$ rotation),
(inductivity)	μ	,,	ρ,	(density),
(permittivity)	c	,,	$1/n$,	(compliancy),
(mag. source)	$f = \text{curl } e$	,,	f_1,	(impd. force),
(mag. energy)	$\frac{1}{2}\mu H^2$	,,	$\frac{1}{2}\rho q^2$,	(kin. energy).

This is, as far as it goes, an excellent analogy, particularly on account of its directness and ease of application, and on account of the similarity of sources.   The disturbances of H generated by f in the dielectric and propagated away at speed $(\mu c)^{-\frac{1}{2}}$ are precisely represented by the velocity generated by similarly distributed impressed Newtonian force, which is propagated in the same manner through the solid at the equivalent speed $(n/\rho)^{\frac{1}{2}}$.   Of course the correspondence at similar moments also applies to the other quantities compared.

But it is imperative (in general) that the motions in the solid should be small.   The disturbances should therefore be of the fluctuating or alternating character, so that Z, the time-integral of H, does not mount up.   For, correspondingly, if this were allowed, then G would mount up, and the elastic solid get too much out of shape to allow us to treat d/dt at a fixed point and $\delta/\delta t$ for a moving particle as identical.

Observe also, that the electric energy $\frac{1}{2}cE^2$ is not matched by the potential energy of distortion; although their total amounts are equal, their distributions are entirely different. This is a fatal failure in detail.   Nor have we any right to expect that a scheme like Maxwell's, depending upon rotations, can be perfectly matched by one which depends on shears.

Second Example: Same as last, but Electric Force compared with Velocity.

§ 147. Another analogy of the same class is got by making the displacement in the solid represent the time-integral of $\mathbf{E}$ in the dielectric. Thus let $e = 0$ in (323), and $-\operatorname{curl} \mathbf{h} = +\mathbf{g}$. It is now $\mathbf{g}$ that is the source. Eliminate $\mathbf{H}$ between the two equations (323). We get

$$cp\mathbf{E} + \mathbf{g} = \operatorname{curl}(-\operatorname{curl}\mathbf{E})/\mu p. \quad . \quad . \quad (327)$$

Or, since $\operatorname{div} \mathbf{E} = 0$ (when there is no electrification)

$$\mathbf{g} = \left(cp - \frac{\nabla^2}{\mu p}\right)\mathbf{E} = \left(cp^2 - \frac{\nabla^2}{\mu}\right)\mathbf{A}, \quad . \quad . \quad (328)$$

if $\mathbf{A} = \mathbf{E}/p$, making $\mathbf{E} = p\mathbf{A}$.

Comparing (327) with (324) for the solid, we see that we have $\mathbf{E}$ representing velocity, μ compliancy, c density, and so on, being a general turning over of all the quantities. Thus

$\mathbf{A} = \mathbf{E}/p$ corresponds to $\mathbf{G}$ (displacement),

μ	,,	n^{-1},
c	,,	ρ,
(el. force) $\mathbf{E}$	,,	$\mathbf{q}$ (velocity),
$\frac{1}{2}c\mathbf{E}^2$	,,	$\frac{1}{2}\rho\mathbf{q}^2$,
$\mathbf{H}$	,,	$n \operatorname{curl} \mathbf{G}$,
$-\operatorname{curl}\mathbf{h}$	,,	$\mathbf{f}_1$ (circuital impd. force).

The conclusions are similar as regards the generation and propagation of disturbances of the two kinds compared. There is a similar failure to before as regards the energy of distortion. It ought to be the magnetic energy $\frac{1}{2}\mu\mathbf{H}^2$ now, but it is not. This want of correspondence as regards one of the energies will be removed (and can only be removed) by doing away with the rigidity and substituting something else, which must, however, be equivalent in its results as regards the propagation of disturbances from place to place.

Third Example: A Conducting Dielectric compared with a Viscous Solid. Failure.

§ 148. As we see from the preceding that the propagation of electromagnetic disturbances through a non-conductor can be imitated in an incompressible solid, possessing the usual distortional rigidity, let us next introduce electric conductivity on the one hand, and examine what changes are needed in the analogies of §§ 146, 147 to keep them working, if it be possible. Since there is waste of energy associated with the conduction current, it is clear that some sort of frictional force requires to be introduced into the elastic solid.

First try distortional friction. The equation of motion of an incompressible viscous solid is

$$\mathbf{f}_1 = \left[\rho p^2 - (n_0 + n_1 p)\nabla^2\right]\mathbf{G}, \quad . \quad . \quad (329)$$

where $\mathbf{f}_1$ is the impressed force per unit volume, $\mathbf{G}$ the displacement, ρ the density, n_0 the rigidity, and n_1 the associated frictionality. Also, p stands for d/dt, for practical convenience in the operations, whether direct or inverse. Thus, $p^{-1}\mathbf{q}$ or $\mathbf{q}/p$ means the time-integral of $\mathbf{q}$, which would in the ordinary notation of the integral calculus be expressed by $\int_0^t \mathbf{q}\,dt$, a notation which is not convenient for manipulative purposes in investigations of this class. We may derive (329) from (324) by changing n to $n_0 + n_1 p$; or by (306), (313). [*See also* p. 229.]

Now the circuital equations of $\mathbf{E}$ and $\mathbf{H}$ in a conducting dielectric (homogeneous, isotropic, stationary), are

$$\operatorname{curl}(\mathbf{H} - \mathbf{h}) = (k + cp)\mathbf{E}, \qquad \operatorname{curl}(\mathbf{e} - \mathbf{E}) = \mu p\mathbf{H}, \quad . \quad (330)$$

comparing which with (323) we see that the effect of introducing conductivity is to change cp to $k + cp$, so that the equation of $\mathbf{H}$ is got by making this change in (325), giving

$$\mathbf{f} = \left(\mu p - \frac{\nabla^2}{k + cp}\right)\mathbf{H} = \left(\mu p^2 - \frac{p\nabla^2}{k + cp}\right)\mathbf{Z}, \quad . \quad . \quad (331)$$

where $\mathbf{Z} = \mathbf{H}/p$. But on comparing this equation with (329), we see that the viscosity in the one case and the conductivity in the other enter into the equations in different ways. We can only produce a proper correspondence when the coefficients of ∇^2 are equal; or, in another form, when the operators $p(k + cp)^{-1}$ and $(n_0 + n_1 p)$ are equivalent in effect. This is not generally possible. But there are two extreme cases of agreement. One we know already, viz., when there is no conduc-

tivity and no viscosity, as in § 146. The other is when there is
no permittivity and no rigidity, which is sufficiently important
to be separately considered.

Fourth Example: A Pure Conductor compared with a Viscous Liquid. Useful Analogy.

§ 149. This is the extreme case of a pure conductor, or a
conductor which cannot support elastic displacement. It dissipates energy, but does not store it electrically; though, on the
other hand, its magnetic storage capacity is retained. It
is compared with an incompressible viscous solid with the
rigidity abolished; that is to say, with an incompressible
viscous liquid. Put $c = 0$ in (331), and we have the equation

$$\mathbf{f} = (\mu p - k^{-1}\nabla^2)\mathbf{H} ; \qquad . \quad . \quad (332)$$

and putting $n_0 = 0$ in (329), with $\mathfrak{q}$ the velocity substituted for
$p\mathfrak{G}$, gives us the equation

$$\mathbf{f}_1 = (\rho p - n_1\nabla^2)\mathfrak{q} \quad . \quad . \quad . \quad . \quad (333)$$

in the liquid. These admit of immediate comparison. Observe,
however, the curious fact that whereas in § 146 the permittivity was the reciprocal of the rigidity, so that the vanishing
of one means the infinitude of the other, yet now both the
permittivity and the rigidity vanish together, as if they were
equivalent. This emphasises the incompatibility of (329), (331).
Our present analogy must stand by itself, all idea of permittivity and rigidity being thrown away.

We compare magnetic force $\mathbf{H}$ in a pure conductor with the
velocity $\mathfrak{q}$ in a viscous liquid; the inductivity μ with the
density of the liquid, so that the magnetic energy and the
kinetic energy are compared, and the current is twice the
spin. Also the source $\mathbf{f} = \text{curl } \mathbf{e}$ in the case of the conductor is
compared with circuital impressed force in the liquid. So far is
the same as in § 146. But now, in addition, we have the electric
resistivity k^{-1} represented by the liquid viscosity. We conclude
that the disturbances $\mathfrak{q}$ generated by $\mathbf{f}_1$ in the liquid are similar
to the disturbances $\mathbf{H}$ in the conductor generated by $\mathbf{f}$, and
that the propagation of electrical disturbances in a conductor
is like that of motion in a viscous incompressible liquid. It

takes place by the process called diffusion. It is the limiting case of elastic wave-propagation, with distortion and dissipation of energy by friction.

We are usually limited to small motions in this analogy, so that the impressed forces should not act in one sense continuously, but should fluctuate about the mean value zero, for the reason mentioned before. But we are not always limited to small motions. In certain symmetrical distributions of magnetic force and current we may remove this restriction altogether. In the liquid case (333) p strictly means $\delta/\delta t$, whilst in (332) it means d/dt. In the former a moving particle is followed; in the latter we keep to one place. But in the cases of laminar flow referred to, $\delta/\delta t$ reduces to d/dt readily enough, so that the analogy is carried on to cases of steadily acting impressed forces.

The analogy is an important one, owing to the readiness with which the setting in motion of water by sliding friction can be followed, as when a wind blows over its surface. Two interesting cases are the penetration of magnetic induction and (with it) electric current into a core enveloped by a solenoidal coil, in whose circuit an impressed force, variable or steady, acts; and the penetration of magnetic induction and electric current into a straight wire when an impressed force acts in its circuit. Considering the latter we may state the analogy thus. Understanding, in the first place, that the circumstances should be such that the influence of electric displacement outside the wire is negligible, we may replace the wire by a similar tube of water, and then replace the impressed voltage in the electric circuit by a uniform tangential drag upon the surface of the water in the direction of the length of the pipe. The current of water in the pipe and the electric current in the wire, will vary similarly under the action of similarly varying impressed forces. Thus, a steadily applied force on the water will first pull the outermost layer into motion; this, by the viscosity, will pull the next layer, and so on up to the axis. The initial current is purely superficial; a little later we have a central core in which the water is practically motionless, surrounded by a tubular portion whose outermost layer is moving rapidly, and innermost very slowly; later still, the whole mass is in motion, though less rapidly at the axis than at the boundary;

finally, the whole mass of water acquires a state of uniform motion. Substituting electric current for current of water, we obtain a representation of the way the electric current is set up in a wire, passing through the various stages from the initial surface current to the final uniformly distributed current. If the impressed force be rapidly oscillatory, we stop the penetration as above described in its early stage, so that if the frequency be great enough, there is a practical confinement of the current (in sensible amount) to the skin of the wire (or of the water respectively). But if the mean value of the oscillatory force be not zero, it is the same as having two impressed forces, one steady, the other periodic, without any bias one way or the other. Then we have finally a steady current throughout the wire, *plus* the oscillatory current with superficial concentration.

When a straight core is magnetised in a solenoid, it is the magnetic induction which is longitudinal, or parallel to the axis. Comparing this with the current of water in the pipe, we have the same state of things as before described, as regards the penetration of magnetic induction into the core.

One effect is to increase the resistance of a wire when the frequency is great enough. It will be remembered that Prof. Hughes (in 1886) brought forward evidence of this increase, and, therefore, in the opinion of others, of the truth of the theory of surface conduction along wires under certain circumstances which was advanced by me a year previous. There has since been plenty of confirmatory evidence of a more complete nature; that is, not merely of an approximation towards, but of an almost complete attainment of surface-conduction.

This action of conductors is sometimes referred to as magnetic screening. It should, however, be noted that the screening is done by the conductor itself, not by the currents "induced" in the outer part of the wire or core, which are, as it were, merely a sign that the screening is taking place. The screening action often seems to be merely superficial; but this is accidental, from the disturbances being communicated to the conductor from without. In reality, any and every part of a conductor is screened from the rest by the portion immediately surrounding it, that is, by its skin, and by that only. It is primarily the conductivity that causes the screening, so that it is rather conductive screening than magnetic screening. The result is that disturbances can only pass through a conductor by the (relatively) very slow process of diffusion, so unlike that by which they are transferred through a non-conductor. But the property of conductivity acts conjointly with the inductivity, so that although copper would be the best screener on account of its high conductivity, the great inductivity of iron usually far more than compensates for its inferior conductivity, and makes iron take the first place as a screener.

Fifth Example: Modification of the Second and Fourth.

§ 150. As we were not able to make an analogy with c and k both finite, likewise n_0 and n_1, when H was taken to be velocity, let us try the effect of taking E to be velocity, as in § 147. Change the source from f to $g = -\operatorname{curl} h$; then we have merely to alter cp in (328) to $k + cp$ to obtain the required modification. Thus,

$$g = \left((k + cp) - \frac{\nabla^2}{\mu p} \right) E = \left((kp + cp^2) - \frac{\nabla^2}{\mu} \right) A, \quad . \quad (334)$$

if $A = E/p$. Here, again, on comparison with (329), we see that distortional friction will not furnish a proper analogy in the general case. There seem to be but two special cases possible; first, with the conductivity and frictionality both zero (not the resistivity and frictionality), which reproduces the case in § 147; and, secondly, the permittivity and the rigidity both zero (not permittivity and compliancy), which brings us to the viscous liquid again. Thus, putting $c = 0$ in (334) brings us to

$$g = \left(kp - \frac{\nabla^2}{\mu} \right) A, \quad . \quad . \quad . \quad . \quad . \quad (335)$$

which compares with (333). We now compare conductivity with density, and inductivity with the reciprocal of the viscosity. This is by no means so useful as the previous form of the viscous fluid analogy, although it is fundamentally of the same nature, involving propagation by diffusion.

As in §§ 146, 147, we had a failure of representation of one of the energies by the energy of distortion, so now, in the viscous fluid analogies, we fail to represent the dissipation of energy properly. The latter depends, as shown before, on the velocity of distortion, so we might expect a similar failure.

Sixth Example: A Conducting Dielectric compared with an Elastic Solid with Translational Friction.

§ 151. But we can get a working analogy to suit the conducting dielectric in another way. Thus, still keeping the distortional elasticity or rigidity, give up the viscosity or distortional friction, and substitute translational friction. For the equation of motion of an incompressible solid with the usual constants, rigidity n and density ρ_2, and with translational frictionality ρ_1 is

$$f_1 = (\rho_1 p + \rho_2 p^2 - n\nabla^2)G, \quad . \quad . \quad . \quad (336)$$

where f_1 is circuital impressed force. Now operate on (331) by $1 + k/cp$, producing the equation

$$\left(1 + \frac{k}{cp}\right)f = \left(\frac{\mu k}{c}p + \mu p^2 - \frac{\nabla^2}{c}\right)Z. \quad . \quad . \quad (337)$$

Comparing this with (336) we see that there is a proper correspondence of form on the right sides, though not on the left. The circuital impressed force f_1 is replaced, not by f, but by $f + kf/cp$. This is awkward for the analogy as regards the generation of disturbances. But away from the sources, there is a useful correspondence. As in § 146, Z corresponds to spacial displacement in the solid, H to velocity, μ to density, c to compliancy (distortional); whilst now, in addition, the translational frictionality in the solid has the complex representative $\mu k/c$ in the dielectric; or $k(\mu v)^2$, if v is the speed of propagation of disturbances. So disturbances of H are propagated in the same way through a conducting dielectric, as are disturbances in an incompressible solid with the usual rigidity, with frictional resistance to displacement superadded. But we do not correctly localise the energy dissipated. The rate of waste per unit volume is kE^2 in the dielectric, whilst it is $\rho_1 q^2$ in the solid, which corresponds to $k(\mu v)^2 H^2$ in the dielectric. This being entirely against Joule's law renders this form of analogy unsatisfactory. But we may easily make a substantial improvement by taking E to be velocity in the solid.

Seventh Example: Improvement of the Sixth.

§ 152. Thus, let $g = -$ curl h be the source of disturbance in the dielectric, so that we have the equations (334) to consider.

Take the second form and compare it with (336), and we see at once that there is a considerably more simple correspondence than in the case last treated. Thus, the source f_1 of circuital displacement G corresponds to the source g of circuital A (the time-integral of E); the electric force now means velocity; μ is compliancy; c is density; k itself is the frictionality; the electric energy $\frac{1}{2}c\mathbf{E}^2$ is kinetic energy; and the waste (Joulean) is properly localised in the solid by the frictional waste of energy $\rho_1\mathbf{q}^2$ per second. The induction $\mu\mathbf{H}$ is twice the rotation. In fact, this case is the same as in § 147, with translational friction added without destroying the analogy. It is clear that the correct localisation of the waste, and the correspondence of sources, makes the analogy be more readily followed. Of course, there is the same failure as regards the distortional energy as in § 147.

Eighth Example: A Dielectric with Duplex Conductivity compared with an Elastic Solid with Translational Elasticity and Friction. The singular Distortionless Case.

§ 153. As a last example in which use is made of distortional elasticity, let us generalise the non-conducting dielectric by the introduction of magnetic conductivity. We have

$$\text{curl}\,(\mathbf{H} - \mathbf{h}) = (k + cp)\mathbf{E}, \quad . \quad . \quad (338)$$

$$\text{curl}\,(\mathbf{e} - \mathbf{E}) = (g + \mu p)\mathbf{H}, \quad . \quad . \quad (339)$$

where g is the magnetic conductivity. So, if curl $\mathbf{h} = 0$, and we eliminate E, we have

$$\text{curl}\,\mathbf{e} = \left((g + \mu p) - \frac{\nabla^2}{k + cp} \right)\mathbf{H}. \quad . \quad . \quad (340)$$

Or, if curl $\mathbf{e} = 0$ and we eliminate H, we have

$$-\text{curl}\,\mathbf{h} = \left((k + cp) - \frac{\nabla^2}{g + \mu p} \right)\mathbf{E}. \quad . \quad . \quad (341)$$

We see that we cannot, when k and g are both finite, get a satisfactory analogy as regards the generation of disturbances by f_1, circuital impressed force in a solid. Consider, therefore, only the free propagation, subject to the equation

$$0 = \left((k + cp)(g + \mu p) - \nabla^2 \right)\mathbf{Z}, \quad . \quad . \quad (342)$$

R

where for Z we may substitute H or E or other electrical quantities. Since, in (342), we have a term proportional to Z itself (or its substitute), we must introduce elastic resistance to displacement in the elastic solid to make an analogy. Thus,

$$f_1 = (\rho_0 + \rho_1 p + \rho_2 p^2)\, G - n\nabla^2 G. \quad . \quad . \quad (343)$$

is the equation of motion in the incompressible solid if there be elastic and frictional as well as inertial resistance to translation, by (307), (313), § 145.

Now expand (342) and divide by c. We get

$$0 = \left(\frac{kg}{c} + \left(g + \frac{k\mu}{c}\right)p + \mu p^2\right) Z - \frac{\nabla^2}{c} Z, \quad . \quad (344)$$

which is suitable for comparison with (343). When Z is the spacial displacement, H is velocity, c compliancy, μ density, as in § 146 ; but in other respects the correspondence is complex, for the elastic resistance to translation ρ_0 is kg/c, whilst the frictionality ρ_1 is $g + k\mu/c$.

If we take E to be velocity there will be another set of correspondences, c being density and μ compliancy.

However unsatisfactory the analogy may be in details, we have still the result that disturbances in a dielectric with duplex conductivity (electric and magnetic) are propagated similarly to motions in an elastic solid constrained in the manner above decribed. The effect of the two kinds of distortion and waste of energy involved in the existence of the two conductivities are therefore imitated by one kind of frictionality in the elastic solid, assisted by elastic resistance to translation.

The singular distortionless case comes about when

$$\left.\begin{array}{l} 4\rho_0\rho_2 = \rho_1{}^2 \quad \text{in the solid,} \\ k/c = g/\mu \quad \text{in the dielectric.} \end{array}\right\} \quad . \quad . \quad (345)$$

In either case the effect of the frictional resistance on the disturbances is made merely attenuative. Disregarding the attenuation with the time, the propagation takes place in the dielectric in the same way as if it were non-conducting, and in the solid in the same way as if it were devoid of frictional and elastic resistance to translation. In the electromagnetic case we may follow this into detail with ease in a symmetrical manner. It is, however, rather troublesome in the elastic solid, on account of the want of correspondence in detail, due to the employment of distortional elasticity, and for other reasons.

The Rotational Ether, Compressible or Incompressible.

§ 153A All the above analogies, however good or bad they may be in other respects, are deficient in one vital property, inasmuch as they involve the elasticity of distortion, and with it, the energy of distortion. But the electromagnetic equations, especially when put in the duplex form symmetrical with respect to the electric and magnetic sides, have nothing in them suggestive of distortional forces, nor can we represent either the electric or the magnetic energy as the energy of distortion. On the other hand, the equations are fully suggestive of rotation. If, then, the elastic solid has still to do duty for purposes of analogy, and yet not fail upon the point mentioned, it is clear that the rigidity must be done away with, at least as an electromagnetically active influence. Its place must then be taken by elastic resistance to rotation, by using a rotational instead of a shearing stress. This brings us to the medium invented by Lord Kelvin, called by him simply Ether, and contrasted with Jelly, which means an incompressible elastic solid with ordinary rigidity. This rotational ether has mass of course, for one thing, which brings in kinetic energy (of translation); and, on the other hand, it possesses, by means of internal arrangements with which we are not immediately concerned, the property of elastically resisting rotation, and consequently stores up energy of the potential kind, to balance the kinetic. If this were all, there would be little more to say. But Lord Kelvin found that his Ether enabled him to complete a long-delayed work, viz., to produce a satisfactory elastic-solid analogy to suit the problem of magnetic induction ("Mathematical and Physical Papers," Vol. I., art. 27, and Vol. III., art. 49); and I have pointed out that this rotational ether enables us to construct good all-round analogies for the propagation of disturbances in a stationary dielectric (Appendix to chap. II., *ante*), with correct localisation of both energies and of the flux of energy. As these are the only all-round elastic-solid analogies yet known, we may here consider them again,

R 2

especially in relation to the previously given analogies, and
to show where the changes come in.

Go back to equation (313), giving the translational force $\mathbf{F}$
due to the stress in an elastic solid when there is rotational
elasticity as well as rigidity.   We have

$$\mathbf{F} = n \, (\nabla^2 \mathbf{G} + \tfrac{1}{3}\nabla \operatorname{div} \mathbf{G}) + k\nabla \operatorname{div} \mathbf{G} - \nu \operatorname{curl}^2 \mathbf{G}, \quad (346)$$

where $\mathbf{G}$ is the displacement, n the rigidity, k the compressive
resistivity, and ν the new elastic constant connected with the
rotation.   Now, we have $\operatorname{curl}^2 = \nabla \operatorname{div} - \nabla^2$, so we may rearrange
(346) thus,

$$\mathbf{F} = (n + \nu)\nabla^2\mathbf{G} + (k + \tfrac{1}{3}n - \nu)\nabla \operatorname{div} \mathbf{G}, \quad . \quad (347)$$

and this may be split up into circuital and divergent parts, viz.,

$$\mathbf{F}_1 = (n + \nu)\nabla^2\mathbf{G}_1, \quad \text{(circuital)}, \ . \quad (348)$$

$$\mathbf{F}_2 = (k + \tfrac{4}{3}n)\nabla^2\mathbf{G}_2, \quad \text{(divergent)}. \quad (349)$$

Now observe that in the circuital equation, n and ν occur
additively, so that equal parts of each have the same effect,
and either may be replaced by an equal amount of the other.
This is irrespective of the compressibility, which is not concerned in circuital disturbances.   We see, then, that the
rigidity may be done away with altogether and equal ν substituted, without affecting the propagation of circuital disturbances.   This is a very remarkable property.

Observe, also, that in the divergent equation ν does not
enter at all, although n does, and in the same way as in the
ordinary compressible rigid solid.   In fact, the coefficient $k + \tfrac{4}{3}n$
is the same as the former $m + n$, of § 142 and later.   So the
propagation of divergent disturbances is the same as if the
rotational elasticity were done away with.

Now, we have already noticed the case in which the speed
of the divergent wave is brought down to zero—the contractile
ether of §§ 142, 143.   In an ordinary rigid solid, which was
then referred to, this requires negative compressibility.   But
now, if n be abolished and ν substituted, so that the circuital
propagation is unaffected, the divergent equation will contain
only k, so that to make the speed of the divergent wave zero
we need only abolish the resistance to compression, or make
$k = 0$.   The medium does not now resist either change of

shape or of size, but resists rotation only, and the potential energy is the energy of rotation.   This is one extreme form of the rotational ether, made by combining it with the contractile, by the evanescence of the rigidity and of the resistance to compression.   The medium is quite neutral as regards expansion and compression.

Now, it is clear, from equation (348) and matter in previous paragraphs, that we may construct analogies to suit the non-conducting dielectric by abolishing n and using ν instead, without concerning ourselves with k at all, which may have any value from zero to infinity.   But Lord Kelvin's ether is got by going to the other extreme from the neutral case just mentioned.

First Rotational Analogy : Magnetic Force compared with Velocity.

§ 154.   Take $k = \infty$ (with $n = 0$) making the medium incompressible, and therefore making disturbances in an unbounded medium be necessarily circuital, whether the impressed forcive be circuital or not.   We now have

$$\mathbf{F}_1 = \nu \nabla^2 \mathbf{G}, \quad \cdots \cdots \quad (350)$$

by (348), whilst the other equation (349) will merely assert that $\mathbf{F}_2$ is balanced by difference of pressure.   This we are not concerned with, since our impressed forcive should be circuital in the electromagnetic comparisons.   The stress (310) is now

$$\mathbf{P}_N = - \nu \mathbf{VN} \operatorname{curl} \mathbf{G}, \quad \cdots \cdot \quad (351)$$

where we ignore the uniform pressure for the reason mentioned. The torque accompanying this stress is, by (311),

$$\mathbf{S} = 2\nu \operatorname{curl} \mathbf{G}, \quad \cdots \cdots \quad (352)$$

and the translational force is

$$\mathbf{F} = - \operatorname{curl} \tfrac{1}{2}\mathbf{S} = \nu \nabla^2 \mathbf{G}, \quad \cdots \quad (353)$$

agreeing with (350).   So the equation of motion is, by (306),

$$\mathbf{f}_1 = (\rho p^2 - \nu \nabla^2)\mathbf{G} = (\rho p - \frac{\nu}{p}\nabla^2)\mathbf{q}, \quad \cdots \quad (354)$$

where $\mathbf{f}_1$ is the circuital impressed force, and $\mathbf{q}$ is the velocity. As before, we keep to small motions in general.

Equation (354) may be at once compared with the electro-magnetic

$$f = \left(\mu p - \frac{\nabla^2}{cp}\right)H = \left(\mu p^2 - \frac{\nabla^2}{c}\right)Z, \quad . \quad . \quad (355)$$

as in (325), (326), our first example, and we therefore deduce the following correspondences :—

	$Z = H/p$ stands for G,		(spacial displ.)
(mag. force)	H	,, ,, q,	(velocity).
(el. current)	curl H	,, ,, curl q,	(2 × spin).
(el. displ.)	D	,, ,, curl G,	(2 × rotation).
(inductivity)	μ	,, ,, ρ,	(density).
(permittivity)	c	,, ,, $1/\nu$,	(rotl. compliancy)
(induction)	B	,, ,, ρq,	(momentum).
(el. force)	E	,, ,, $\frac{1}{2}S$,	($\frac{1}{2}$ × torque).
(mag. energy)	$\frac{1}{2}\mu H^2$	,, ,, $\frac{1}{2}\rho q^2$,	(kin. energy).
(el. energy)	$\frac{1}{2}cE^2$	,, ,, $\frac{1}{2}\nu(\text{curl}G)^2$,	(rotl. energy).
(source of H)	$f = $ curl e	,, ,, f_1,	(impd. force).
(energy flux)	VEH	,, ,, $V\frac{1}{2}Sq$,	(energy flux)

Referring to the similar list in § 146, it will be seen that the correspondences there given are repeated here, except that the elastic constant changes its meaning, and that there are now additional correspondences. The electric energy is correctly localised by the potential energy of the rotation. Since the kinetic energy is also correctly localised, we need not be surprised to find that the energy-flux has the same distribution.

The activity of the stress P_N is $P_N q$ per unit area, which is the same as qNP_q, if P_q be the stress on the plane whose normal is q, provided the stress is irrotational. But if ro-tational, as at present, it is qNQ_q, where Q is conjugate to P. This activity means energy transferred from the side of the plane when the stress vector is reckoned, to the other side, or against the motion. The vector expressing the energy flux is therefore $-qQ_q$, not counting the convective flux. In our present case, by (351), Q is the negative of P, because it is purely a rotational stress, so $W = qP_q$ is the vector flux of energy ; or, by (351),

$$W = -\nu q V q_1 \text{ curl } G = -\nu V q \text{ curl } G, \quad . \quad (356)$$

which is, by the above table, the same as **VEH**, the electromagnetic flux of energy. (In the above, q and q_1 are the tensor and unit vector of **q**.)

The above reasoning is applied directly to the unit element of volume. But it may be easier to follow by taking any volume into consideration. If **N** is the normal outwards from the surface enclosing it, then P_N is the pull, per unit area, of the matter outside, on the matter inside the surface, across the unit area of the interface; and $P_N q$ is its activity. The total activity of P_N all over the surface is therefore $\Sigma P_N q$, which is the same as $-\Sigma N q P_q$, and expresses the work done per second by the matter outside on the matter inside the surface, or the rate of transfer of energy from the outside to the inside; and its equivalent is the rate of increase of the stored energy, potential and kinetic, within the surface. The convergence of the vector $q P_q$ expresses the same property for the unit volume, so that $q P_q$ is the flux of energy (per unit area) itself. (The small convective flux of energy is ignored here.)

Circuital Indeterminateness of the Flux of Energy in general.

§ 155. That *any* circuital flux of energy may be superadded, without making any difference in the transformations of energy, is a fact which is of importance as evidence against the objectivity of energy, but is of no moment whatever in the practical use of the idea of a flux of energy for purposes of reasoning. We should only introduce an auxiliary circuital flux when some useful purpose is served thereby.

I may here remark that (speaking from memory) when Professor J. J. Thomson first objected to the **VEH** formula as representative of energy-flux, by reason of the circuital indeterminateness, he added that a knowledge of the real flux could only be determined from an actual knowledge of the real dynamical connections involved, that is to say, by a knowledge of the mechanism. But this is surely an inconclusive argument, because the circuital indeterminateness applies even then. In fact, it is universal in its application, as energy is at present understood. For there to be an absolutely definite flux of energy through space seems to require energy to have objectivity in the same sense as matter, which is a very difficult notion to grasp, and still more difficult to accept. But even if it be ac-

cepted, the argument of circuital indeterminateness remains in
action, when it is desired to find what *is* the flux in a given case.

In a recent number of the *Phil. Mag.*, Mr. Macaulay, by the
addition of a certain circuital flux to **VEH**, brings out the
result that in stationary states the modified flux reduces to
p**C**, where p is potential and **C** current-density. I am, however (unless I forget much of what I have learnt in the last
15 years), unable to see that the auxiliary flux proposed serves
any useful purpose. In the first place, it greatly complicates
the flux of energy in general, and is entirely against the more
simple ideas to which we are naturally led in the study of
electromagnetic waves, whether in dielectrics or conductors.
Besides that, it is implied that the function p, or electric
potential, is a determinate quantity specifying some definite
state of the medium. This seems to me to be an idea which
has no place at all in Maxwell's theory. It may have place in
other theories, but that is not to the point. To exemplify,
put a closed circuit (with battery) supporting a steady current
inside a metal box, and electrify the latter from outside.
According to the p**C** formula, the flux of energy in the wire
suffers a very remarkable change, by reason of the raising of
the potential of the box and its interior. But, according to the
interpretation of Maxwell's scheme which I expound, there is
no change whatever in the electrical state of the interior of the
box in the steady state (though there may and must be a
transient disturbance in the act of charging it, which is a
separate question), and the energy distribution and its flux are
the same as before, because **E** and **H**, which settle the state of
affairs, electric and magnetic, are unchanged. There are,
however, I believe, some electricians who will be much gratified
with the p**C** flux in steady states.

The total flux of energy through a wire will be p**C**, where by
C we mean the total current, obtained from p**C** by integration
over the cross-section. Now this is somewhat suggestive of
my expression V**C** for the total flux of energy along a circuit of
two conductors. But there are radical differences. For V**C**
is the integral form of **VEH** in the dielectric. Although C in
V**C** means the same as in p**C**, being the circuitation of **H**, or
the gaussage, V has no connection with p the potential, for it
(V) is the line-integral of **E** across the circuit from one wire

to the other—*i.e.*, the transverse voltage.   Again, the VC formula holds good in variable as well as in steady states, whilst the proposed pC holds for steady states only.   Even in steady states, when V specializes itself and becomes difference of potential, it remains different from p.

But although I cannot see the utility of the proposed change, which seems to be a retrograde step, I think that Mr. Macaulay's mathematics, which is of a strong kind, may be of value in the electromagnetic field in other ways, especially when cleared of useless and treacherous potentials.

Second Rotational Analogy : Induction Compared with Velocity.

§ 156.   Returning to the suppositional Ether, the conception is of an incompressible medium possessing mass, which involves translational inertia, and therefore kinetic energy, and which elastically resists rotation, and so stores potential energy ; and we have supposed that the velocity of the medium means magnetic force, and that its density means μ, so that the kinetic energy and the magnetic energy are compared.   But we may equally well have this comparison of the energies combined with a different interpretation of μ.   Take, for instance, the induction to mean velocity.   Then, since the magnetic energy is $\frac{1}{2}\mu^{-1}\mathbf{B}^2$, we see that it is now, not μ, but μ^{-1} that is the density, whilst $\mathbf{H}$ is momentum.   Also, since $\mu c v^2 = 1 = \rho \nu^{-1} v^2$, we see that the new interpretation of c is νv^{-4}.   That is, the change of μ from density to its reciprocal involves the change of c from compliancy to its reciprocal multiplied by v^{-4}, making ρ^2/ν or ρ/v^2. But it is better to write them out side by side, thus :—

(mag. energy)	$\frac{1}{2}\mu^{-1}\mathbf{B}^2$ stands for		$\frac{1}{2}\rho\mathbf{q}^2$,	(kin. energy).
(induction)	$\mathbf{B}$	,,	,,	$\mathbf{q}$,	(velocity).
(reluctivity)	μ^{-1}	,,	,,	ρ,	(density).
(mag. force)	$\mathbf{H}$	,,	,,	ρq,	(momentum).
(el. current)	curl $\mathbf{H}$	,,	,,	ρ curl $\mathbf{q}$,	
(el. displ.)	$\mathbf{D}$	,,	,,	ρ curl $\mathbf{G}$,	
(permittivity)	c	,,	,,	$\nu^{-1}\rho^2$,	
(el. force)	$\mathbf{E}$	,,	,,	$\nu\rho^{-1}$ curl $\mathbf{G}$,	
(el. energy)	$\frac{1}{2}c\mathbf{E}^2$	,,	,,	$\frac{1}{2}\nu(\text{curl } \mathbf{G})^2$	(potl. energy).

The potential energy still correctly localising the electric energy, in spite of the other changes, we may expect the flux of energy to be correct. We have

$$\mathrm{VEH} = \nu\rho^{-1}\mathrm{V}\,\mathrm{curl}\,\mathbf{G}.\,\rho\mathbf{q} = -\nu\mathrm{V}\mathbf{q}\,\mathrm{curl}\,\mathbf{G},$$

as before, equation (356).

Similarly, we may assume that μ is not the density, nor its reciprocal, but an unstated function of the density, say $\mu = f(\rho)$, and work out the various correspondences that this necessitates. We shall, for example, by the formulæ for magnetic and kinetic energy, require

$$\mathbf{q} = \mathbf{H}\{\rho^{-1}f(\rho)\}^{\frac{1}{2}}.$$

Of course, in thus comparing magnetic with kinetic energy, we shall always have $\mathbf{H}$ compared with $\mathbf{q}$, or with some multiple of $\mathbf{q}$, which may change its "dimensions," as, for example, in the above detailed cases, where $\mathbf{H}$ is changed from velocity to momentum.

Probability of the Kinetic Nature of Magnetic Energy.

§ 157. If it be asked why, in the previous analogies, a preference has usually been shown for the representation of magnetic force (or a constant multiple thereof) by velocity, the answer would be, substantially, that it has been done in order to make the magnetic energy be kinetic. Now, it is true that in a perfectly abstract electromagnetic scheme, arranged in duplex form—in which every electric magnitude has its magnetic representative—and, therefore, including a magnetic conduction-current with waste of energy, there would be a perfect balance of evidence as regards the kinetic nature of either the electric or the magnetic energy, when the other is to be potential energy, as of a state of strain in an elastic medium. For, if it were argued from a certain set of relations that the magnetic energy was kinetic, the force of the argument could be at once destroyed by picking out an analogous set of relations tending to show (in the same way as before) that the electric energy was kinetic.

But, as matters actually stand, with an imperfect correspondence between the electric and magnetic sides of the electromagnetic scheme, there seems to be a considerable preponderance of evidence in favour of the kinetic nature of the magnetic energy. We may refer, in particular, to the laws of linear electric circuits, which were shown by Maxwell to be simply deducible by the ordinary equations of motion (generalised) of a dynamical system, on the assumption that magnetic energy is kinetic; there being one degree of freedom for every circuit, and the variables being such that every linear electric current is a (generalised) velocity. In this theory, the dielectric in which the conducting circuits are immersed is regarded as unyielding, so that electric displacement cannot occur in it, and the currents are confined entirely to the conductors.

Now, we could construct a precisely similar theory of conductive magnetic circuits immersed in a medium permitting displacement, but destitute of magnetic inductivity. The circuital flux of magnetic induction in the former case would now be replaced by a similar circuital flux of electric displacement; the former electric currents becoming magnetic currents, and the magnetic energy becoming electric energy. But this electric energy, when expressed in terms of the linear magnetic currents, would possess the property of allowing us to deduce from it the laws of the magnetic circuits, by using the generalised equations of motion, on the assumption that electric energy is kinetic, in a manner resembling Maxwell's deduction of the laws of electric circuits. Therefore, supposing the state of things mentioned to really exist, we might become impressed with the idea that electric energy is kinetic; just as, at present, it seems hardly possible to avoid entertaining the idea of the kinetic nature of the magnetic energy.

We do not, however, need to go to generalised dynamics to arrive at this probable conclusion. It is sufficient to start with a sound general knowledge of dynamical facts and principles, not necessarily mathematical, but such as may be acquired in practical experience by an intelligent and thoughtful mind, involving clear ideas about inertia, momentum, force, and work, and how they are practically connected (the Act of Parliament notwithstanding). On then proceeding to the experimental study of electrokinetics, including the phenomena of self-induction in particular, the dynamical ideas will be found to come in quite naturally. Lastly, the generalised theoretical dynamics will serve to clinch the matter. Although adding little that is novel, it will corroborate former conclusions, and co-ordinate the facts in a compact and systematic manner, suitable to a dynamical science.

On the other hand, there is little that is suggestive of kinetic ideas in electrostatics, whilst there is much that is suggestive of the potential energy of a strained state. This fact, combined with the kinetic suggestiveness of the facts (and the equations embodying them) in which magnetic induction and electric currents are concerned, explains why the association of magnetic with kinetic, and electric with potential energy becomes natural. It is, therefore, somewhat a matter of surprise, as well as rather vexing, to find that in order to extend the last-considered rotational analogy to a conducting dielectric, we must, if we wish to do it simply, give up the comparison of magnetic force with velocity and electric force with torque, and adopt the converse system, making the electric energy be kinetic, and the magnetic energy the potential energy of the rotation.

Unintelligibility of the Rotational Analogue for a Conducting Dielectric when Magnetic Energy is Kinetic.

§ 158. We can, perhaps, most easily see that this plan should be adopted by writing down the two circuital equations of electromagnetism, and then, immediately under them, the corresponding proposed circuital equations of the rotational ether. Thus, taking H to be velocity, as before, we have, to express the first circuital law in the dielectric and its mechanical companion,

$$g + \operatorname{curl} H = cp E, \quad \ldots \quad (357)$$

$$g_1 + \operatorname{curl} q = \frac{1}{\nu} p(\tfrac{1}{2} S). \quad \ldots \quad (358)$$

Here the terms in vertical are to be compared. Magnetic force H becoming velocity q, the electric force E becomes $\tfrac{1}{2}S$, or half the torque; c the permittivity is the compliancy ν^{-1}, and g_1 is a newly-introduced companion to g, which is the source $- \operatorname{curl} h$. It will be interpreted later.

The other circuital equations are

$$f - \operatorname{curl} E = \mu p H, \quad \ldots \quad (359)$$

$$f_1 - \operatorname{curl} (\tfrac{1}{2} S) = \rho p q. \quad \ldots \quad (360)$$

We now have the additional analogues of inductivity and density, and f, which is the source curl e, is compared with f_1, which is impressed translational force per unit volume.

So far relating to a non-conducting dielectric, and giving an intelligible dynamical analogy, if we wish to extend it to a conducting dielectric, according to Maxwell's scheme (including, of course, a pure conductor which has, or is assumed to have, no permittivity), we require to change cp in the first circuital law (357) to $k + cp$, whilst the second circuital law needs no change. But the right member of (358) needs to be changed to match the modified (357). Thus,

$$g + \text{curl } H = (k + cp)E, \quad . \quad . \quad . \quad . \quad (361)$$

$$g_1 + \text{curl } q = (k_1 + v^{-1}p)(\tfrac{1}{2}S), \quad . \quad . \quad (362)$$

where k_1 is the new coefficient, to match k. But what is its interpretation in the rotational ether, and how is the latter to be modified to make k_1 as intelligible as the other constants?

Now, since we use rotational elasticity to obtain the potential energy, and we associate the rotation with the electric displacement, it is suggested that rotational friction should be introduced to cause the waste of energy analogous to that of Joule. But if there were frictional resistance proportional to the spin, we should have

$$S = 2(v + v_1p) \text{ curl } G, \quad . \quad . \quad . \quad (363)$$

instead of (352), connecting the torque with the rotation. But this will not harmonise at all with (362). We cannot do what we want by rotational friction, but require some special arrangement, whose nature does not appear, in order to interpret (362) intelligibly.

The Rotational Analogy, with Electric Energy Kinetic, extended to a Conducting Dielectric by means of Translational Friction.

§ 159. But, by changing the form of the analogy, choosing the electric energy to be kinetic, the extension to a conducting dielectric can be made in a sufficiently obvious manner. Put (360) under (361), thus

$$g + \text{curl } H = (k + cp)E, \quad . \quad . \quad . \quad (364)$$

$$f_1 + \text{curl } (-\tfrac{1}{2}S) = (\rho_1 + \rho p)q, \quad . \quad . \quad (365)$$

where we introduce ρ_1 to match k. Also put (358) under (359) to get the other pair, thus

$$\mathbf{f} - \text{curl } \mathbf{E} = \mu p \mathbf{H}, \quad \cdot \quad \cdot \quad \cdot \quad (366)$$

$$-\mathbf{g}_1 - \text{curl } \mathbf{q} = \frac{1}{\nu} p (-\tfrac{1}{2}\mathbf{S}). \quad \cdot \quad \cdot \quad \cdot \quad (367)$$

We have now a fit, with an intelligible meaning to be given to the new coefficient that brings in waste of energy. For (365), which is compared with the first circuital law, is the translational equation of motion in the rotational ether when there is frictional resistance to translation, expressed by $\rho_1 \mathbf{q}$, so that ρ_1 is the frictionality, and $\rho_1 \mathbf{q}^2$ the rate of waste.

The correspondences in detail are as follows :—

		$\mathbf{A} = \mathbf{E}/p$ stands for	$\mathbf{G}$,	(spacial displ.).
(el. force)	$\mathbf{E}$	,, ,,	$\mathbf{q}$,	(velocity).
(permittivity)	c	,, ,,	ρ,	(density).
(el. displ.)	$\mathbf{D}$	,, ,,	$\rho \mathbf{q}$,	(momentum).
(el. energy)	$\tfrac{1}{2}c\mathbf{E}^2$	,, ,,	$\tfrac{1}{2}\rho \mathbf{q}^2$,	(kin. energy).
(mag. force)	$\mathbf{H}$	,, ,,	$-\tfrac{1}{2}\mathbf{S}$,	($-\tfrac{1}{2}$ torque).
(inductivity)	μ	,, ,,	ν^{-1},	(compliancy).
(induction)	$\mathbf{B}$	,, ,,	$-\tfrac{1}{2}\nu^{-1}\mathbf{S}$,	
(mag. energy)	$\tfrac{1}{2}\mu\mathbf{H}^2$	,, ,,	$\tfrac{1}{2}\nu(\text{curl }\mathbf{G})^2$,	(rotl. energy).
(conductivity)	k	,, ,,	ρ_1,	(frictionality).
(cond. current)	$\mathbf{C} = k\mathbf{E}$	,, ,,	$\rho_1 \mathbf{q}$,	
(Joule-heat)	$k\mathbf{E}^2$	,, ,,	$\rho_1 \mathbf{q}^2$,	(rate of waste).
(true current)	$\mathbf{C} + p\mathbf{D}$	,, ,,	$(\rho_1 + \rho p)\mathbf{q}$.	

These should be studied in connection with the circuital equations (364), (366), and their matches underneath them, if it be desired to obtain "an intelligent comprehension" of the true nature of the analogy, and to correct any errors that may have crept in.

If we keep away from the sources of energy, there is little difficulty in understanding the analogy in a broad manner. But the sources require special attention before the electromagnetic and rotational analogues are intelligible. We know that activity is the product of two factors, a "force" and a "velocity." Newton knew that. In modern dynamics, too, where the velocity is not a primitive velocity, but has a generalised meaning—the time-rate of change of some variable —the corresponding force generalised (not a primitive force) is still such that the product of "force" and "velocity" is activity, or activity per unit volume, &c., according to convenience. In our electromagnetic equations, for instance, $E\dot{C}$ and $H\dot{B}$ and $E\dot{D}$ are activities (per unit volume), and we call E and H the "forces" (intensities), and the other factors the fluxes, the corresponding "velocities." Similarly, $e\dot{C}$, $e\dot{D}$, and $h\dot{B}$ are activities (per unit volume understood), where e and h are intrinsic, communicating energy to the system.

Now, since the "variables" may be variously chosen in a dynamical system, we need not be surprised if it should sometimes happen that the (generalised) force turns out to be a (primitive) velocity, and the (generalised) velocity to be a (primitive) force. Here is food for the scoffer, for one thing. At any rate, we should be careful not to confound distinct ideas, and remember the meaning of the activity product. Our present rotational analogy furnishes an illustration. We have the equation of activity,

$$e(\dot{C} + \dot{D}) + h\dot{B} = Q + \dot{U} + \dot{T} + \operatorname{div} W, \quad . \quad (368)$$

where $$W = V(E - e)(H - h). \quad . \quad . \quad (369)$$

W is the energy-flux, Q the rate of waste, U the electric, and T the magnetic energy per unit volume, whilst the left member represents the rate of supply of energy by the intrinsic sources e and h, being the sum of their activities.

Now what are the analogues of e and h? Remember that only their curls appear in the circuital equations, and that f and g are the sources of disturbances. By inspection of (365), (367), we see that f_1 is the curl of a torque, and g_1 the curl of a velocity, say

$$2f_1 = \operatorname{curl} S_0, \qquad g_1 = - \operatorname{curl} q_0. \quad . \quad . \quad (370)$$

Then, just as the analogue of H is $-\tfrac{1}{2}S$, the analogue of h is $-\tfrac{1}{2}S_0$. Impressed magnetic force is, therefore, represented by an impressed torque per unit volume, and its activity is

$$(-\tfrac{1}{2}S_0)v^{-1}p(-\tfrac{1}{2}S), \quad . \quad . \quad . \quad (371)$$

which is, of course, plain enough.   But the other activity is

$$q_0 (\rho_1 + \rho p) q, \quad \ldots \ldots \quad (372)$$

where q_0 is the analogue of e.   Here the supposed "force" has
become a velocity, and the "velocity" a force.   For the factor
of q_0 in (372) is the analogue of electric current-density, and
means the (primitive) force per unit volume in the rotational
ether, partly employed in increasing momentum, partly working against friction.   Of course this perversion is rather
extreme.

The rotational analogues of (368), (369) may be readily
written down by proper translations in accordance with the above,
remembering (370).

Mr. W. Williams, who has recently (Physical Society, 1892)
published a very close study of the theory of the "dimensions"
of physical magnitudes, on applying his views to the electromagnetic equations and the rotational ether, has arrived at the
conclusion that the representation of either **E** or **H** by velocity
leads to the only two systems that are dynamically intelligible.
It should be remembered, however, that his conclusion is
subject to certain limitations (mentioned in his paper) regarding
dimensions, for otherwise the conclusion might seem to be of
too absolute a nature, as if one of **E** or **H** *must* be velocity.   In
any case, I cannot go further myself at present than regard the
rotational ether as furnishing a good analogy, which may lead
later to something better and more comprehensive.   The present limitation to small motions (in general) is a serious one,
and there are other difficulties.

Symmetrical Linear Operators, direct and inverse, referred to the principal Axes.

§ 160.   Linear vector functions of a vector play a very
important part in vector-algebra and analysis, just as simple
equations do in common algebra.   Thus, in the theory of
elasticity the strain vector is a linear function of the direction
vector, and the stress vector is another linear function of the
direction vector.   In electromagnetism the electric displacement is a linear function of the electric force, and so is the
conduction current-density ; whilst the magnetic induction is a
linear function of the magnetic force, when its range is small

enough. There are, therefore, at least five examples of linear vector functions to be considered in electromagnetism. In other sciences too, the linear function often turns up, and frequently in pure geometry, when treated algebraically. Finite rotations may be treated by the method, and in the algebra of surfaces of the second order the linear connection between two vectors is prominent. It would be inexcusable not to give some account of linear operators in this chapter on vector-analysis and its application to electromagnetism. The subject, however, is such a large one that it is only possible to deal with its more elementary parts in a somewhat brief manner—not, however, so brief as to be useless.

The simplest kind of linear connection is that of Ohm's law in isotropic conductors. We have $C = kE$, where E is the intensity of electric force and C the current-density, whilst k is a constant, the conductivity. No specification of direction is here made. But when vectorised, we have the equation $\mathbf{C} = k\mathbf{E}$ instead, k being as before. We now assert that **E** and **C** are parallel, whilst their tensors are in a constant ratio, for all directions in space.

Suppose, however, we alter the conductivity of the body in a certain direction, say that of i (*e.g.*, by means of compression parallel to i), making it k_1, whilst its conductivity is k_2 in all directions transverse to i. We now have $C_1 = k_1 E_1$, and also $C_1 = k_1 E_1$, when the electric force is parallel to i. But if it be transverse to i, then we have $C = k_2 E$. There is, in both cases, parallelism of **E** and **C**, but the ratio of the tensors changes from k_1 to k_2. What, then, is the current when **E** is neither parallel nor transverse to i? The answer is to be obtained by decomposing **E** into two components, one parallel to i, the other transverse; then reckoning the currents to match by the above, and finally combining them by addition. Thus

$$\mathbf{C} = \mathbf{i}.k_1 E_1 + \mathbf{j}.k_2 E_2 + \mathbf{k}.k_2 E_3, \quad . \quad . \quad . \quad (1)$$

where E_1, E_2, E_3 are the i, j, k scalar components of **E**. We have here made use of the linear principle. It is that the sum of the currents due to any two electric forces is the current due to the sum of the electric forces. This principle (suitably expressed) applies in all cases of linear connection between two vectors, and is the ultimate source of the simplicity of treatment that arises.

s

More generally, let the conductivity be different in the three co-perpendicular directions of **i**, **j**, **k**, being k_1, k_2, k_3 respectively. Then we have

$$\mathbf{C} = k\mathbf{E} = \mathbf{i}.k_1 \mathrm{E}_1 + \mathbf{j}.k_2 \mathrm{E}_2 + \mathbf{k}.k_3 \mathrm{E}_3. \quad . \quad . \quad (2)$$

The current is now coincident with the electric force in three directions only, namely, along the principal axes of conductivity. When **E** has any other direction than that of a principal axis, we have no longer parallelism of **C** and **E**, and their relation is expressed by equation (2). It is the general type of all linear relations of the symmetrical kind when referred to the principal axes, if we remove the restriction which obtains in the electrical example that the k's must be all positive. Given, then, that $\mathbf{C} = k\mathbf{E}$, with the understanding that **C** is a symmetrical linear function of **E**, so that k must represent the linear operator connecting them (the conductivity operator in the above example) the full answer to the question, What is the **C** corresponding to a given **E**?—is obtainable in the above manner, viz., by first finding the axes of parallelism of **E** and **C** and the values of the principal k's, and next, by adding together the three **C**'s belonging to the three component **E**'s parallel to the axes.

The inverse question, Given **E**, find **C**, is similarly answerable. For if $\mathbf{C} = \rho\mathbf{E}$, where ρ is the operator inverse to k (or the resistivity operator in the case of conduction-current), we know that $k_1\rho_1 = 1$, if ρ_1 is the constant resistivity parallel to **i**, and similarly $k_2\rho_2 = 1$ and $k_3\rho_3 = 1$ for the **j** and **k** axes, so that in full,

$$\mathbf{E} = \rho\mathbf{C} = \mathbf{i}.\rho_1 \mathrm{C}_1 + \mathbf{j}.\rho_2 \mathrm{C}_2 + \mathbf{k}.\rho_3 \mathrm{C}_3, \quad . \quad . \quad (3)$$

where C_1, C_2, C_3 are the scalar components of current.

Precisely similar remarks apply to the permittivity operator c in $\mathbf{D} = c\mathbf{E}$, connecting the displacement **D** with the electric force **E**, and to the inductivity operator μ in $\mathbf{B} = \mu\mathbf{H}$, connecting the induction **B** with the magnetic force **H**, when the connection is a linear one. It is not without importance to constantly bear in mind the above process of passing from **E** to **C** through k, and the process of inverting k to ρ, because when treated in a general manner, with reference to any axes, the relations become much more complicated, whilst their intrinsic and resultant meaning is identically the same.

Geometrical Illustrations.   The Sphere and Ellipsoid. Inverse Perpendiculars and Maccullagh's Theorem.

§ 161. Next let us obtain some geometrical illustrations of the previous. The transition from E to C is like that from a sphere to an ellipsoid. If a solid be uniformly subjected to what is termed a homogeneous strain, any initially spherical portion of it becomes an ellipsoid. If this be done without rotation, the state is such that all lines parallel to a certain axis, say that of i, in the unstrained solid, are lengthened or shortened in a certain ratio in passing to the strained state, without change of direction. The same is true as regards lines parallel to two other axes, say j and k, perpendicular to each other and to the first axis, with different values given to the ratio of lengthening or shortening in the three directions of preservation of parallelism.

The equation of the ellipsoidal surface itself is

$$1 = \frac{x^2}{a^2} + \frac{y^2}{b^2} + \frac{z^2}{c^2}, \quad \cdots \quad (4)$$

when referred to the centre and the principal axes. Here a, b, c, are the lengths of the principal semi-axes, whilst x, y, z, are the scalar components of r, the radius vector from the centre to any point of the surface. It may be written

$$1 = \left(i \frac{x}{a} + j \frac{y}{b} + k \frac{z}{c} \right)^2 = N^2 \text{ say,} \quad \cdots \quad (5)$$

from which we see that when r belongs to the ellipsoid, N is a unit vector, the vector radius of a certain sphere from which the ellipsoid r may be obtained by homogeneous strain, the three ratios of elongation being a, b, c (that is, the co-ordinate x/a in the sphere becomes x in the ellipsoid, and so on).

Now, if we square (3) and divide by E^2, we may write it

$$1 = \frac{C_1^2}{(k_1 E)^2} + \frac{C_2^2}{(k_2 E)^2} + \frac{C_3^2}{(k_3 E)^2}, \quad \cdots \quad (6)$$

where the C's are the components of C, and E is the tensor of E. Now suppose E is constant, and let E take all directions in succession. Its extremity will range over a spherical surface, whilst the end of the corresponding C or kE will range over an ellipsoidal surface. For we may take $k_1 E$, $k_2 E$, $k_3 E$ in (6) to be a, b, c in (4); when C_1, C_2, C_3 will simultaneously be x, y, z. The semi-axes of this current-ellipsoid are the principal currents in the directions of parallelism of electric force and current.

Similarly to (6), we have, by squaring (2),

$$1 = \frac{E_1{}^2}{(\rho_1 C)^2} + \frac{E_2{}^2}{(\rho_2 C)^2} + \frac{E_3{}^2}{(\rho_3 C)^2}, \quad . \quad . \quad . \quad (7)$$

from which we see that when C is constant, so that the vector C ranges over a spherical surface, the corresponding E simultaneously ranges over the surface of an ellipsoid whose principal semi-axes are the principal E's.

There are other ways of illustration than the above. Thus we may see from the form of the right member of (4) that it is the scalar product of $\mathbf{r}$ and another vector $\mathbf{s}$, thus

$$\mathbf{rs} = 1, \quad . \quad . \quad . \quad . \quad . \quad . \quad (8)$$

where $\mathbf{r}$ and $\mathbf{s}$ are given by

$$\mathbf{r} = \mathbf{i}x + \mathbf{j}y + \mathbf{k}z, \quad . \quad . \quad . \quad . \quad (9)$$

$$\mathbf{s} = \mathbf{i}\frac{x}{a^2} + \mathbf{j}\frac{y}{b^2} + \mathbf{k}\frac{z}{c^2} \quad . \quad . \quad . \quad . \quad (10)$$

Here we see that $\mathbf{s}$ is a linear function of $\mathbf{r}$, and such that if $\mathbf{N} = \phi \mathbf{r}$, then $\mathbf{s} = \phi \mathbf{N} = \phi \phi \mathbf{r} = \phi^2 \mathbf{r}$. See equation (5). The interpretation of the new vector $\mathbf{s}$ is easily to be found. It is the reciprocal of the vector perpendicular from the centre upon the tangent plane to the ellipsoid at the extremity of $\mathbf{r}$. For if $\mathbf{p}$ be this perpendicular, the projection of $\mathbf{r}$ upon $\mathbf{p}$ is evidently $\mathbf{p}$ itself, because the three vectors $\mathbf{r}$, $\mathbf{p}$, and the line in the tangent plane joining their extremities form a right-angled triangle whose longest side is $\mathbf{r}$. Therefore

$$\mathbf{rp} = \mathbf{p}^2, \quad . \quad . \quad . \quad . \quad . \quad . \quad (11)$$

or, dividing by $\mathbf{p}^2$,

$$\mathbf{rp}^{-1} = 1, \quad . \quad . \quad . \quad . \quad . \quad (12)$$

comparing which with (8), which is true for all $\mathbf{r}$'s of the ellipsoid, we see that $\mathbf{s} = \mathbf{p}^{-1}$, as stated

Now we have

$$\frac{\mathbf{EC}}{\mathbf{EC}} = 1 = \frac{E_1{}^2}{\rho_1 \mathbf{EC}} + \frac{E_2{}^2}{\rho_2 \mathbf{EC}} + \frac{E_3{}^2}{\rho_3 \mathbf{EC}}, \quad . \quad . \quad (13)$$

from which we see that if we construct the ellipsoid whose principal semi-axes are $(\rho_1 \mathbf{EC})^{\frac{1}{2}}$, and so on, the radius vector will be $\mathbf{E}$ and the corresponding reciprocal perpendicular will be $\mathbf{C}/\mathbf{EC}$. Here $\mathbf{EC}$ is understood to be constant.

Similarly, we have

$$1 = \frac{C_1{}^2}{k_1 EC} + \frac{C_2{}^2}{k_2 EC} + \frac{C^2{}_3}{k_3 EC}. \quad . \quad . \quad . \quad (14)$$

so that in the ellipsoid with principal semi-axes $(k_1 EC)^{\frac{1}{2}}$, &c., the radius vector is C and the reciprocal perpendicular is E/EC. In both these methods of representation neither E nor C has constant tensor, but their product, the activity, is constant.

In the case (6), where it is E that is considered constant, the reciprocal of the perpendicular on the tangent plane to the ellipsoid of C is

$$\frac{iC_1}{(k_1 E)^2} + \ldots = \frac{i\rho_1 E_1}{E^2} + \ldots = \frac{\rho E}{E^2}. \quad . \quad . \quad . \quad (15)$$

This is also an ellipsoid. For kE is an ellipsoid, and ρE only differs in the principal constants being the reciprocals of those in the former case.

Similarly, in the case (7), where C is constant, the reciprocal perpendicular is

$$\frac{iE_1}{(\rho_1 C)^2} + \ldots = \frac{ik_1 C_1}{C^2} + \ldots = \frac{kC}{C^2}, \quad . \quad . \quad . \quad (16)$$

and again the extremity of kC ranges over an ellipsoidal surface.

Thus the sphere of E with constant tensor is associated with an ellipsoid kE, whose reciprocal perpendicular is $\rho E/E^2$ or $k^{-1}E/E^2$, another ellipsoid. It follows that the reciprocal perpendicular of the latter gives us the former ellipsoid again.

This reciprocal relation of two ellipsoids through their inverse perpendiculars is an example of Maccullagh's extraordinary theorem of reciprocal surfaces. It may be stated thus Let

$$r_1 s_1 = 1, \quad \text{and} \quad r_2 s_2 = 1, \quad . \quad . \quad . \quad (17)$$

be the equations of two surfaces referred to the same origin, r_1 and r_2 being the radius vectors, and s_1, s_2 the reciprocals of the vector perpendiculars from the origin on their tangent planes. If, then, we choose r_2 to be s_1, we shall simultaneously make s_2 be r_1. That is, starting with the surface r_1, construct the surface whose radius vector r_2 is s_1, the reciprocal perpendicular of the r_1 surface. Then the reciprocal perpendicular s_2 of the second surface (s_1 or r_2) will be the radius vector r_1 of the first surface.

This admits of simple vectorial proof.   For if $r_1 s_1 = 1$ be the equation of a surface according to the above notation, we obtain, by differentiation,

$$r_1 ds_1 + s_1 dr_1 = 0, \quad . \quad . \quad . \quad . \quad (18)$$

if dr_1 and ds_1 are simultaneous variations of r_1 and s_1.   But dr_1 is in the tangent plane to r_1, and therefore at right angles to s_1, so that $s_1 dr_1 = 0$.   This leaves $r_1 ds_1 = 0$ also.   But ds_1 or dr_2 equivalently is in the tangent plane of the surface whose vector radius is s_1, or r_2 ;   therefore r_1 is parallel to the perpendicular on the same, or parallel to s_2, say, $r_1 = x s_2$, from which $r_1 s_1 = x s_1 s_2$ or $x r_2 s_2$.   But $r_1 s_1 = 1$ and $r_2 s_2 = 1$, so $x = 1$, and $r_1 = s_2$, which completes the connections.

In terms of the perpendiculars, let p_1 and p_2 be their lengths and r_1, r_2 the corresponding radius vectors, then

$$r_1 : r_2 = p_1 : p_2, \quad . \quad . \quad . \quad . \quad (19)$$

and r_1 is parallel to p_2, and r_2 to p_1.

Internal Structure of Linear Operators.   Manipulation of several when Principal Axes are Parallel.

§ 162.   Leaving now the geometrical illustrations connected with the ellipsoid, consider the symmetrical linear operator by itself.   If $D = cE$, where c is the linear operator, we know by the above exactly how it operates through the principal c's. We can, however, write c in such a manner that it shall state explicitly its meaning.   Thus, if c_1, c_2, c_3 be the principal c's belonging to the axes of i, j, k, we have

$$cE = i.c_1 E_1 + j.c_2 E_2 + k \ c_3 E_3, \quad . \quad . \quad (20)$$

by § 160.   Now put the E's in terms of E, producing

$$cE = i.c_1 iE + j \ c_2 jE + k.c_3 kE. \quad . \quad . \quad (21)$$

In this form, the operand may be separated from the operator, thus—

$$cE = (i.c_1 i + j.c_2 j + k.c_3 k)E, \quad . \quad . \quad (22)$$

so that the expression for c itself is

$$c = i.c_1 i + j \ c_2 j + k.c_3 k. \quad . \quad . \quad . \quad (23)$$

In this form, with a vector to operate upon implied, the nature of c is fully exhibited.   But the operand need not follow the operator. If it precedes it the result is just the same. Thus by (23),

$$\mathbf{E}c = \mathbf{E}\mathbf{i}.c_1\mathbf{i} + \mathbf{E}\mathbf{j}.c_2\mathbf{j} + \mathbf{E}\mathbf{k}.c_3\mathbf{k}$$
$$= c_1\mathbf{E}_1.\mathbf{i} + c_2\mathbf{E}_2.\mathbf{j} + c_3\mathbf{E}_3.\mathbf{k} = c\mathbf{E}. \quad . \quad . \quad (24)$$

When, however, the operator is unsymmetrical, the vectors $c\mathbf{E}$ and $\mathbf{E}c$ are not identical. They are said to be conjugate to one another, so that in the symmetrical case of identity, c is also called a self-conjugate operator. The distinction between symmetrical and skew operators will appear later.

There are six vectors concerned in c, viz., $\mathbf{i}, \mathbf{j}, \mathbf{k}$, and $c_1\mathbf{i}, c_2\mathbf{j}, c_3\mathbf{k}$, which call $\mathbf{I}, \mathbf{J}, \mathbf{K}$. Thus

$$c = \mathbf{i}.\mathbf{I} + \mathbf{j}.\mathbf{J} + \mathbf{k}.\mathbf{K} \quad . \quad . \quad . \quad . \quad (25)$$

is the type of a symmetrical operator referred to the principal axes. Now the general form of c referred to any axes and with the symmetrical restriction removed is obtained by turning these six vectors to any six others; thus,

$$\phi = \mathbf{a}.\mathbf{l} + \mathbf{b}.\mathbf{m} + \mathbf{c}.\mathbf{n}, \quad . \quad . \quad . \quad . \quad . \quad (26)$$

so that $\qquad \phi\mathbf{E} = \mathbf{a}.\mathbf{l}\mathbf{E} + \mathbf{b}.\mathbf{m}\mathbf{E} + \mathbf{c}.\mathbf{n}\mathbf{E}, \quad . \quad . \quad . \quad . \quad (27)$

$$\mathbf{E}\phi = \mathbf{E}\mathbf{a}.\mathbf{l} + \mathbf{E}\mathbf{b}.\mathbf{m} + \mathbf{E}\mathbf{c}.\mathbf{n}, \quad . \quad . \quad . \quad . \quad (28)$$

show the linear function and its conjugate explicitly. This is (with a changed notation, however) Prof. Gibbs's way of regarding linear operators. The arrangement of vectors in (26) he terms a dyadic, each term of two paired vectors (a . l, &c.,) being a dyad. Prof. Gibbs has considerably developed the theory of dyadics.

Returning to the simpler form (25) or (22), we may note the effect of the performance of the operation symbolised by c a number of times, directly or inversely. Thus by (22) or (21), or (20), we have

$$\left.\begin{array}{l} c^2\mathbf{E} = cc\mathbf{E} = \mathbf{i}.c_1{}^2\mathbf{E}_1 + \mathbf{j}.c_2{}^2\mathbf{E}_2 + \mathbf{k}.c_3{}^2\mathbf{E}_3, \\[4pt] c^3\mathbf{E} = cc^2\mathbf{E} = \mathbf{i}.c_1{}^3\mathbf{E}_1 + \mathbf{j}.c_2{}^3\mathbf{E}_2 + \mathbf{k}.c_3{}^3\mathbf{E}_3, \\[4pt] c^{-1}\mathbf{E} = \mathbf{i}.\dfrac{\mathbf{E}_1}{c_1} + \mathbf{j}.\dfrac{\mathbf{E}_2}{c_2} + \mathbf{k}.\dfrac{\mathbf{E}_3}{c_3}, \end{array}\right\} \quad . \quad . \quad (29)$$

and so on. Thus c^n means the operator whose principals are the nth powers of those of the primitive c, with the same axes. We therefore get a succession of ellipsoids with similarly directed principal axes, only altering their lengths, starting from initial **E** with constant tensor. (That is, provided the principal c's are all positive. If some be negative, we have other surfaces of the second order to consider.)

We may also manipulate in the same simple manner **any** number of operators, a, b, c, &c., which have the same principal axes, though they may differ in other respects. Thus, using the same kind of notation,

$$
\left.
\begin{aligned}
a\mathbf{E} &= \mathbf{i} \cdot a_1 \mathbf{E}_1 + \mathbf{j} \cdot a_2 \mathbf{E}_2 + \mathbf{k} \cdot a_3 \mathbf{E}_3, \\
ba\mathbf{E} = ab\mathbf{E} &= \mathbf{i} \cdot a_1 b_1 \mathbf{E}_1 + \mathbf{j} \cdot a_2 b_2 \mathbf{E}_2 + \mathbf{k} \cdot a_3 b_3 \mathbf{E}_3, \\
cba\mathbf{E} = abc\mathbf{E} &= \mathbf{i} \cdot a_1 b_1 c_1 \mathbf{E}_1 + \mathbf{j} \cdot a_2 b_2 c_2 \mathbf{E}_2 + \mathbf{k} \cdot a_3 b_3 c_3 \mathbf{E}_3,
\end{aligned}
\right\} \quad . \quad (30)
$$

and so on. That is, the successive action of any number **of** symmetrical operators with common principal axes is equivalent to the action of a single operator of the same kind, whose principals are the products of the similar principals of the set of operators. The resulting ellipsoids have their axes parallel throughout.

Next, multiply the equation (20) by any other vector **F**, producing

$$
\mathbf{F}c\mathbf{E} = c_1 \mathbf{E}_1 \mathbf{F}_1 + c_2 \mathbf{E}_2 \mathbf{F}_2 + c_3 \mathbf{E}_3 \mathbf{F}_3. \quad . \quad . \quad (31)
$$

By symmetry we see that this is identically the same as **E**c**F**. We might also conclude this from the fact that c**E** and **E**c are the same vector; or, thirdly, by forming c**F**, and then multiplying it by **E**. We may therefore regard **E**c**F** as the scalar product of **E** and c**F** (or **F**c), or, as the equal scalar product of **E**c (or c**E**) and **F**. This reciprocity is the general characteristic of symmetrical operators, by which they can be distinguished from the skew operators. The electrical meaning in terms of displacement and electric force is that the component displacement in any direction **N**, due to an electric force acting in any other direction **M**, equals the component displacement along **M**, due to an electric force with the same tensor acting along **N**.

Theory of Displacement in an Eolotropic Dielectric. The Solution for a Point-Source.

§ 163. We see from the preceding that when it is the electric force that is given in a dielectric there is no difficulty in finding the displacement; and conversely, when the displacement is given, the electric force similarly becomes known through the values of the principal permittivities. Suppose, however, that the data do not include a knowledge of either the electric force or the displacement, both of which have to be found to suit other data. It is then sufficient to find either, the linear connection settling the other. Thus, to take an explicit case which has, for a reason which will appear, a special interest, suppose the dielectric medium to have uniform permittivity transverse to the axis of **k**, but to be differently permittive to displacement parallel to this axis. Further, put a point-charge q at the origin, and enquire what is the equilibrium distribution of displacement.

The statement that there is a charge q at the origin means nothing more than that displacement diverges or emanates from that place, to the integral amount q. We also understand, of course, that the displacement has no divergence anywhere else. But the manner of emanation is, so far, left quite arbitrary. Certain general results may, however, be readily arrived at. To begin with, if the medium is isotropic, the emanation of the displacement must not favour one direction more than another, from which it follows that the displacement at distance r from the source is spread uniformly over the area $4\pi r^2$, so that $q/4\pi r^2$ expresses its density. Now if we alter the permittivity to displacement parallel to **k** only, we favour displacement in that direction if the permittivity is increased, at the expense of the transverse displacement, because the total amount, which measures the strength of the source, is the same. Conversely, if we reduce the permittivity parallel to **k**, we favour the transverse displacement. The lines of displacement, originally spread equably, therefore separate themselves about the **k** axis (both ways), and concentrate themselves transversely, or about the equatorial plane, the plane passing through the charge which is cut by the **k** axis perpendicularly.

If we carry this process so far as to reduce the permittivity along **k** to zero, there can be no displacement at all parallel to **k**, so that the displacement must be entirely confined to the equatorial plane itself. In this plane it spreads equably from the source, because of the transverse isotropy assumed at the beginning. The amount q therefore spreads uniformly over the circle of circumference $2\pi r$ in the displacement sheet, so that the density (linear) is now $q/2\pi r$. The law of the inverse square of the distance has been replaced by the law of the inverse distance, with confinement to a single plane, however.

Next, if we introduce planar eolotropy, say by reducing the permittivity in direction $\mathbf{j}$, we cause the displacement in the sheet to concentrate itself about the $\mathbf{i}$ axis, at the expense of the displacement parallel to $\mathbf{j}$. Finally, if we abolish altogether the permittivity parallel to $\mathbf{j}$, the whole of the displacement will be confined to the $\mathbf{i}$ axis, half going one way and half the other.

We may go further by introducing heterogeneity. Thus, reduce the permittivity on one side of the equatorial plane. The effect will be to favour displacement on the other side; and, in the limit, when the permittivity is altogether abolished on the first side, the whole of the displacement goes unilaterally along the $\mathbf{k}$ axis on the other side.

In the case of the planar distribution of displacement, the surface-density is infinite, but the linear-density is finite, except at the source. We may, however, spread out the source along a finite straight line (part of the $\mathbf{k}$ axis) when the displacement will simultaneously spread out in parallel sheets, giving a finite surface-density. Similarly, in the case of the linear distribution of displacement along the $\mathbf{i}$ axis, we may spread out the source upon a portion of the $\mathbf{j}$, $\mathbf{k}$ plane, when the former line of displacement will become a tube, with finite surface-density of displacement inside it.

In the above manner, therefore, we obtain a general knowledge of results without mathematics. Except, however, in the extreme cases of spherical isotropy, of planar isotropy with zero permittivity perpendicular to the plane, and of purely linear displacement due to the vanishing of the transverse permittivity, and other extreme cases that may be named, we do not obtain an exact knowledge of the results, except in one respect, that the total displacement must always be the same. Return, therefore, to the case of initial isotropy upset by changed permittivity parallel to the $\mathbf{k}$ axis, or, if need be, with three different principal permittivities, so that the displacement is

$$\mathbf{D} = \mathbf{i}.c_1 F_1 + \mathbf{j}.c_2 F_2 + \mathbf{k}.c_3 F_3 = c\mathbf{F}, \quad . \quad . \quad . \quad (32)$$

if $\mathbf{F}$ be the electric force. We know that in the state of equilibrium, there must be no voltage in any circuit, or the curl of **F** must be zero, so that **F** must be the slope $-\nabla P$ of a scalar P, the potential. Furthermore, the divergence of **D** is the electrification-density ρ. Uniting, then, these results, we obtain

$$\operatorname{div}(-c\nabla P) = \rho, \quad \ldots \quad \ldots \quad (33)$$

or $$(c_1\nabla_1{}^2 + c_2\nabla_2{}^2 + c_3\nabla_3{}^2)P = -\rho, \quad \ldots \quad (34)$$

which is the characteristic of P when referred to the principal axes. Now when there is a point-source at the origin, and the three c's are equal, we know that the potential is $q/4\pi rc$ at distance r. From this, remembering that r^2 is the sum of the squares of the components of **r**, it is easy to see that the potential in the case of eolotropy is

$$P = \frac{fq/4\pi}{(x^2/c_1 + y^2/c_2 + z^2/c_3)^{\frac{1}{2}}} = \frac{fq/4\pi}{(\mathbf{r}c^{-1}\mathbf{r})^{\frac{1}{2}}}, \quad \ldots \quad (35)$$

where f is some constant, because this expression (35) satisfies (34) with $\rho = 0$, that is to say, away from the source.

The constant f may be evaluated by calculating the displacement passing through the surface of any sphere $r = $ constant, according to (35), and equating it to q. The result is $f = (c_1 c_2 c_3)^{-\frac{1}{2}}$. The complete potential in the eolotropic medium due to the point-source q at the origin is therefore

$$P = \frac{q/4\pi}{(c_1 c_2 c_3)^{\frac{1}{2}}(\mathbf{r}c^{-1}\mathbf{r})^{\frac{1}{2}}}. \quad \ldots \quad \ldots \quad (36)$$

The equipotential surfaces are therefore ellipsoids centred at the charge, the equation of any one being

$$\mathbf{r}c^{-1}\mathbf{r} = \frac{x^2}{c_1} + \frac{y^2}{c_2} + \frac{z^2}{c_3} = \text{constant}, \quad \ldots \quad (37)$$

so that the lengths of the principal axes are proportional to the square roots of the principal permittivities. The lines of electric force cut through the equipotential surfaces at right angles. But the lines of displacement do not do so, on account of the eolotropy. To find their nature, first derive the electric force from the potential by differentiation. Then (36) gives

$$\mathbf{F} = \frac{(q/4\pi)c^{-1}\mathbf{r}}{(c_1 c_2 c_3)^{\frac{1}{2}}(\mathbf{r}c^{-1}\mathbf{r})^{\frac{3}{2}}}, \quad \ldots \quad \ldots \quad (38)$$

and the displacement is obtained by operating by c on this, which cancels the c^{-1}, and gives

$$D = \frac{(q/4\pi)\mathbf{r}}{(c_1 c_2 c_3)^{\frac{1}{2}}(\mathbf{r}c^{-1}\mathbf{r})^{\frac{3}{2}}} \quad . \quad . \quad (39)$$

The displacement is therefore radial, or parallel to $\mathbf{r}$ itself. That is, the lines of displacement remain straight, only altering their distribution as the medium is made eolotropic.

Observe, in passing, that the scalar product $\mathbf{FD}$ varies as P^4; that is, the density of the energy varies as the fourth power of the potential. Also note that $\mathbf{F}r$ is proportional to P; that is, the radial component of $\mathbf{F}$, or the component parallel to the displacement, varies as P/r.

If we select any pair of equipotential surfaces between which the electric force and displacement are given by (38), (39), we may, if we please, do away with the rest of the electric field. That is, we may let the displacement and electric force terminate abruptly upon the two ellipsoidal surfaces. What is left will still be in equilibrium. For the voltage remains zero in every circuit possible. This is obviously true between the surfaces, because no change has been made there. It is also true in any circuit beyond the surfaces, because of the absence of electric force. And lastly, it is true for any circuit partly within and partly beyond the region of electric force, because the electric force is perpendicular to the surface. So the electric field is self-contained. The electrification is on the surfaces, where the normal component of the displacement measures the surface-density. That is, there is a total electrification q on the inner and $-q$ on the outer surface. The elastance of the condenser, or "leyden," to use Lord Rayleigh's word, formed by the two surfaces, is the ratio of the voltage between them to q. That is, the coefficient of q in (36) is the elastance between any equipotential surface and the one at infinity.

In thus abolishing the electric field beyond the limited region selected, we may at the same time change the nature of the medium, making it isotropic, for example, or conducting, &c. Any change is permissible that does not introduce sources that will disturb the equilibrium of the electric field between the equipotential surfaces.

Starting from isotropy, when the equipotential surface is a sphere, if we keep $c_1 = c_2$ constant, and reduce c_3, the sphere will become an oblate spheroid, like the earth (and other bodies)—flattened at the poles. Let the common value of c_1

and c_2 be denoted by c_0, then the potential due to the central charge q is

$$P = \frac{q/4\pi}{c_0 c_3{}^{\frac{1}{2}}(x^2/c_0 + y^2/c_0 + z^2/c_3)^{\frac{1}{2}}}, \quad \cdot \quad \cdot \quad \cdot \quad (40)$$

and the displacement and electric force are given by

$$c\mathbf{F} = \mathbf{D} = \frac{(q/4\pi)\mathbf{r}}{c_0 c_3{}^{\frac{1}{2}}(x^2/c_0 + y^2/c_0 + z^2/c_3)^{\frac{3}{2}}}. \quad \cdot \quad \cdot \quad (41)$$

As c_3 is continuously reduced the oblate spheroid of equilibrium becomes flattened more and more, and is finally, when $c_3 = 0$, reduced to a circular disc. The displacement is now entirely in the equatorial plane, so that if we terminate the displacement at distance r from the centre, we obtain a circular line of electrification (that is, on the edge of the disc). We may have the displacement going from the central charge to an equal negative charge spread over the circle, or from one circle to another in the same plane ; and so on.

Similarly, when c_1 and c_2 are unequal, but $c_3 = 0$, the circular disc is replaced by an elliptical disc. But, without introducing transverse eolotropy, we see that the abolition of the permittivity in a certain direction has the effect of converting the usual tridimensional solutions relating to an isotropic medium to bidimensional solutions, in which the displacement due to sources situated in any plane perpendicular to the axis of symmetry is also in that plane. We may have any number of such planar distributions side by side, with the axis of symmetry running through them. But they are quite independent of one another. If we restore the permittivity parallel to the axis of symmetry, the displacement will usually spread out to suit tridimensional isotropy. The exception is when every plane perpendicular to the axis has identically similar sources, similarly situated. Then the restoration of the permittivity will produce no change.

Theory of the relative Motion of Electrification and the Medium. The Solution for a Point-Source in steady rectilinear motion. The Equilibrium Surfaces in General.

§ 164. Let us now pass to what is, at first sight, an entirely different problem, but which involves essentially the same mathematics. As before, let there be a point-source of displace-

ment at the origin, but let the dielectric medium be homogeneous and fully isotropic.   The displacement will be radially
distributed, without bias one way or another.   Now, keeping
the charge fixed in space, imagine the medium to move steadily
from right to left past the charge.   How will this affect the
displacement ?

If we desire the solution expressing how the displacement
changes from its initial distribution, as the medium is brought
from rest into steady motion, we can obtain it by going the
right way to work.   But it is complex, and will, therefore, not
be given here, especially as it is not intelligible without close
study.   But if we only wish to know the finally-assumed distribution of displacement when the initial irregularities have
subsided, we may obtain and express the result in a comparatively simple manner.   Thus, to begin from the foundation, we
have the two fundamental circuital laws,

$$\text{curl} (\mathbf{H} - \mathbf{h}) = cp\mathbf{E} + \rho\mathbf{u}, \quad . \quad . \quad . \quad . \quad (42)$$

$$\text{curl} (\mathbf{e} - \mathbf{E}) = \mu p\mathbf{H}, \quad . \quad . \quad . \quad . \quad (43)$$

where $\mathbf{E}$ and $\mathbf{H}$ are the electric and magnetic forces, μ and c
the inductivity and permittivity, ρ the density of electrification,
supposed to have velocity $\mathbf{u}$; and $\mathbf{e}$, $\mathbf{h}$ are the motional electric
and magnetic forces given by

$$\mathbf{e} = V\mathbf{wB}, \qquad \mathbf{h} = V\mathbf{Dw}, \quad . \quad . \quad . \quad (44)$$

$\mathbf{B}$ and $\mathbf{D}$ being the induction and displacement, and $\mathbf{w}$ the
velocity of the medium.   (See equations (1), (2), or (3), (4),
§ 66, and equations (5), (6), § 44, for the motional forces. Also
§§ 33 to 70 generally.)

Now in our present case, $\mathbf{E}$ and $\mathbf{H}$ are steady, and the electrification is at rest.   The right sides of (42), (43) are
therefore zero, which is a great simplification.   By (42) and
the second o. (44) we obtain

$$\text{curl } \mathbf{H} = \text{curl } \mathbf{h} = \text{curl } V\mathbf{Dw}. \quad . \quad . \quad (45)$$

But $\mathbf{B}$ is circuital, and therefore so is $\mathbf{H}$.   It follows that

$$\mathbf{H} = V\mathbf{uD}, \quad . \quad . \quad . \quad . \quad . \quad (46)$$

if we substitute $-\mathbf{u}$ for $\mathbf{w}$, for future convenience.   This gives
$\mathbf{H}$ explicitly in terms of the displacement, when that is known,
so we have no further trouble with the magnetic force.

To find the displacement, (43) gives, along with the first of (44),

$$\text{curl } (\mathbf{E} + V u \mathbf{B}) = 0, \quad \ldots \quad (47)$$

or say,

$$\text{curl } \mathbf{f} = 0, \quad \text{if} \quad \mathbf{f} = \mathbf{E} + V u \mathbf{B}. \quad \ldots \quad (48)$$

Now put the value of $\mathbf{B}$ in $\mathbf{f}$ by (46), thus,

$$\mathbf{f} = \mathbf{E} + \mu c V u V u \mathbf{E} = \mathbf{E} + \frac{u^2}{v^2} V k V k \mathbf{E}, \quad \ldots \quad (49)$$

since $\mu c v^2 = 1$, v being the speed of propagation of disturbances, and u the tensor of $\mathbf{u}$, supposed to be parallel to $\mathbf{k}$.

Expanding by the fundamental formula (52), § 114, we obtain

$$\mathbf{f} = \mathbf{E} + \frac{u^2}{v^2}(\mathbf{k}.\mathbf{E}_3 - \mathbf{E})$$

$$= (1 - \frac{u^2}{v^2})\mathbf{i}\mathbf{E}_1 + (1 - \frac{u^2}{v^2})\mathbf{j}\mathbf{E}_2 + \mathbf{k}\mathbf{E}_3. \quad \ldots \quad (50)$$

Now let $\mathbf{f} = (1 - u^2/v^2)\mathbf{F}$. Insert in (50), and divide by $(1 - u^2/v^2)$. The result is

$$\mathbf{F} = \mathbf{i}\mathbf{E}_1 + \mathbf{j}\mathbf{E}_2 + \frac{\mathbf{k}\mathbf{E}_3}{1 - u^2/v^2}. \quad \ldots \quad (51)$$

This vector $\mathbf{F}$ must also have no curl. Therefore, if P is the potential whose slope is $\mathbf{F}$, we have

$$E_1 = - \frac{dP}{dx}, \quad E_2 = - \frac{dP}{dy}, \quad E_3 = - (1 - \frac{u^2}{v^2})\frac{dP}{dz}, \quad (52)$$

and the components of displacement are given by

$$D_1 = - c_1\frac{dP}{dx}, \quad D_2 = - c_2\frac{dP}{dy}, \quad D_3 = - c_3\frac{dP}{dz}, \quad (53)$$

if

$$c_1 = c_2 = c, \quad \text{and} \quad c_3 = c(1 - u^2/v^2). \quad \ldots \quad (54)$$

Now observe that (53) express the displacement in terms of the electric force in an eolotropic medium at rest, whose principal permittivities are given by (54), when P represents the electrostatic potential. Our present problem of a moving isotropic medium is therefore reduced to one relating to a stationary eolotropic medium. Thus, to state the comparison fairly, when the isotropic medium moves bodily past a stationary charge, the displacement distribution becomes identically the same as if the medium were at rest, but had its permittivity in the direction of motion reduced from c to $c(1 - u^2/v^2)$. The solution is therefore given by (40), (41) in § 163, by taking $c_0 = c$, and c_3 as in (54). The displacement concentrates itself about the equatorial plane, or plane through the charge perpendicular to the axis of motion in the one problem, or axis of reduced permittivity in the other. In the limit, when the speed of motion reaches v, the speed of light, the displacement is wholly planar, as in the eolotropic case when the permittivity c_3 vanishes.

Observe, in (52), that the square of u occurs. It does not matter, therefore, which way the motion takes place, so far as the displacement is concerned. But it makes a great difference in the magnetic force which accompanies the lateral concentration of the displacement. Being given by (46), we see that H reverses itself when the direction of motion is reversed.

It should be carefully noted that P, which is the electrostatic potential in the eolotropic stationary medium, is not the electrostatic potential in the moving medium, although it is the same function precisely. In the eolotropic medium its slope is the electric force F. In the moving medium its slope is the same force F, but this is no longer the electric force, which is E instead, obtained from F by (51) or (52). Again, whilst the displacement and electric force are not parallel in the stationary eolotropic medium; on the other hand, in the moving medium they are parallel, because the medium is isotropic. To complete the differences, there is no magnetic force in the eolotropic case, but there is in the moving medium, and it is its existence which allows the parallel to exist as regards the identical distributions of displacement.

Instead of moving the medium past a fixed charge, we may fix the medium, and move the charge through t the reverse way at the same speed. This is why we put $w = -u$. The above results apply when the charge q moves with velocity u through the stationary medium. But it is necessary now to travel with the charge in order to see the same results, because the origin is taken at the charge, and it is now in motion. The preliminary mathematics is therefore less easy.

The above theory of convection-currents and the eolotropic comparison I have given before.* I have now to add a somewhat important correction. For the opportunity of making

---

* "Electrical Papers," Vol. II., pp. 492-499, 504-516.

this correction I am indebted to Mr. G. F. C. Searle, of Cambridge, who has been at the trouble of working his way through my former calculations. Whilst fully confirming my results for a point-charge, and therefore all results obtained therefrom by integration, he has recently (in a private communication) cast doubt upon the validity of the extension of the solution to the case of a moving charged conducting sphere, and has asked the plain question (in effect) :—What justification is there for taking distributions of displacement calculated in the above manner, and cutting them short by surfaces to which the displacement is perpendicular, which surfaces are then made equilibrium surfaces? For example, in the solution for a point-charge, I assumed that the solution was the same for a charged sphere, because the lines of electric force met it perpendicularly. On examination, however, I find that there is no justification for this process. The electric force should not be normal to a surface of equilibrium, exceptions excepted.

The true boundary condition for equilibrium, however, needs no fresh investigation, being implicitly contained in the above. In the stationary case of eolotropy, the vector F must have no curl, and since it is the electric force, it is the same as saying that there must be no voltage in any circuit, so that F must be normal to an equilibrium surface. Now, in the problem of a moving medium with charge fixed the same condition formally obtains, viz., that F shall have no curl. But as it is not the electric force E, it is clear that the displacement (which is parallel to E) cannot be perpendicular to an equilibrium surface. The voltage calculated by the fictitious electric force F must come to zero in every circuit. The boundary condition is, therefore, that F is perpendicular to an equilibrium surface. That is, P = constant is the equation to a surface of equilibrium (no longer equipotential in the electrostatic sense), where P is, however, the same function as in the eolotropic stationary problem of electrostatics.

Therefore, by § 163, we see that the sphere of equilibrium when the medium (isotropic) is stationary, becomes an oblate spheroid when the medium is set into a state of steady motion, the shorter axis being parallel to the line of motion. In the limit, when the speed is v, the spheroid is flattened to a circular disc, on whose edge only is the electrification.

T

This applies when the whole dielectric medium is moving past a fixed spheroidal surface of electrification. It equally applies when the electrification moves the other way through a fixed medium. There will be no electric force in its interior. We may therefore fill it up with conducting matter, provided we do not interfere with the free motion of the external medium through it in the one case, or with the rest of the external medium when the conductor moves through it. But in the limiting case of motion at the speed of light, when only a single circular line of electrification is concerned, it would seem to be immaterial whether the conductor be a flat disc or a sphere, subject to the reservation of the last sentence.

Whether it is possible to move matter through the ether without disturbing it forms an entirely different question, to which no definite answer can be given at present. We can, in any case, fall back upon the more abstract theory of electrification moving through the ether, without having conductors to interfere.

Theory of the relative Motion of Magnetification and the Medium.

§ 165. From the preceding, relating to the magnetic effects produced by moving the medium bodily past stationary electrification; or equivalently, by moving the electrification the other way through the medium; or, more generally, by relative motion of the sources of displacement and the medium supporting it, we may readily deduce the corresponding results when the sources are of the flux induction, instead of displacement. Thus, in (42), (43), do away with the convective electric current ρu, and substitute convective magnetic current σu, where σ is volume-density of magnetification (suppositional). Our circuital equations are then

$$\operatorname{curl} (\mathbf{H} - \mathbf{h}) = cp\mathbf{E}, \quad \cdots \quad (55)$$

$$\operatorname{curl} (\mathbf{e} - \mathbf{E}) = \mu p\mathbf{H} + \sigma u, \quad \cdots \quad (56)$$

with the auxiliary equations of motional electric and magnetic force

$$\mathbf{e} = V\mathbf{wB}, \qquad \mathbf{h} = V\mathbf{Dw}, \quad \cdots \quad (57)$$

as before, where $\mathbf{w}$ is the velocity of the medium, and with the further auxiliary conditions that the divergence of the displacement is zero, whilst that of the induction is σ. Now we really do not need to employ the formal mathematics. Knowing the results relating to the motion of electrification, we may infer the corresponding ones concerning the motion of magnetification, by making use of the analogies between the electric and magnetic sides of electromagnetism, translating results in the appropriate manner. But as our object is to illustrate the working of vectors as well as electromagnetic principles, it will not be desirable to leave out the mathematics entirely, especially as the safe carrying out of the analogies requires that they should be thoroughly understood first, which requires some practice.

First, if we convert the problem to one of stationary waves, by keeping the sources at rest, and letting only the medium move bodily past them, we have, when the stationary state of induction and displacement is reached, disappearance of the right members of (55) and (56). Then (56) gives, when united with the first of (57),

$$\text{curl } \mathbf{E} = \text{curl } \mathbf{e} = \text{curl } V\mathbf{wB}. \quad . \quad . \quad (58)$$

In the result for a moving point-charge, curl $\mathbf{E}$ is perpendicular to $\mathbf{w}$, so now we may write

$$\mathbf{E} = V\mathbf{wB} = V\mathbf{Bu}, \quad . \quad . \quad . \quad . \quad (59)$$

where $\mathbf{w} = -\mathbf{u}$, because, similarly, curl $\mathbf{H}$ will be perpendicular to $\mathbf{w}$, and so make $V\mathbf{Bu}$ circuital. Here $\mathbf{u}$ is the equivalent velocity of the sources, when the medium is stationary. Thus $\mathbf{E}$, and therefore $\mathbf{D}$, are fully known in terms of $\mathbf{B}$. Comparing with (46), we see that the electric force set up when a magnetic charge moves is related to the induction from the charge in the same way as the magnetic force set up when an electric charge moves is related to the displacement from it, with, however, a change of sign, or reversal of direction.

Using the second of (57) in the other circuital equation (55) produces

$$\text{curl } (\mathbf{H} - V\mathbf{uD}) = 0 = \text{curl } \mathbf{f}, \text{ say}, \quad . \quad . \quad (60)$$

where, by (59),

$$\mathbf{f} = \mathbf{H} - cV\mathbf{u}V\mathbf{Bu} = \mathbf{H} + \mu cV\mathbf{u}V\mathbf{uH}, \quad . \quad . \quad (61)$$

or, if $\mathbf{u}$ is parallel to $\mathbf{k}$,

$$\mathbf{f} = \mathbf{H} + \frac{u^2}{v^2}\mathrm{V}\mathbf{k}\mathrm{V}\mathbf{k}\mathbf{H} = \mathbf{H} + \frac{u^2}{v^2}(\mathbf{k}\mathrm{H}_3 - \mathbf{H})$$

$$= (1 - u^2/v^2)\mathbf{i}\mathrm{H}_1 + (1 - u^2/v^2)\mathbf{j}\mathrm{H}_2 + \mathbf{k}\mathrm{H}_3; \quad \text{. .} \quad (62)$$

from which we see that the vector $\mathbf{F}$ derived from $\mathbf{f}$ by $\mathbf{f} = (1 - u^2/v^2)\mathbf{F}$ has no curl, or is derived from a scalar potential Ω thus, $\mathbf{F} = -\nabla\Omega$, and that the induction is given by

$$\mathrm{B}_1 = -\mu\frac{d\Omega}{dx}, \quad \mathrm{B}_2 = -\mu\frac{d\Omega}{dy}, \quad \mathrm{B}_3 = -\mu(1 - u^2/v^2)\frac{d\Omega}{dz}. \quad (63)$$

That is, the induction distributes itself in the same way as if the medium were at rest, but had its inductivity reduced from μ to $\mu(1 - u^2/v^2)$ in the line of motion. In fact, the theory of the effects produced by relative motion of magnetic sources and the medium is essentially the same as that of moving electric sources ; translating displacement in the latter case to induction in the former, permittivity to inductivity, electric potential (real or fictitious, as the case may be) to magnetic potential, and the magnetic induction which accompanies moving electrification to electric displacement set up by moving magnetification, remembering, however, the change of sign.

We know, therefore, the result of the steady rectilinear motion of any distribution of magnetic sources. If the same motion is common to all the elementary sources, we reduce the problem to that of magnetic eolotropy without relative motion of the medium and sources, and there is a definite system of surfaces of equilibrium ($\Omega = $ constant) where we may, if we please, terminate the electric and magnetic fields, thereby substituting new arrangements of magnetic sources for the old. In this case it is most convenient to keep the sources at rest, and move the medium alone.

But should the relative motion of source and medium not be the same for every source, then, since we cannot move the medium bodily more than one way at a time, we may imagine it to be stationary, and obtain the effect due to the motion of the magnetification by superimposing the separate effects of the elementary sources, which are known by the above.

Theory of the relative Motion of Magnetisation and the Medium. Increased Induction as well as Eolotropic Disturbance.

§ 166. There is, however, no such thing as magnetification, the magnetic analogue of electrification. The induction is always circuital. If it were not circuital, we should have unipolar magnets. We must, therefore, somewhat modify the conditions assumed to prevail in the above, in order to come closer to reality. Consider, therefore, instead of magnetification, a distribution of intrinsic magnetisation. This quantity is $I = \mu h_0$, where h_0 is the equivalent intrinsic magnetic force. Abolish σ in (56), and introduce h_0 in (55). The circuital equations are now

$$\text{curl } (H - VDw - h_0) = cpE, \quad . \quad . \quad . \quad (64)$$

$$\text{curl } (VwB - E) = \mu pH, \quad . \quad . \quad . \quad (65)$$

where we have introduced the motional forces (57). In the steady state, the right members vanish as before. Now assume that

$$E = VwB, \quad . \quad . \quad . \quad . \quad . \quad (66)$$

so that the theory is unchanged so far, the displacement depending on the induction and velocity of the medium in the same way as when the source was magnetification. We shall see the limitation of application later. Next put (66) in (64) and we find, if the fraction $1 - u^2/v^2$ be denoted by s,

$$\text{curl } f = \text{curl } h_0, \quad . \quad . \quad . \quad . \quad (67)$$

$$\text{curl } F = \text{curl } h_0/s, \quad . \quad . \quad . \quad (68)$$

where f and F are the same vectors (in terms of H) as in the last case, such that $sF = f$. We see that the vector F has now definite curl, and that the induction is derived from F by

$$B_1 = \mu F_1, \qquad B_2 = \mu F_2, \qquad B_3 = \mu s F_3; \quad . \quad (69)$$

that is to say $B = \lambda F$, where λ is the inductivity operator whose scalar principals are μ, μ, and μs. We therefore have to find the steady state from these complete connections,

$$\text{div } B = 0, \qquad B = \lambda F, \qquad \text{curl } F = \text{curl } h_0/s . \quad (70)$$

That is, the induction is the same as in a stationary eolotropic medium (according to λ), provided the intensity of the intrinsic magnetic force be increased from h_0 to h_0/s. This is an important point.

We know that there are certain respects in which the theory of induction due to intrinsic magnetic force is identical with that due to magnetification. For instance, the steady induction outside a magnet due to h_0 is the same as that due to a distribution of σ, measured by the convergence of the intrinsic magnetisation. From this we might hastily conclude that the disturbance from the steady distribution produced by moving the medium would be the same for h_0 as for the equivalent σ, provided we keep outside the region of magnetisation. But the above investigation only partly confirms this conclusion. It shows a likeness and a difference. The likeness is in the eolotropic peculiarity brought in by the motion. In both cases we may do away with the motion provided we simultaneously reduce the inductivity parallel to the (abolished) motion from μ to $s\mu$, so as to cause the induction to retreat from this direction and concentrate itself transversely. The difference is in the reckoning of the strength of sources. In the case of magnetification (as of electrification) we do not alter the strength of the source. But in the case of magnetisation we do, or, at any rate, produce an equivalent result. For, along with the reduction of inductivity in the direction of motion, we require to increase the intensity of the intrinsic magnetic force from h_0 to h_0/s. Remember that s is a proper fraction, going from unity to zero as the speed of motion increases from 0 to v, the speed of propagation of disturbances.

To exemplify this, put

$$\mathbf{F} = h_0/s - \nabla\Omega, \quad \ldots \quad \ldots \quad (71)$$

as we see we may do, by the third of (70). Using this in the second and first of (70) we obtain

$$\operatorname{div} \lambda\nabla\Omega = \operatorname{div} \lambda h_0/s, \quad \ldots \quad \ldots \quad (72)$$

the equation of the potential Ω. Or, in terms of Cartesians,

$$\mu(\nabla_1^2 + \nabla_2^2 + s\nabla_3^2)\Omega = \mu s^{-1}(\nabla_1 h_{01} + \nabla_2 h_{02} + s\nabla_3 h_{03}) . \ (73)$$

The equivalent magnetification is the convergence of $\lambda h_0/s$, not of μh_0 the real intensity of magnetisation, nor yet of $\mu h_0/s$, the same increased in a constant ratio.

There is, therefore, a remarkable difference between the two cases of motion of the medium parallel to and transverse to the lines of real magnetisation. Thus, first let h_0 be parallel to w or k. Then the s^{-1} outside the brackets in (73) cancels the s inside, so that the right member becomes simply $\mu \nabla_3 h_{03}$, where h_{03} is now the tensor of h_0. That is, the effective magnetification is the convergence of the real magnetisation, just as when the medium is at rest. But if h_0 is perpendicular to k, say parallel to i, the right member of (73) becomes $\mu s^{-1} \nabla_1 h_{01}$, or the effective magnetification is s^{-1} times the convergence of the magnetisation, and is therefore increased.

Suppose, for example, our magnet is a straight filamentary magnet. It has two poles, of strength m and $-m$ say, so that induction to the amount m diverges from one and converges to the other pole, whilst continuity is made between them in the filament itself. Now let the medium move past the filament, the direction of motion being parallel to it. Then the strength of the poles is unchanged, but the induction outside is diverted laterally by the effective reduced inductivity parallel to the filament. This case, then, resembles that of a pair of oppositely signed point-charges, if we keep outside the filament. In the limit, therefore, when the speed is raised to v, we have a pair of parallel plane induction-sheets whose cores are joined by the straight filament of induction, to make continuity.

For (66) to be valid, B should be such as to make E circuital. This requires that w and curl B should be perpendicular to one another. They are perpendicular in the example just mentioned, on account of the symmetry of B with respect to the axis or line of motion. But if the same filament be held transversely across the line of motion the property stated is no longer true, so that (66) is not true, and this will, to an unknown extent, upset the later equations. We can, however, still employ them provisionally to obtain a first approximation to the results. On this understanding, then, the strength of the poles (effective) is made m/s and $-m/s$, with unimpaired effective inductivity parallel to the filament, whilst that perpendicular to the filament is reduced, as before. So there is now an increased total induction as well as concentration about the plane through the filament perpendicular to the motion.

We may, however, arrange matters in such a way that the equation (66) shall be still valid when the line of motion is perpendicular to the magnetisation, and so obtain an exact solution showing the increased induction, that is, exact in the absence of working errors. Let the region of magnetisation be confined between two infinite parallel planes, and the magnetisation $I = \mu h_0$ be uniformly distributed, parallel to the boundaries. We thus do away with the poles. Now when the medium is at rest the induction is $B = I$ within the plate (which may be of any thickness), and zero outside, whilst there is no displacement. The intrinsic magnetic force h_0 produces the greatest effect that it can produce unaided. But if the medium be made to move steadily straight across the magnetised region, then, after certain transient effects have passed away (which may be readily calculated, because they form simply plane electromagnetic waves), although there will still be no induction (or displacement) outside the plate, that within it will be increased (without change of direction) from I to I/s. That is, the motional magnetic force comes in to assist the intrinsic magnetic force. Here we have the explanation of the previous result relating to the increased effective strength of poles.

Along with this increased induction, there will be electric force, according to (66). It is entirely within the magnetised region. As the speed increases up to v, the induction and the accompanying displacement in the plate go up to infinity.

The theory of an electrised plate is similar. The displacement due to intrinsic electrisation $J = ce_0$, where e_0 is the equivalent electric force, the electrisation being uniform and parallel to the sides of the plate, will be increased from $D = J$ to $D = J/s$ by setting the medium moving straight through it, it being now the motional electric force that assists the intrinsic. The accompanying magnetic force is accordingly given by

$$H = h = VDw. \quad . \quad . \quad . \quad . \quad . \quad (74)$$

Outside the plate there is no disturbance in the steady state.

It will, of course, be understood that if, instead of suppositional impressed or intrinsic forces in a uniform medium, we employ actual material plates, magnetised or electrised as the case may be, they must, in the first place, be non-conductors, and next, they must not interfere with the supposed uniform motion of the medium.   In short, the reservation is similar to that mentioned in connection with moving charged conductors. Of course, in our present case, the plates may be moved, whilst the medium is supposed to be at rest.

It will be readily seen that the full investigation of the effects of moving practicable electromagnetic arrangements of conductors presents considerable difficulties.   We can, however, get some information relating to the motion of a linear circuit.

Theory of the Relative Motion of Electric Currents and the Medium.

§ 167. Thus, let us first abolish the intrinsic magnetisation of § 166, and substitute equivalent electric current.   Here, by equivalent electric current, we mean a distribution of electric current which produces the same induction as the intrinsic magnetisation ; so that if both were to exist together, the induction would be everywhere doubled ; and then, if either of them were negatived, the resulting induction would be nil. That this is possible, whatever may be the distribution of inductivity, is a very remarkable property.   The induction due to magnetisation is conditioned solely by the curl of the intrinsic magnetic force, and the "poles" are essentially quite a secondary matter.   That is, we may vary the poles as we like, provided we do not alter the curl of the intrinsic force, without affecting the induction or the associated energy.

The question then presents itself whether this equivalence continues to hold good when the medium is set in motion past the stationary magnetisation or electric current respectively. In the circuital equation (64), put $h_0 = 0$, and introduce a term C_0 on the right side, producing

$$\text{curl } (H - VDw) = C_0 + cpE, \quad . \quad . \quad . \quad (75)$$

expressing the first circuital law ; whilst (65), or

$$\text{curl } (VwB - E) = \mu pH \quad . \quad . \quad . \quad . \quad (76)$$

expresses the second.   Now (75) only differs from (64) in the substitution of C_0 here, for curl h_0 there.   That is, the new C_0 and the old curl h_0 are equivalent.   We may, therefore,

dismiss the idea of magnetisation, and let the source of the
induction be any distribution of intrinsic electric current C_0.
With it is associated a certain distribution of induction, identical with that due to any distribution of intrinsic magnetisation
for which we have curl $h_0 = C_0$, and the equivalence persists
when the medium is moving, or when the sources are moving.
Corresponding to (70) we shall have

$$\text{div } B = 0, \qquad B = \lambda F, \qquad \text{curl } F = C_0/s, \quad . \quad (77)$$

for the determination of the induction in the eolotropic manner
previously pursued, when the steady state of affairs is reached;
subject also to the previous reservation that w and curl B are
perpendicular. That is to say, the induction is affected by the
motion in the same way as if the medium had its inductivity
decreased from μ to μs in the direction of motion, without any
change transversely, whilst at the same time the source C_0 is
effectively increased to C_0/s.

Equations (77) are suitable when the electric currents are in
planes perpendicular to the direction of motion. For instance,
let there be (to take a very easy example) two parallel plane
sheets of electric current of surface-density C_0 and $-C_0$ respectively. The induction between them, which is $B = \mu C_0$ when
at rest, becomes $\mu C_0/s$ when they are in motion with velocity u
perpendicular to their planes. Here s is the fraction $(1 - u^2/v^2)$
as before, and the electric force accompanying the changed in
duction is simply the motional electric force.

There is another way of making the eolotropic comparison,
namely, by employing the vector f instead of F. Then, instead
of (77) we shall have

$$\text{div } B = 0, \qquad B = (\lambda/s)\,f, \qquad \text{curl } f = C_0. \quad . \quad (78)$$

In this way of looking at the matter we regard C_0 as suffering
no change of effective strength, whilst the eolotropic operator
λ/s is such as to indicate increased inductivity (from μ to μ/s)
transverse to the motion, and unchanged inductivity parallel
to it. The vector f is the excess of the magnetic force H of
the flux over the motional magnetic force. Inasmuch as this
way emphasises the fact that the intrinsic sources are really
constant, it is not without its advantages. But the idea of
increased transverse inductivity is rather an unmanageable one
in general, especially in the case of moving electrification.

If the sources and the medium have a common uniform translational motion, it may be seen from the general circuital equations that the steady distribution of the fluxes with respect to their sources is unaffected by the motion. That is, if we travel with the medium there is no change observable. This applies in the case of circuital sources (as $\operatorname{curl} h_0$ above), as well as divergent sources (as of electrification). The naturalness of the result is obvious, when the relativity of motion is remembered.

The General Linear Operator.

§ 168. It is now necessary to leave these special cases of eolotropy for fear of being carried away too far from the main subject, which is, the nature of linear vector operators, in termination of this chapter on vector analysis. Up to the present only the symmetrical operator has been under consideration. Three rectangular axes are concerned, each of which is identified with parallelism of the force and flux, or of a vector and a linear function thereof. Now, if we associate the algebraical ratio of the force to the flux with the three directions of parallelism, we see that the symmetrical linear operator depends upon three vectors. If they were arbitrary, this would involve nine scalar specifications ; but, being coperpendicular, there are really only six independent specifications ; and, moreover, if we transform to any other system of axes (independent or non-coplanar) there can still be no more than six independent data.

But it is easy to see that the general linear vector operator must involve nine scalar specifications, viz., three for each of the three independent axes of reference that may be chosen. Thus, let **D** be *any* linear function of **E**. We must then have, in terms of the **i**, **j**, **k** components, the following set of equations :—

$$
\left.
\begin{aligned}
D_1 &= c_{11}E_1 + c_{12}E_2 + c_{13}E_3, \\
D_2 &= c_{21}E_1 + c_{22}E_2 + c_{23}E_3, \\
D_3 &= c_{31}E_1 + c_{32}E_2 + c_{33}E_3,
\end{aligned}
\right\} \quad . \quad . \quad . \quad (79)
$$

where the nine c's may have any values we please. The axes of reference are perpendicular only for convenience ; any set of three independent axes may be used, with nine properly deter-

mined c's to match. The relation between the vectors **D** and
E, which is fully exhibited in (79), may be conveniently
symbolised by

$$D = c\mathbf{E}, \quad \cdots \quad (80)$$

where c is the general linear operator.

If we exchange c_{12} and c_{21}, &c., in the set (79) we shall
usually change **D**, of course. Let it become **D**′. It is then
fully given by the set

$$\begin{aligned}
\mathbf{D}'_1 &= c_{11}\mathbf{E}_1 + c_{21}\mathbf{E}_2 + c_{31}\mathbf{E}_3, \\
\mathbf{D}'_2 &= c_{12}\mathbf{E}_1 + c_{22}\mathbf{E}_2 + c_{32}\mathbf{E}_3, \\
\mathbf{D}'_3 &= c_{13}\mathbf{E}_1 + c_{23}\mathbf{E}_2 + c_{33}\mathbf{E}_3,
\end{aligned} \quad \cdots \quad (81)$$

and these relations between **D**′ and **E** may be symbolised, like
(80), by the single equation

$$\mathbf{D}' = c'\mathbf{E}. \quad \cdots \quad (82)$$

The manipulation of sets of equations like (79) and (81) is
lengthy and laborious. By the use of the linear operators,
however, with proper attention to the laws governing them,
the vectors may be manipulated with facility. Thus, to give
the first and easiest example that presents itself, we may
instantly turn **AB** to **A**cc^{-1}**B**. This is obvious enough, inasmuch as the operation indicated by c^{-1} will be precisely cancelled by the operation c, so that $cc^{-1}\mathbf{B} = \mathbf{B}$. But there is much
more in it than that. For **A**cc^{-1}**B** is not merely the scalar product of the vectors **A** and cc^{-1}**B**, that is, of **A** and **B**, but is
the scalar product of the vectors **A**c and c^{-1}**B**, which are quite
different. The vector **A**c is the same as the vector c'**A**, whilst
c^{-1}**B** is the vector which, when operated upon by c, gives the
vector **B**.

The constituents of the inverse operator c^{-1} may be found
by solution of (79). Since **D** is a linear function of **E**, it follows
that **E** is a linear function of **D**; or the **E**'s may be expressed
in terms of the **D**'s by means of a set of equations like (79),
with new coefficients instead of the c's there. Similarly as
regards the inverse operator c'^{-1} belonging to (81) and (82).

The vector formed by taking half the sum of the vectors **D**
and **D**′ is a symmetrical function of **E**, say

$$\tfrac{1}{2}(\mathbf{D} + \mathbf{D}') = \lambda\mathbf{E}. \quad \cdots \quad (83)$$

The constituents of λ with two equal suffixes, namely, c_{11}, c_{22}, c_{33}, are the same as those of c and of c'. But the other constituents are half the sum of the corresponding ones of c and c'; for instance $\frac{1}{2}(c_{12} + c_{21})$ in place of c_{12} and c_{21}. From this we see that the operator λ is its own conjugate, or is self-conjugate, making $\lambda\mathbf{E}$ and $\lambda'\mathbf{E}$ identical.

On the other hand, the vector formed by taking half the difference of $\mathbf{D}$ and $\mathbf{D}'$ is a simple vector product. By inspection of (79) and (81), and remembering the structure of a vector product, we may see that

$$\tfrac{1}{2}(\mathbf{D} - \mathbf{D}') = V a\mathbf{E}, \quad \ldots \ldots \quad (84)$$

where a is the vector given by

$$a = \tfrac{1}{2}(c_{32} - c_{23})\mathbf{i} + \tfrac{1}{2}(c_{13} - c_{31})\mathbf{j} + \tfrac{1}{2}(c_{21} - c_{12})\mathbf{k}. \quad (85)$$

We see, therefore, that any linear function, if it be not already of the symmetrical kind, may be represented as the sum of a symmetrical function and of a vector product. Thus, by addition and subtraction of (83) and (84) we obtain

$$c\mathbf{E} = \mathbf{D} = \lambda\mathbf{E} + V a\mathbf{E}, \quad \ldots \ldots \quad (86)$$

$$c'\mathbf{E} = \mathbf{D}' = \lambda\mathbf{E} - V a\mathbf{E}. \quad \ldots \ldots \quad (87)$$

That is, in terms of the operators alone,

$$c = \lambda + V a, \quad \ldots \ldots \quad (88)$$

$$c' = \lambda - V a, \quad \ldots \ldots \quad (89)$$

where c (and therefore c') is general, whilst λ is symmetrical.

The vector a is quite intrinsic, and independent of axes of reference. The symmetrical operator λ involves six scalars, the vector a three more, thus making up the full nine.

Notice, by (86), (87), that

$$\mathbf{E}c\mathbf{E} = \mathbf{E}\lambda\mathbf{E} = \mathbf{E}c'\mathbf{E}, \quad \ldots \ldots \quad (90)$$

or the scalar product of the force and the flux does not involve the vector a at all.

The Dyadical Structure of Linear Operators.

§ 169. Now go back to the equations (79). Observe that the right members, being the sums of three pairs of simple products, are themselves scalar products. Thus, let three new vectors defined by

$$\left.\begin{aligned}
\mathbf{c}_1 &= \mathbf{i}c_{11} + \mathbf{j}c_{12} + \mathbf{k}c_{13}, \\
\mathbf{c}_2 &= \mathbf{i}c_{21} + \mathbf{j}c_{22} + \mathbf{k}c_{23}, \\
\mathbf{c}_3 &= \mathbf{i}c_{31} + \mathbf{j}c_{32} + \mathbf{k}c_{33},
\end{aligned}\right\} \quad \ldots \quad (91)$$

be introduced. Then we see that equations (79) are simply

$$D_1 = \mathbf{c}_1\mathbf{E}, \qquad D_2 = \mathbf{c}_2\mathbf{E}, \qquad D_3 = \mathbf{c}_3\mathbf{E}, \ . \ . \ (92)$$

and, therefore, by combining them, we produce the *one* vector equation

$$\mathbf{D} = \mathbf{i}.\mathbf{c}_1\mathbf{E} + \mathbf{j}.\mathbf{c}_2\mathbf{E} + \mathbf{k}.\mathbf{c}_3\mathbf{E}, \quad \ldots \quad (93)$$

thus proving that every linear operator may be exhibited in the dyadical form

$$c = \mathbf{i}.\mathbf{c}_1 + \mathbf{j}.\mathbf{c}_2 + \mathbf{k}.\mathbf{c}_3, \quad \ldots \ . \ (94)$$

whether it be symmetrical or not. The conjugate operator is, similarly,

$$c' = \mathbf{i}.\mathbf{c}'_1 + \mathbf{j}.\mathbf{c}'_2 + \mathbf{k}.\mathbf{c}'_3, \quad \ldots \ . \ (95)$$

where the accented vectors may be obtained from the un-accented given in (91) by changing c_{12} to c_{21}, &c., just as $\mathbf{D}'$ was got from $\mathbf{D}$, in fact.

It is now easily to be proved that

$$c\mathbf{E} = \mathbf{E}c', \qquad\qquad \mathbf{E}c = c'\mathbf{E}, \ . \ . \ . \ (96)$$

that is, if we wish to remove an operator from before to behind a vector, we may do so by simply turning the operator to its conjugate; that is, by putting on the accent, or by removing it if already there. It will, of course, be understood that when a vector $\mathbf{E}$ follows c, and we use the dyadical form of c, as in (94), it is only the three second vectors that unite with $\mathbf{E}$; whereas, if $\mathbf{E}$ precedes c, it is the first vectors in c that unite with it. Thus :—

$$\mathbf{E}c = \mathbf{E}\mathbf{i}.\mathbf{c}_1 + \mathbf{E}\mathbf{j}.\mathbf{c}_2 + \mathbf{E}\mathbf{k}.\mathbf{c}_3, \quad \ldots \ . \ (97)$$

$$c'\mathbf{E} = \mathbf{i}.\mathbf{c}'_1\mathbf{E} + \mathbf{j}.\mathbf{c}'_2\mathbf{E} + \mathbf{k}.\mathbf{c}'_3\mathbf{E}. \quad \ldots \ . \ (98)$$

These are identical vectors always. In the symmetrical case, the accents may be dropped, of course; then $\mathbf{E}\lambda = \lambda\mathbf{E}$. It will be seen that the above way of writing and working dyadics fits in precisely with all the previous notation employed in the vector algebra. The dots in (94) are merely

separators, not signs of multiplication. When we put on a
vector, before or behind, it unites with the vectors it is *not*
separated from by dots; we therefore obtain the forms (97),
(98), in agreement with the principles of notation described in
the early part of this chapter.

We have already pointed out that $E\lambda F$, where λ is symmetrical, may be regarded as the scalar product of E and λF,
or of $E\lambda$ (the same as λE), and F. The corresponding general
property is the same; that is, c being general, EcF is the
scalar product of Ec and F, or of E and cF. But Ec is the
same as $c'E$, and cF the same as Fc', so we have

$$EcF = c'EF = EFc' = Fc'E = FEc. \quad . \quad . \quad (99)$$

The operator c must be united with one or other of the two
vectors E and F, but it is immaterial which it is, provided we
always change to the conjugate properly, as exemplified.

The same applies when there are many linear operators.
Let there be three, a, b, c. Then we have, to illustrate how
the transformations are made,

$$cE[abF] = Ec'[abF] = Ec'[a(Fb')] = Ec'[Fb'a'], \quad . \quad (100)$$

and so on. Here the brackets are introduced to show the
association of the vectors and the operators.

We may also introduce anywhere a new operator, accompanied, of course, by the reciprocal operator; thus

$$cE[aF] = cE[ab^{-1}bF] = cE[ab^{-1}(Fb')] = \&c. \quad . \quad (101)$$

The above will be sufficient to show the great power gained
by the use of the operators, enabling one to do almost at sight
algebraical work which would, in Cartesians, cover pages.

Hamilton's Theorem.

§ 170. The following little theorem, due to Hamilton, will
serve to illustrate the working of vectors, as well as another
purpose which will appear later.

Let m and n be a pair of vectors, then their vector product
Vmn is perpendicular to both, by our definition of a vector product. That is,

$$0 = mVmn, \qquad 0 = nVmn, \quad . \quad . \quad . \quad (102)$$

by our definition of a scalar product, and the parallelepipedal property. Now introduce cc^{-1} betwen the V's and the vectors preceding them in (102). Thus

$$0 = mcc^{-1}Vmn, \qquad 0 = ncc^{-1}Vmn. \quad . \quad . \quad (103)$$

These assert (and we cannot but believe it) that the vector $c^{-1}Vmn$ is perpendicular to the vectors mc and nc, or $c'm$ and $c'n$; it is, therefore, parallel to their vector product. The last statement is expressed by

$$xc^{-1}Vmn = Vc'mc'n, \quad . \quad . \quad . \quad (104)$$

where x is an unknown (or so far undetermined) scalar. Or, operating on both sides by c,

$$xVmn = cVc'mc'n. \quad . \quad . \quad . \quad (105)$$

To find the value of x, we have merely to multiply (105) by a third vector, say 1. For this gives

$$x1Vmn = 1cVc'mc'n = c'1Vc'mc'n, \quad . \quad . \quad (106)$$

which gives the value of x explicitly, namely,

$$x = \frac{c'1Vc'mc'n}{1Vmn}, \qquad . \quad . \quad . \quad (107)$$

to be used in (105), which, with x thus settled, is the statement of the little (but important) theorem. The quantity x appears to depend upon 1, m, n. These vectors, however, enter into both the numerator and denominator in such a way that they can be wholly eliminated from x. This may be seen by expanding the numerator and denominator. In fact, x is a pure constant, depending upon the operator c alone.

In the symmetrical case we may find its value by taking 1. m, n to be i, j, k, and the latter to be the principal axes. Thus

$$c'1 = ci = c_1i, \qquad c'm = c_2j, \qquad c'n = c_3k,$$

if c_1, c_2, c_3 are the principal scalar c's. So, by (107),

$$x = \frac{c_1iVc_2jc_3k}{iVjk} = c_1c_2c_3, \quad . \quad . \quad . \quad (108)$$

the continued product of the principal c's.

In the general case we may put, as we know,

$$c = \lambda + Va.$$

Using this in (107), on the understanding that c_1, c_2, c_3 are the principals of λ now, we shall get, similarly,

$$x = c_1 c_2 c_3 + \mathbf{a}\lambda\mathbf{a}, \quad . \quad . \quad . \quad . \quad (109)$$

which it will be a useful exercise to verify.

Since c is any linear operator, (105) remains true when for c we substitute its conjugate c', producing

$$x' \mathrm{V} mn = c' \mathrm{V} cmcn, \quad . \quad . \quad . \quad . \quad (110)$$

where x' is got from the expression for x by changing c' to its conjugate c. But c and c' only differ in the changed sign of $\mathbf{a}$, as in (88), (89). On the other hand (109) contains a quadratically, so that a reversal of its sign makes no difference. Therefore, x' is the same as x, which might not have been anticipated at the beginning of the evaluation.

Hamilton's Cubic and the Invariants concerned.

§ 171. The reader of Prof. Tait's profound treatise on Quaternions will probably stick at three places in particular, to say nothing of the numerous minor sticking-points that present themselves in all mathematical works of any value, and which may be readily overcome by the reader if he be a real student as well. First, there is the fundamental Chapter II., wherein the rules for the multiplication of vectors are made to depend upon the difficult mathematics of spherical conics, combined with versors, quaternions and metaphysics. Next, Chapter IV., where the reader may be puzzled to find out why the usual simple notion of differentials is departed from, although the departure is said to be obligatory. Thirdly, Chapter V., where the reader will be stopped nearly at the beginning by a rather formidable investigation of Hamilton's cubic. Only when the student is well acquainted with the nature of linear operators and how to work them can he tackle such an investigation. He should, therefore, pass on, to obtain the necessary experience. On then returning to the cubic he may find the investigation not so difficult after all, especially if it be simplified by some changes calculated to bring out the main points of the work more plainly.

The reader is led to think that the object of the investigation is to invert a linear operator—that is, given $\mathbf{D} = c\mathbf{E}$, to find

U

$\mathbf{E} = c^{-1}\mathbf{D}$. But if this were all, it would be a remarkable example of how not to do it. For the inversion of a linear operator can be easily effected by other far simpler and more natural means. The mere inversion is nothing. It is the cubic equation itself that is the real goal. The process of reaching it is simplified by the omission of inverse operations. (It is also simplified by not introducing the auxiliary function called χ.)

The fundamental cubic is derived from equation (105), or the equivalent equation (110) last investigated. I should remark that this equation is frequently useful in advanced vector-analysis as a transformation formula, turning $\mathbf{V}c\mathbf{m}c\mathbf{n}$ to a function of $\mathbf{V}\mathbf{m}\mathbf{n}$. Now use the form (110), or

$$x\mathbf{V}\mathbf{m}\mathbf{n} = c'\mathbf{V}c\mathbf{m}c\mathbf{n}, \quad \ldots \quad \ldots \quad (111)$$

where x is a known function of c, viz.,

$$x = \frac{c\mathbf{1}\mathbf{V}c\mathbf{m}c\mathbf{n}}{\mathbf{1}\mathbf{V}\mathbf{m}\mathbf{n}}. \quad \ldots \quad \ldots \quad (112)$$

Here c is any linear operator. Now if g be a constant, $g\mathbf{m}$ is a linear function of $\mathbf{m}$, and therefore $(c - g)\mathbf{m}$ is any linear function of $\mathbf{m}$; that is, $c - g$ is a general linear operator. Equation (111) therefore remains true when we substitute $c - g$ for c in it, not forgetting to make the change in x, and also the equivalent change in c', viz., from c' to $c' - g$. Equation (111) then becomes

$$x_g\mathbf{V}\mathbf{m}\mathbf{n} = (c' - g)\mathbf{V}(c - g)\mathbf{m}(c - g)\mathbf{n}, \qquad (113)$$

where x_g is what x becomes by the change, that is, by (112),

$$x_g = \frac{(c - g)\mathbf{1}\mathbf{V}(c - g)\mathbf{m}(c - g)\mathbf{n}}{\mathbf{1}\mathbf{V}\mathbf{m}\mathbf{n}}. \quad \ldots \quad (114)$$

Remembering that g is a constant, we may readily expand (113) to the form

$$(x - x_1 g + x_2 g^2 - g^3)\mathbf{V}\mathbf{m}\mathbf{n}$$
$$= c'\mathbf{V}c\mathbf{m}c\mathbf{n} - g(\mathbf{V}c\mathbf{m}c\mathbf{n} + c'\mathbf{V}\mathbf{m}c\mathbf{n} + c'\mathbf{V}c\mathbf{m}\mathbf{n})$$
$$- g^3\mathbf{V}\mathbf{m}\mathbf{n} + g^2(c'\mathbf{V}\mathbf{m}\mathbf{n} + \mathbf{V}\mathbf{m}c\mathbf{n} + \mathbf{V}c\mathbf{m}\mathbf{n}), \quad . \quad . \quad (115)$$

where the cubic function of g on the left side is the expansion of x_g. The new coefficients x_1 and x_2 are, of course, to be found by expanding (114), but we do not want their values immediately. Since x is a function of c only, x_g is a function of c and g only, so that x_1 and x_2 are functions of c only.

Now since (111) is an identity, so is (115); and since g may have any value, the last equation must be identically true for every power of g concerned, taken one at a time. That is, it splits into four identities. Now, on comparing the coefficients of g^0 on the left and right sides, we obtain (111) again. Similarly, we observe that the coefficients of g^3 are the same. So far, then, we have nothing new. But the g and g^2 terms give

$$x_1 \text{Vmn} = \text{V}cm cn + c'(\text{V}mcn + \text{V}cmn), \quad . \quad . \quad (116)$$

$$x_2 \text{Vmn} = c'\text{Vmn} + (\text{V}mcn + \text{V}cmn), \quad . \quad . \quad (117)$$

which are fresh identities. From these various others may be deduced by elimination or combination. The one we are seeking is obtained by operating on the first by c' and on the second by c'^2, and then subtracting the second result from the first. This eliminates the vector in the brackets and leaves

$$(x_1 c' - x_2 c'^2)\text{Vmn} = c'\text{V}cmcn - c'^3\text{Vmn}, \quad . \quad (118)$$

$$= (x - c'^3)\text{Vmn}, \quad . \quad . \quad (119)$$

where the transition from (118) to (119) is made by using (111) again. Rearranging, we have the final result,

$$(x - x_1 c' + x_2 c'^2 - c'^3)\text{Vmn} = 0, \quad . \quad . \quad (120)$$

which is Hamilton's cubic. Note that Vmn may be any vector we please, and may, therefore, be denoted by a single letter. Observe, also, that the cubic function of c' is of the same form as that of g in the expansion of x.

Since (120) is true for all linear operators, we may write c instead of c', giving

$$x - x_1 c + x_2 c^2 - c^3 = 0. \quad . \quad . \quad . \quad (121)$$

This form is what we should have arrived at had we started from (105) instead of (110). The value of x is the same in either case, and it may be inferred from this that x_1 and x_2 do not suffer any change when we pass from a linear operator to its conjugate.

That the complex function x_g is independent of l, m, n, may be seen by inspecting (114), and observing the parallelepipedal

form of numerator and denominator. For, suppose we alter
l to $l + a\mathbf{m}$, where a is any constant. The addition made to the
numerator vanishes by the parallelepipedal property, and similarly in the denominator. The same invariance obtains when
we change l to $l + a\mathbf{n}$. Also, the change of l to any scalar
multiple of itself alters both the numerator and denominator
in the same ratio. But, unless l, $\mathbf{m}$, $\mathbf{n}$ are coplanar, we may
turn l to *any* vector by adding to it vectors parallel to l, $\mathbf{m}$
and $\mathbf{n}$ of the right size. Therefore, x_g is the same whatever
vector l may be. By the same reasoning $\mathbf{m}$ may be any vector,
and so may $\mathbf{n}$. But l, $\mathbf{m}$, $\mathbf{n}$ should be independent vectors (that
is, not coplanar). This is not because we can suppose that an
actual discontinuity occurs when (for example) l is brought
into the plane $\mathbf{m}$, $\mathbf{n}$, but merely because the expression x_g
assumes an indeterminate form in the coplanar case.

From the invariance of x_g follows that of the three functions
it contains, namely, x, x_1, and x_2. That of x, of course, may be
independently seen by itself, readily enough, but that of the others
is less plain, because on expansion they are found to each
involve three parallelepipedal products in the numerator instead
of only one. Thus, expanding (114) we obtain

$$x_1 = \frac{l V c \mathbf{m} c \mathbf{n} + \mathbf{m} V c \mathbf{n} c l + \mathbf{n} V c l c \mathbf{m}}{l V \mathbf{m} \mathbf{n}}, \quad . \quad . \quad (122)$$

$$x_2 = \frac{c l V \mathbf{m} \mathbf{n} + c \mathbf{m} V \mathbf{n} l + c \mathbf{n} V l \mathbf{m}}{l V \mathbf{m} \mathbf{n}}. \quad . \quad . \quad (123)$$

In these we may, by the above, substitute c' for c, that is,
turn the linear operator to its conjugate. This does not mean
that x_g is independent of the vector $\mathbf{a}$ which comes in when
the operator is not symmetrical, but that it only involves $\mathbf{a}$
quadratically.

The value of x being

$$x = c_1 c_2 c_3 + \mathbf{a} c \mathbf{a}, \quad . \quad . \quad . \quad . \quad (124)$$

expressed in terms of the principal constants of the symmetrical
operator and the rotation vector, we may obtain the reduced
values of x_1 and x_2 from it directly instead of from their general
expressions. Thus: turn c to $c - g$ (in 124), making it

$$x_g = (c_1 - g)(c_2 - g)(c_3 - g) + \mathbf{a}(c - g)\mathbf{a}. \quad . \quad (125)$$

On expansion, the coefficients of g and g^2 show that

$$x_1 = c_2c_3 + c_3c_1 + c_1c_2 + \mathbf{a}^2, \quad \cdot \ \cdot \ \cdot \quad (126)$$

$$x_2 = c_1 + c_2 + c_3.$$

Thus x_2 is independent of $\mathbf{a}$, whilst x_1 involves its square, as we concluded previously.

When referred to the principal axes, Hamilton's cubic therefore reduces to

$$0 = (c_1c_2c_3 + \mathbf{a}c\mathbf{a}) - (c_1c_2 + c_2c_3 + c_3c_1 + \mathbf{a}^2)c \left. \atop + (c_1 + c_2 + c_3)c^2 - c^3, \right\} \quad (127)$$

and in the symmetrical case of vanishing $\mathbf{a}$ to

$$0 = (c - c_1)(c - c_2)(c - c_3). \quad \cdot \ \cdot \ \cdot \quad (128)$$

Here the axes are coperpendicular. But we may disregard the principal axes altogether, and write the general cubic in the form

$$0 = (c - g_1)(c - g_2)(c - g_3), \quad \cdot \ \cdot \ \cdot \quad (129)$$

where the g's are the roots of $x_g = 0$. If, then, these roots are all real and different, there are three directions of parallelism of $\mathbf{r}$ and $c\mathbf{r}$, or three $\mathbf{r}$'s such that

$$c\mathbf{r}_1 = g_1\mathbf{r}_1, \qquad c\mathbf{r}_2 = g_2\mathbf{r}_2, \qquad c\mathbf{r}_3 = g_3\mathbf{r}_3. \quad (130)$$

Now multiply the first by $\mathbf{r}_2$ and the second by $\mathbf{r}_1$, giving

$$\mathbf{r}_2 c\mathbf{r}_1 = g_1\mathbf{r}_1\mathbf{r}_2, \qquad \mathbf{r}_1 c\mathbf{r}_2 = g_2\mathbf{r}_1\mathbf{r}_2 \quad \cdot \ \cdot \ \cdot \quad (131)$$

The left members are not usually equal, but in the symmetrical case they are, and then the right members are equalised. Then $\mathbf{r}_1\mathbf{r}_2 = 0$, or $\mathbf{r}_1$ is perpendicular to $\mathbf{r}_2$. Similarly $\mathbf{r}_2$ is perpendicular to $\mathbf{r}_3$. This is the case of three mutually perpendicular axes of parallelism of force and flux we started from.

The Inversion of Linear Operators.

§ 172. If we write c^{-1} for c in the cubic (121) we see that $c^{-1}\mathbf{r}$ becomes expressed as a function of $\mathbf{r}$, $c\mathbf{r}$, and $c^2\mathbf{r}$. This is one way of inverting the operator, but a very clumsy way. The simple way is in terms of dyads, and is fully described by saying that if

$$c = \mathbf{a}.\mathbf{l} + \mathbf{b}.\mathbf{m} + \mathbf{c}.\mathbf{n} \quad \cdot \ \cdot \ \cdot \ \cdot \quad (132)$$

is any linear operator in dyadical form, then its reciprocal is

$$c^{-1} = \mathbf{L.A} + \mathbf{M.B} + \mathbf{N.C}, \quad \ldots \quad (133)$$

where $\mathbf{A}$, $\mathbf{B}$, $\mathbf{C}$ is the set complementary to a, b, c, and $\mathbf{L}$, $\mathbf{M}$, $\mathbf{N}$ the set complementary to l, m, n. We have already used these complementary vectors in the early part of this chapter, equations (54) and (55), § 114. We showed by elementary considerations that any vector r could be expressed in terms of any three independent vectors a, b, c by

$$\mathbf{r} = \mathbf{rA.a} + \mathbf{rB.b} + \mathbf{rC.c}, \quad \ldots \quad (134)$$

where the complementary vectors $\mathbf{A}$, $\mathbf{B}$, $\mathbf{C}$ are got from a, b, c by

$$\mathbf{A} = \frac{\mathbf{V}bc}{a\mathbf{V}bc}, \qquad \mathbf{B} = \frac{\mathbf{V}ca}{a\mathbf{V}bc}, \qquad \mathbf{C} = \frac{\mathbf{V}ab}{a\mathbf{V}bc}. \quad (135)$$

These equations serve to define the set complementary to a, b, c. We also showed at the same time that the vector r could be expressed in terms of the complementary set by

$$\mathbf{r} = \mathbf{ra.A} + \mathbf{rb.B} + \mathbf{rc.C}. \quad \ldots \quad (136)$$

To verify, we have merely to multiply (134) by $\mathbf{A}$, $\mathbf{B}$, $\mathbf{C}$ in turn, and (136) by a, b, c in turn.

These equations may be written

$$\mathbf{r} = (\mathbf{a.A} + \mathbf{b.B} + \mathbf{c.C})\mathbf{r}, \quad \ldots \quad (137)$$

$$\mathbf{r} = (\mathbf{A.a} + \mathbf{B.b} + \mathbf{C.c})\mathbf{r}, \quad \ldots \quad (138)$$

showing that the dyadic in the brackets is of a very peculiar kind, inasmuch as its resultant effect on any vector is to reproduce the vector. That is, taken as a whole, and disregarding its detailed functions, the dyadic is equivalent to unity.

Now, suppose it is given that

$$\mathbf{R} = (\mathbf{a.l} + \mathbf{b.m} + \mathbf{c.n})\mathbf{r}, \quad \ldots \quad (139)$$

so that $\mathbf{R}$ is any linear function of r. We know that

$$\mathbf{R} = (\mathbf{a.A} + \mathbf{b.B} + \mathbf{c.C})\mathbf{R}, \quad \ldots \quad (140)$$

and that

$$\mathbf{r} = (\mathbf{L.l} + \mathbf{M.m} + \mathbf{N.n})\mathbf{r}, \quad \ldots \quad (141)$$

by the property of the complementary vectors explained. Comparing (140) with (139), we see that

$$\mathbf{lr} = \mathbf{AR}, \qquad \mathbf{mr} = \mathbf{BR}, \qquad \mathbf{nr} = \mathbf{CR}. \quad \ldots \quad (142)$$

Using these in (141), we convert it to

$$r = (L.A + M.B + N.C)R, \quad . \quad . \quad . \quad (143)$$

comparing which with (139), we see that the inversion of the dyadic has been effected in a simple and neat manner.

Professor Gibbs calls the vectors A, B, C the reciprocals of a, b, c. They have some of the properties of reciprocals. Thus,

$$aA = 1, \quad bB = 1, \quad cC = 1. \quad . \quad . \quad . \quad (144)$$

But it seems to me that the use of the word reciprocal in this manner is open to objection. It is in conflict with the obvious meaning of the recipocal a^{-1} of a vector a, that it is the vector whose tensor is the reciprocal of that of a with unchanged ort (or with ort reversed in the quaternionic system). It will also be observed that Gibbs's reciprocal of a vector depends not upon that vector alone, but upon two others as well. It would seem desirable, therefore, to choose some other name than reciprocal. I have provisionally used the word " complementary " in the above, to avoid confusion with the more natural use of " reciprocal."

Vector Product of a Vector and a Dyadic. The Differentiation of Linear Operators.

§ 173. In connection with the dyadic, it should be remarked that we have only employed them in the manner they usually present themselves in physical mathematics, namely, so as to make scalar products with the vectors they are associated with. But there is also the vector product to be considered in a complete treatment. Thus,

$$V\phi r = a.Vlr + b.Vmr + c.Vnr, \quad . \quad . \quad (145)$$

$$Vr\phi = Vra.l + Vrb.m + Vrc.n. \quad . \quad . \quad (146)$$

Observe that, as before with the scalar products, the vector r only combines with the vectors nearest to it in the dyadic, and not separated from it by the dots. Notice, too, that whereas the scalar product ϕr of the dyadic ϕ and a vector r is a vector ; on the other hand, the vector products $V\phi r$ and $Vr\phi$ are themselves dyadics, and behave as such in union with other vectors. Thus $sV\phi r$ is a vector, and so is $sVr\phi$, the first being got by making scalar products of s with a, b, c, and the second with Vra, &c. But to go further in this direction would be to go beyond the scope of the present treatment.

The differentiation of linear operators must, however, be mentioned, because the process is of frequent occurrence in electromagnetic investigations. If we think only of the differential coefficient of a scalar, that it is its rate of increase with some variable, it might seem at first sight that the differential coefficient of a linear operator was nonsense. But a little consideration will show that it is a perfectly natural, and by no means a difficult conception. Thus, from

$$\mathbf{D} = c\mathbf{E}, \quad \cdots \quad (147)$$

where c is a linear operator, we obtain, by differentiation with respect to a scalar variable, say the time,

$$\dot{\mathbf{D}} = \dot{c}\mathbf{E} + c\dot{\mathbf{E}}. \quad \cdots \quad (148)$$

Here in the second term on the right we suppose that c is constant, and in the first that $\mathbf{E}$ is constant. The meaning of $\dot{c}\mathbf{E}$, then, is the rate of increase of $\mathbf{D}$ when c alone varies. It is the linear operator whose constituents are the rates of increase of the constituents of c. That this is so will be evident on differentiating the set of equations (79). Therefore, if

$$c = a.\mathbf{l} + b.\mathbf{m} + c.\mathbf{n}, \quad \cdots \quad (149)$$

we shall have

$$\dot{c} = \dot{a}.\mathbf{l} + \dot{b}.\mathbf{m} + \dot{c}.\mathbf{n} + a\dot{\mathbf{l}} + b.\dot{\mathbf{m}} + c.\dot{\mathbf{n}}, \quad \cdots \quad (150)$$

the sum of two dyadics. But it may be that our axes of reference are invariable, as for example when

$$c = \mathbf{i}.c_1 + \mathbf{j}.c_2 + \mathbf{k}.c_3, \quad \cdots \quad (151)$$

$\mathbf{i}, \mathbf{j}, \mathbf{k}$ being a fixed set of rectangular orts. Then we have the one dyadic

$$\dot{c} = \mathbf{i}.\dot{c}_1 + \mathbf{j}.\dot{c}_2 + \mathbf{k}.\dot{c}_3, \quad \cdots \quad (152)$$

where, of course, the dots over the $\mathbf{i}$ and $\mathbf{j}$ do not signify any differentiation.

In an isotropic dielectric of variable permittivity the electric stress leads to the force $-\nabla_c(\frac{1}{2}c\mathbf{E}^2)$ per unit volume, where the scalar c alone is differentiated, so that the result is $-\frac{1}{2}\mathbf{E}^2\nabla c$.

But if the dielectric be eolotropic, the corresponding force is $-\nabla_c(\frac{1}{2}\mathbf{E}c\mathbf{E})$. This is equivalent to

$$-\tfrac{1}{2}\left(\mathbf{i}.\mathbf{E}\frac{dc}{dx}\mathbf{E} + \mathbf{j}.\mathbf{E}\frac{dc}{dy}\mathbf{E} + \mathbf{k}.\mathbf{E}\frac{dc}{dz}\mathbf{E}\right), \quad . \quad . \quad (153)$$

where dc/dx, &c., are differential coefficients of c as above explained. It means the same as

$$-(\nabla_\mathrm{D} - \nabla_\mathrm{E})(\tfrac{1}{2}\mathbf{E}\mathbf{D}), \quad . \quad . \quad . \quad (154)$$

where ∇_D means that $\mathbf{D}$ alone, and ∇_E that $\mathbf{E}$ alone is differentiated.

Summary of Method of Vector Analysis.

§ 174. In the last paragraph I came dangerously near to overstepping the imposed limits of my treatment of vectors, which is meant to present the subject merely in the form it assumes in ordinary physical mathematics. If we were to ignore the physical applications, and treat vector algebra as a branch of pure mathematics, regardless of practical limitations, there would be no bounds to the investigation of the subject. It will now be convenient to wind up with a few remarks on the previous, and on the nature and prospects of vector analysis.

Since we live in a world of vectors, an algebra or language of vectors is a positive necessity. At the commencement of this chapter, §§ 97 to 102, I made some general remarks on the nature of cartesian analysis, vector analysis, and quaternions ; and the reader is recommended to read them again from a more advanced point of view. Then, he was supposed to know next to nothing about vectors. Now, although he need not have absorbed all the special applications of vector algebra that have been given since, it may be presumed that he has acquired a general knowledge of the principles of the subject. There is no longer any question as to the desirability and utility of vectorial analysis. The present question is rather as to the form the vector algebra should take. On this point there is likely to be considerable difference of opinion, according to the point of view assumed, whether with regard to physical applications, or abstract mathematical theory. Let us then, to begin with, summarise the leading points in the algebra given above.

First, there is the idea of the vector as a distinct entity. Nearly everyone nowadays knows and appreciates the idea.

And it is a noteworthy fact that ignorant men have long been in advance of the learned about vectors. Ignorant people, like Faraday, naturally think in vectors. They may know nothing of their formal manipulation, but if they think about vectors, they think of them *as* vectors, that is, directed magnitudes. No ignorant man could or would think about the three components of a vector separately, and disconnected from one another. That is a device of learned mathematicians, to enable them to evade vectors. The device is often useful, especially for calculating purposes, but for general purposes of reasoning the manipulation of the scalar components instead of the vector itself is entirely wrong.

In order to facilitate the reading of vector work, the vector has its special type, thus $\mathbf{E}$. This Clarendon, or any very similar neat black type (not block letters), is meant to mark the vector, always the vector, and never anything else, with the obvious exception of headlines that speak for themselves. Also, to economise letters, and ease the strain on the memory, the same letter suitably modified serves for the tensor, the ort, and the three scalar components. Thus E is the tensor, or size, and $\mathbf{E}_1$ the ort (signifying the orientation) so that $\mathbf{E} = E\mathbf{E}_1$; whilst the scalar components referred to rectangular axes are E_1, E_2, E_3. The physical dimensions may be most conveniently merged in the tensor E. Every vector has its species. The one we are most familiar with is the space vector, or straight line joining two points, or more strictly the displacement from one point to another. But all kinds of vector magnitudes are formally similar in having size and ort, and therefore, when considered vectorially, obey the same laws. The properties of the space vector therefore supply us with the rules or laws of vector algebra.

The addition and subtraction of vectors is embodied in the assertion (whose truth is obvious) that the sum of any number of vectors making (when put end to end) a circuit is zero; or, in another form, all the various paths by which we may pass from one point to another are vectorially equivalent, namely, to the vector straight from point to point. Every vector equation therefore expresses this fact. Every term is a vector; and if all the vectors be put on one side, with zero on the other, we express the circuital property; whilst if we have vectors on both sides of the equation, we assert the vectorial equivalence of two paths.

Any vector **A** and its reciprocal A^{-1} have the same ort, and their tensors are reciprocal to one another.

Coming next to combinations of vectors of the nature of products, we find they are of two kinds, scalar and vector. The scalar product of **E** and **F** is denoted by **EF**, and the vector product by **VEF**. Their tensors are $EF \cos\theta$ and $EF \sin\theta$ respectively, θ being the included angle. The scalar product is directionless; the ort of the vector product is that of the normal to the plane of **E** and **F**. The tensor and ort of **VEF** are V_0EF and V_1EF.

The justification for the treatment of the scalar and vector products as fundamental ideas in vector algebra is to be found in the distributive property they possess, thus,

$$(a + b + ...) (c + d + ...) = ac + ad + bc + bd + ..., \qquad . \quad (155)$$
$$V(a + b + ...) (c + d + ...) = Vac + Vad + Vbc + Vbd + ...,(156)$$

which hold good for any number of vectors. In the first equation the manipulation is as in common algebra, in all respects. In the second, in all respects save one; for the reversal of the order of the vectors in a vector product negatives it, thus, $Vac = -Vca$.

Here we have very complex relations of geometry brought down to elementary algebra. As Prof. Gibbs has remarked, the scalar and vector product, which consist of the cosine and sine of trigonometry combined with certain other simple notions, are "incomparably more amenable" to algebraical treatment than the sine and cosine themselves.

When we go on to combinations in threes, fours, &c., we find it is the same thing over again in various forms. Thus c.ab and cVab and VcVab define themselves by the above, being ab times **c**, and the scalar and vector products of c and Vab respectively.

Similarly with four vectors, as in dc.ab, the product of dc and ab; dc.Vab, which is dc times Vab; d.cVab, which is cVab times **d**; dVcVab and VdVcVab, which are the scalar and vector products of d and VcVab; VdcVab and VVdcVab, which are the scalar and vector products of Vdc and Vab. Combinations of two vectors are universal, and those of three are pretty frequent, but those of four are exceptional. But it should be noted that the scalar and vector product of two vectors are involved throughout, no new idea being introduced when more than two vectors are concerned.

When we pass on to analysis, we find just the same vector algebra to be involved. Vectors are differentiated with respect to scalars to make new vectors in the same way as scalars, and the ideas connected with differentials (infinitesimal) and differential coefficients are essentially the same for vectors as for scalars. We never require to differentiate a vector with respect to a vector, and there is a very good reason for it, because the operation is an indeterminate one.

The fictitious vector ∇ is omnipresent in tridimensional analysis. It only differs from a vector in being a differentiator as well, so that it follows vector rules combined with other functions. With it are associated the ideas of the slope of a scalar, and the divergence and curl of a vector, with corresponding important theorems relating to the transition from line to surface, and from surface to volume summations. The theory of potentials in its broad sense is also involved in the mathematics of ∇, including the relations, direct and inverse, of potential, slope, curl, and divergence.

As for the linear operator, that is only the scalar and vector product system again in a special form. Professor Gibbs's dyadic is a useful idea, and I have modified his notation to suit the rest.

It is neither necessary nor desirable that a student should know all that is sketched out above before he makes practical use of it. For vector analysis, like many other things, is best studied in the concrete application. The principles may be very concisely stated, being little more than explanations of the scalar and vector product, with some conventions about notation. A student might learn this by heart, and be little the better for it. He must sit down and work if he wants to assimilate it usefully. He may work with scalar products only to begin with, for quite a large ground is covered thereby. When familiarised with the working of vectors by practice so far as the scalar product goes, the further introduction of the vector product will come comparatively easy. The former knowledge is fully utilised, and also receives a vast extension

of application. For the vector product is a powerful engine,
which, with its companion the scalar product, is fully capable
of working elaborate mathematical analysis in a concise and
systematic manner. The student need not trouble about linear
operators at first. He will grow into them. It is far more
important that he should understand the operation of ∇ as a
vector, and its meaning in the space variation of functions.

Unsuitability of Quaternions for Physical Needs.   Axiom :— Once a Vector, always a Vector.

§ 175. Now, a few words regarding Quaternions. It is
known that Sir W. Rowan Hamilton discovered or invented a
remarkable system of mathematics, and that since his death
the quaternionic mantle has adorned the shoulders of Prof.
Tait, who has repeatedly advocated the claims of Quaternions.
Prof. Tait in particular emphasises its great power, simplicity,
and perfect naturalness, on the one hand; and on the other
tells the physicist that it is exactly what he wants for his physical purposes. It is also known that physicists, with great
obstinacy, have been careful (generally speaking) to have
nothing to do with Quaternions ; and, what is equally remarkable, writers who take up the subject of Vectors are (generally
speaking) possessed of the idea that Quaternions is not
exactly what they want, and so they go tinkering at it, trying
to make it a little more intelligible, very much to the disgust
of Prof. Tait, who would preserve the quaternionic stream pure
and undefiled. Now, is Prof. Tait right, or are the defilers
right? Opinions may differ. My own is that the answer all
depends upon the point of view.

If we put aside practical application to Physics, and look
upon Quaternions entirely from the quaternionic point of
view, then Prof. Tait is right, thoroughly right, and Quaternions furnishes a uniquely simple and natural way of treating
quaternions. Observe the emphasis.

For consider what a quaternion is. It is the operator which
turns one vector to another. For instance, we have

$$B = A . A^{-1} B + B - A^{-1} . AB$$
$$= A^{-1} B . A + V A V B A^{-1},$$

identically. Or,

$$B = (A^{-1}B + V.VA^{-1}B)A. \quad \cdots \quad (157)$$

Here the operator in the brackets, say q, depends upon **A** and **B** in such a way that it turns **A** to **B** ; thus, $B = qA$. It is a quaternion. It has a scalar and a vector part. The general type of a quaternion is

$$q = w + Va, \quad \cdots \quad \cdots \quad (158)$$

where w is any scalar, and a is any vector. Since we know the laws of scalar and vector products we may readily deduce the laws of quaternions should we desire to do so. But we shall find that the above notation, though so well suited for vectors, is entirely unfitted for displaying the merits of quaternions. To do this we must follow Hamilton, and make the quaternion itself the master, and arrange the notation and conventions to suit it, regardless of the convenience of the vector and the scalar. The result is the singularly powerful algebra of quaternions. To give a notion of its power and essential simplicity we may remark that the product of any number of quaternions, p, q, r, s, say, in the order named, is independent of the manner of association in that order. That is,

$$pqrs = p(qrs) = (pq)(rs) = (pqr)s. \quad (159)$$

This remarkable property is the foundation of the simplicity of the quaternionic algebra. It is uniquely simple. No algebra of vectors can ever match it, and Prof. Tait is quite right in his laudations from the quaternionic point of view.

But when Prof. Tait vaunts the perfect fitness and natural-ness of quaternions for use by the physicist in his inquiries, I think that he is quite wrong. For there are some very serious drawbacks connected with quaternions, when applied to vectors. The quaternion is regarded as a complex of scalar and vector, and as the principles are made to suit the quaternion, the vector itself becomes a degraded quaternion, and behaves as a quaternion. That is, in a given equation, one vector may be a vector, and another be a quaternion. Or the same vector in one and the same equation, may be a vector in one place, and a quaternion (versor, or turner) in another. This amalgamation of the vectorial and quaternionic functions is very puzzling. You never know how things will turn out.

Again, the vector having to submit to the quaternion, leads to the extraordinary result that the square of every vector is a *negative* scalar. This is merely because it is true for quadrantal versors, and the vector has to follow suit. The reciprocal of a vector, too, goes the wrong way, merely to accommodate versors and quaternions.

And yet this topsyturvy system is earnestly and seriously recommended to physicists as being precisely what they want. Not a bit of it. They don't want it. They have said so by their silence. Common sense of the fitness of things revolts against the quaternionic doctrines about vectors. Nothing could be more unnatural.

Are they even convenient ? Not to the physicist. He is very much concerned with vectors, but not at all, or at any rate scarcely at all, with quaternions. The vector algebra should satisfy his requirements, not those of the quaternion. Let the quaternion stand aside. The physicist wants, above all, to have clear ideas, and to him the double use of vectors, as vectors and versors, combined with unnatural properties of vectors to suit quaternions, is odious. Once a vector, always a vector, should be a cardinal axiom.

If the usual investigations of physical mathematics involved quaternions, then the physicist would no doubt have to use them. But they do not. If you translate physical investigations into vectorial language, you do not get quaternions ; you get vector algebra instead. Even Prof. Tait's treatise teaches the same lesson, for in his physical applications the quaternion is hardly ever concerned. It is vector algebra, although expressed in the quaternionic notation.

Vectors should be treated vectorially. When this is done, the subject is much simplified, and we are permitted to arrange our notation to suit physical requirements. This is a very important matter. Most calculations are, and always will be, scalar calculations, that is, according to common algebra. The special extensions to tridimensional space involving vectors should therefore be done so as to harmonise with the scalar calculations, and so that only one way of thinking is required, instead of two discrepant ways, and so that mutual conversion of scalar and vector algebra is facilitated. The system which I expound I believe to represent what the physicist wants, at least to begin with. It is what I have been expounding, though rather by example than precept, since 1882. There is only one point where I feel inclined to accept a change, viz., to put a prefix before the scalar product, say Sab instead of ab, to balance Vab. Not that the S prefix is of any use in the algebra as I do it, but that mathematical writers on vectors seem to want to put the scalar and vector parts together to make a general or complete product, say Qab = Sab + Vab. The function Qab, however, does not occur in physical applications, which are concerned with the scalar and vector products separately. If I could omit the V as well as the S I would do so ; but some sign is necessary, and the V of Hamilton is not very objectionable.

Gibbs denotes the scalar and vector products by $a.\beta$ and $a \times \beta$, using Greek letters. In my experience this is not a good way when worked out. It is hard to read, and is in serious conflict with the common use of dots and crosses in algebra, which use has to be given up. I use dots in their common meaning, either as multipliers or separators.

Prof. Macfarlane, who is the latest vectorial propagandist, denotes the scalar and vector products by cos ab and Sin ab. The trigonometrical origin is obvious. But, on this point, I would refer the reader to the equations (155) (156) above, and to the very trenchant remark of Prof. Gibbs which I have already quoted, as to the incomparably greater amenity of the scalar and vector product to algebraical treatment than the trigonometrical functions themselves. The inference is that vector algebra is far more simple and fundamental than trigonometry, and that it is a mistake to base vectorial notation upon that of a special application thereof of a more complicated nature. I should rather prefer sca ab and vec ab. Better still, Sab and Vab, with the understanding that the S may be dropped unless specially wanted.

Macfarlane has also a peculiar way of treating quaternions, about which I will express no opinion at present, being doubtful whether, if the use of quaternions is wanted, *the* quaternionic system should not be used, with, however, a distinction well preserved between the vector and the quaternion, by special type or otherwise.

I am in hopes that the Chapter which I now finish may serve as a stopgap till regular vectorial treatises come to be written suitable for physicists, based upon the vectorial treatment of vectors.   The quaternionists want to throw away the " cartesian trammels," as they call them.   This may do for quaternions, but with vectors would be a grave mistake.   My system, so far from being inimical to the cartesian system of mathematics, is its very essence.

CHAPTER IV.

___

THEORY OF PLANE ELECTROMAGNETIC WAVES

Action at a Distance versus Intermediate Agency. Contrast of New with Old Views about Electricity.

§ 176. It has often been observed that the universe is in an unstable condition. Nothing is still. Nor can we keep motion, once produced, to a particular quantity of matter. It is diffused or otherwise transferred to other matter, either immediately or eventually. The same fact is observed in the moral and intellectual worlds as in the material, but this only concerns us so far as to say that it underlies the communication of knowledge to others when the spirit moves, even though the task be of a thankless nature.

The laws by which motions, or phenomena which ultimately depend upon motion, are transferred, naturally form an important subject of study by physicists. There are two extreme main views concerning the process. There is the theory of instantaneous action at a distance between different bodies without an intervening medium; and on the other hand there is the theory of propagation in time through and by means of an intervening medium. In the latter case the distant bodies do not really act upon one another, but only seem to do so. They really act on the medium directly, and between the two are actions between contiguous parts of the medium itself. But in the former case the idea of a medium does not enter at all. We may, however, somewhat modify the view so that action, though seemingly instantaneous and direct, does take place through an intervening medium, the speed of transmission being so great as to be beyond recognition. Thus, the two ideas of direct action, and through a yielding medium, may be somewhat harmonised, by being made extreme cases of one theory. For example, if we know that there *is* a yielding medium and a finite speed, but that in a certain case under examination the influence of the yielding is insensible, then we may practically assume the speed to be infinite. But although this comes to the same thing as action at a distance, we need not go further and do away with the medium altogether.

There is another way of regarding the matter. We may explain propagation in time through a medium by actions at a distance. But this is useless as an explanation, being at best merely the expression of a mathematical equivalence.

Now consider the transmission of sound. This consists, physically, of vibratory motions of matter, and is transmitted by waves in the air and in the bodies immersed in it. The speed through air is quite small, so small that it could not even escape the notice of the ancients, who were, on the whole, decidedly unscientific in their mental attitude towards natural phenomena. Suppose, however, that the speed of transmission of sound through air was a large multiple of what it is, so that the idea of a finite speed, or of speed at all, did not present itself. We should then have the main facts before us, that material bodies could vibrate according to certain laws, and that they could, moreover, set distant bodies vibrating. This would be the induction of vibrations, and it might be explained, in the absence of better knowledge, by means of action at a distance of matter upon matter, ultimately resolvable into some form of the inverse square law, not because there is anything essentially acoustical about it, but because of the properties of space. Furthermore, we should naturally be always associating sound with the bodies bounded by the air, but never with the air itself, which would not come into the theory. A deep-minded philosopher, who should explain matters in terms of an intervening medium, might not find his views be readily accepted, even though he pointed out independent evidence for the existence of his medium—physiological and mechanical—and that there was a harmony of essential properties entailed. Nothing short of actual proof of the finite speed of sound would convince the prejudiced.

Now, although there is undoubtedly great difference in detail between the transmission of sound and of electrical disturbances, yet there is sufficient resemblance broadly to make the above suppositional case analogous to what has actually happened in the science of electromagnetism. There were conductors and non-conductors, or insulators, and since the finite speed of propagation in the non-conducting space outside conductors was unknown, attention was almost entirely concentrated upon the conductors and a suppositional fluid which was supposed to reside upon or in them, and to move about upon or through them. And the influence on distant conductors was attributed to instantaneous action at a distance, ignoring an intermediate agency. Again, a very deep-minded philosopher elaborated a theory to explain these actions by the intermediate agency of a yielding medium transmitting at finite speed. But although in doing so he utilised the same medium for whose existence there was already independent evidence, viz., the luminiferous ether, and pointed out the consistency of the essential properties required in the two cases, and that his electromagnetic theory made even a far better theory of light than the old, yet his views did not spread very rapidly. The old views persisted in spite of the intrinsic probability of the new, and in spite of the large amount of evidence in support of the view that some medium outside conductors, and it may be also inside them as well, but not particularly conducting matter itself, was essentially concerned in the electrical phenomena. The value and validity of evidence varies according to the state of mind of the judge. To some who had seriously studied Maxwell's theory the evidence in its favour was overwhelming; others did not believe in it a bit. I am, nevertheless, inclined to think that it would have prevailed before very long, even had no direct evidence of the finite velocity been forthcoming. It was simply a question of time. But the experimental proof of the finite speed of transmission was forthcoming, and the very slow influence of theoretical reasoning on conservative minds was enforced by the common-sense appeal to facts. It is now as much a fact that electromagnetic waves are propagated outside conductors as that sound waves are propagated outside vibrating bodies. It is as legitimate a scientific inference that there is a medium to do it in one case as in the other, and there is independent evidence in favour of both media, air for sound and ether for electrical disturbances. It is also so excessively probable now that light vibrations are themselves nothing more than very rapid electromagnetic vibrations, that I think this view will fully prevail, even if the gap that exists between Hertzian and light vibrations is not filled up experimentally, through want of proper appliances—provided some quite new discovery of an unanticipated character is not made that will disprove the possibility of the assumed identity.

Now, the immediate question here is how to propagate a knowledge of the theory of electromagnetic waves. If we could assume the reader to have had a good mathematical training, that would greatly ease matters. On the other hand, it is hardly any use trying to do it for those who have no mathematical knowledge—least of all for those anti-mathematical attackers of the theory of the wave propagation of electrical disturbances who show plainly that they do not even know intelligibly what a wave means, and what is implied by its existence. But, between the two, I think it is possible to do a good deal in the way of propagation by means of a detailed exposition of the theory of plane waves, especially in dielectrics. By the use of plane waves, the mathematical complexity of waves in general in a great measure disappears, and common algebra may be largely substituted for the analysis which occurs in more advanced cases. Most of the essential properties and ideas may be assimilated by a thinking reader who is not advanced in his mathematics, and what is beyond him he can skip. Besides this, the treatment of plane waves is itself the best preliminary to more general cases.

The problems that present themselves by the artificial limitation to plane waves are often of an abstract nature. There are, however, some important exceptions; the most notable being that of propagation along straight wires, of which the theory is essentially that of plane waves, modified by the resistance of the wires. Before, however, proceeding to the details of plane waves, which will form the subject of this chapter (although it will be needless to altogether exclude connected matters), it will be desirable to prepare the mind by an outline of some general notions concerning electromagnetic waves, irrespective of their precise type.

General Notions about Electromagnetic Waves. Generation of Spherical Waves and Steady States.

§ 177. Consider a non-conducting dielectric, to begin with. The two properties it possesses of supporting electric displacement and magnetic induction, which we symbolise by μ and c, the inductivity and permittivity, are, independently of our actual ignorance of their ultimate nature, so related that the speed of propagation v depends upon them in the way expressed by the equation $\mu c v^2 = 1$, or $v = (\mu c)^{-\frac{1}{2}}$. Thus, an increase either of the permittivity, or of the inductivity, lowers the speed of transmission. In transparent bodies we cannot materially alter the inductivity, but we can very considerably increase the permittivity, and so lower the speed. If on the other hand, we wish to have infinite speed, for some practical purpose of calculation, we may get it by assuming either $\mu = 0$ or else $c = 0$. In the latter case, for example, we destroy the power of supporting electric displacement, whilst preserving the magnetic induction. This is what is done in magnetic problems (the theory of coils, self and mutual induction) when we ignore the existence of electric displacement. On the other hand, in the theory of condensers connected up by inductionless resistances, and in the electrostatic theory of a submarine cable, it is the magnetic induction that is ignored, in a manner equivalent to supposing that $\mu = 0$. In either case we have infinite speed or apparent instantaneous action.

Now, consider some of the consequences of the property of propagation with finite speed. Let there be a source of disturbance at a point for simplicity. It may be either an electric source or a magnetic source. By an electric source we mean a cause which will, if it continue steadily acting, result in setting up a state of electric displacement, whilst a magnetic source steadily acting would result in a state of magnetic induction. In the former case there will be no magnetic induction along with the displacement, and in the latter case no displacement along with the induction. Now, if the speed of propagation were infinite in the former case of an electric source, by the non-existence of μ, the steady state would be

set up instantly, and consequently all variations of the intensity
of the source would be immediately and simultaneously accompanied by the appropriate corresponding distribution of displacement in the dielectric.   Similarly, with a magnetic source,
if the speed be infinite by the non-existence of c, all variations in
the source will be simultaneously accompanied by the distribution of magnetic induction appropriate to the instantaneous
strength of the source.

Now do away with the artificial assumption made, and let
both μ and c be finite, and v therefore also finite.   At the time
t after starting a source, the extreme distance reached by the
disturbance it produces is vt.   That is to say, beyond the sphere
of radius vt, whose centre is at the source, there is no disturbance, whilst within it there is.   The wave-front is thus a spherical surface of radius vt, increasing uniformly with the time.
Along with this continuous expansion of the range of action
there are some other things to be considered.   The mere spreading causes attenuation, or weakening of intensity as the disturbance travels away from the source.   Besides this, the intensity
of disturbance is not the same at a given distance in all directions from the source, and since the displacement and induction
are vectors, their directions are not everywhere the same. They
are distributed in the circuital manner, so that we have expanding rings or sheets of displacement or induction.   Whether the
source be of the electric or the magnetic kind, it produces both
fluxes initially and when varying, and generally speaking to an
equal degree as regards energy.   This is a main characteristic
of pure electromagnetic waves, a coexistence of the electric and
magnetic fluxes with equal energies ; and if the source be in a
state of sufficiently rapid alternating variation this state of
things continues.   But if the source, after varying in any way,
finally become steady, the generation of electromagnetic disturbances will speedily cease, and the production of a steady
state will begin round the source—of displacement if the source
be electric, and of induction if the source be magnetic—whilst
the previously generated electromagnetic disturbances will pass
away to a great distance.   In the end, therefore, we have
simply the steady state of the flux corresponding to the
impressed force, whilst the electromagnetic disturbances are
out of reach, still spreading out, however, and in doing so,

leaving behind them the outside portion of the residual steady flux due to the source, which steady flux, however, requires an infinite time to become quite fully established.

We may take a specially simple case in illustration of the above general characteristics. Let the source of energy be contained within a small spherical portion of the dielectric, of radius a, and let it be of the simplest type, viz., a uniform distribution of impressed electric force within the sphere. We know by static considerations alone what the nature of the final displacement due to such a source is. It is a circuital distribution, out from the sphere on one half, and in on the other, connection being made through the sphere itself by a uniform distribution; being, in fact, of the same nature as the induction due to a spherical body uniformly magnetised in a medium of the same inductivity as its own. Now, we can describe the setting up of the final steady state due to the source, when it is suddenly started and kept constant later, thus:—At the first moment an electromagnetic wave is generated on the surface of the sphere, which immediately spreads both ways and becomes a spherical shell, whose outer boundary goes outward at speed v, whilst the inner boundary goes inward. At the time $t = a/v$, therefore, the disturbance fills the sphere of radius $2a$, the centre being just reached. The steady state then begins to form, commencing at the centre and expanding outwards thereafter at speed v. At the time $t = 2a/v$, therefore, we have the steady state fully formed within the sphere of radius a, whilst just outside it is an electromagnetic shell of depth $2a$. Up to this moment the impressed force has been continuously working—not, indeed, in all parts of the sphere it occupies, but in all parts passed by the front of the inward wave from the beginning until the centre was reached, and after that, in the parts not occupied by the already formed steady state. Consequently, at the moment $t = 2a/v$, when the steady state is fully formed throughout the region occupied by the impressed force, the latter ceases to work. It has wholly done its work. The amount done is twice the energy of the final complete steady state, say $2U$. The rest of the work is done by the electromagnetic wave itself. For the subsequent course of events is that the fully-formed electromagnetic shell of depth $2a$ runs

out to infinity, of course expanding on the way, and in doing
so it leaves the steady state behind it.   That is to say, it
drops a part of its contents as it moves on, so that the steady
state is always fully formed right up to the rear of the expanding
shell during the whole of its passage to infinity.   This shell is
not a pure electromagnetic wave, with equal electric and
magnetic energies.   The magnetic energy is constant, of
amount $\frac{1}{2}$U, on the whole journey, but the electric energy
is in excess.   The excess is employed in forming the steady
state, so that when the shell has reached a great distance it
becomes appreciably a pure electromagnetic wave, having the
amount U of energy, half magnetic, half electric, with a slight
excess in the latter, to be later left behind in forming the
remainder of the steady state.   The energy U of the shell is
wholly wasted if the dielectric be unbounded, for there is
nothing to stop the transference to an infinite distance.

If we wish to form the steady state without this waste of
energy, we must bring the impressed force into action very
gradually—infinitely slowly, in fact.   In this way the energy
wasted in the very weak electromagnetic waves generated will
tend to become infinitely small, and the work done by the
impressed force will tend to the value U, the energy of the
steady state.

When the sphere of impressed force is not finite, but is
infinitely small, so that it may be regarded as a special kind of
point-source, we simplify matters considerably in some respects.
For, by the above, the result is that the moment the source
starts, say at full strength, the steady state also immediately
begins to grow, so that at the time t it occupies the sphere of
radius vt.   Outside it there is no disturbance, but on its surface is
an infinitely thin electromagnetic shell, which performs the same
functions as the previous finite shell.   That is, it lays down
the steady state as it expands, and then carries out to infinity
in itself as much energy as it leaves behind.

Notice, however, this peculiarity, that if we had started with
what appears at first sight to be the simpler problem, viz., a
point-source, we should be quite unable to see and understand
the functions of the electromagnetic shell on the extreme verge
of the steady field.   The reader may compare this case with
that of the establishment of the steady state of displacement

due to another kind of point-source, viz., electrification suddenly
brought to rest at a point after previous motion at the speed of
light, as discussed in § 55.   There is a perfect similarity as respects the uniform growth of the steady state.   But the nature
of the bounding electromagnetic shell is not the same.   In the
case of electrification it belongs to the zero degree of spherical
harmonics ; in our present problem it is of the first degree.

If, after keeping on the impressed force at a point for, say,
an interval τ, we suddenly remove it, this is equivalent to
keeping it on, but with the addition of an impressed force which
is the negative of the former.   From this we see that the result of putting on the impressed force for an interval τ only is to
generate a shell of depth $v\tau$, which runs out to infinity.
Within this shell is the steady electric displacement due to the
source appropriate to the instantaneous position of the shell.
On its outer surface is an electromagnetic wave moving out to
infinity, and generating or laying down the steady electric displacement as it goes, whilst on the inner surface is another
wave running after the first, and undoing its effects.   That is,
it takes up the displacement laid down by the first wave, so
that inside the inner wave (just as outside the outer) there is
no disturbance.

Intermittent Source producing Steady States and Electro-magnetic Sheets.   A Train of S.H. Waves.

§ 178. Using the same kind of point-source, let it act inter-
mittently and alternatingly ; say, on positively for an interval
a, then off for an interval β, then on negatively for an interval
a, followed by off for an interval β, and so on recurrently, like
positive and negative applications of a battery with dead intervals.   The result is, by the last case, to divide the spherical
space occupied by the disturbances at a given moment into
concentric shells, of depths va and $v\beta$ respectively.   In the
latter is no disturbance, since they correspond to the dead intervals.   In the former are the steady displacements appropriate to their distance from the source, consecutive shells
being positive or negative according to the state of the source
when they started from it.   On their boundaries are electromagnetic sheets generating these steady states continuously on
one side, and destroying them on the other.

Next, do away with the dead intervals, so that the applied force is like that of a common reversing key, first positive of full strength, and then negative of full strength, but without any interval. The result is to completely fill up the sphere of disturbance, the successive shells of depth va containing steady fields—alternately positive and negative—being brought into contact, and only separated from one another by the electromagnetic sheets. This may be considered rather an unexpected result. Our impressed force is periodic, but expresses the extremest form of variation, namely discontinuities, and is only expressible simple-harmonically by an infinite series of simple-harmonic forces of all frequencies from zero to infinity. But only when the impressed force varies is electromagnetic disturbance generated, so that when in the above manner we confine its variations to be momentary, the electromagnetic disturbances are also momentary, with the consequent result that we have our dielectric occupied by quite steady states of displacement separated by infinitely-thin electromagnetic shells. The latter travel. The steady states also appear to travel, but do not really do so. They are being continuously generated and destroyed at their boundaries, but are quite steady elsewhere.

If the impressed force vary simple-harmonically instead of discontinuously, the result is less simple. We find that there is no clean separation into steady states and electromagnetic waves, the two being mixed inextricably. We can see this by substituting for the simply periodic variation of the force a very great number of constant forces of small duration and of different strengths, so as to roughly imitate the simple harmonic variation. Every discontinuity in the force will behave in the way described, and the final result of the superimposition of effects, when we proceed to the limit and have continuous simple-harmonic variation of the force, is a train of simple-harmonic waves proceeding from the source. They are not pure electromagnetic waves, and there is no steady state anywhere. The wave-length λ is given by $\lambda = v\tau$, where τ is the period of the impressed force, or the reciprocal of the frequency. The electric and magnetic forces are simple-harmonic functions of the time everywhere, right up to the extreme wave-front of the disturbance, where, of course, is an electromagnetic sheet following a different law. But should the source occupy a finite spherical space, then the bounding sheet becomes a shell, of depth equal to the diameter of the sphere, as before described. Now the steady fields become insensible at a great distance in the discontinuous case ; to correspond with this we have the fact that the simple harmonic waves tend to become pure as they expand. The greatest departure from the pure state is round about the source. Here the frequency becomes a matter of importance. When very low, the wave-length is great, and we should, there-fore, have to go a great distance to find any sign of wave-propagation. Moreover, the waves themselves would be exceedingly weak. Only round the source is there sensible disturbance, and it is practically the steady displacement appropriate to the momentary state of the source, accompanied by a weak state of magnetic force connected with the displace-ment nearly in the instautaneous manner. That is, the time-variation of the displacement is the electric current, and the displacement itself is sensibly that of the static theory. Under these circumstances there is next to no waste of energy.

But by increasing the frequency we can completely alter this state of things. For we increase the rate of change of the impressed force, and therefore the strength of the electromagnetic waves generated, for one thing. Besides this, we can bring the region of electromagnetic waves nearer to the source, and so contract the region in which the displacement was approxi-mately static as much as we please. In the limit with very great frequency we have nearly pure electromagnetic waves close up to the source. The waste of energy increases very rapidly with the frequency, varying as its fourth power with simple-harmonic waves. The extremest case of waste is that of a discontinuity, already mentioned, as when setting up the steady state we waste half the work done, and then, when the force is removed, waste the other half.

Self-contained Forced Electromagnetic Vibrations. Contrast with Static Problem.

§ 179. When the sphere of impressed force is of finite size, another effect comes into view of a somewhat striking character, of which the theory of a point-source gives no information. By increasing the frequency sufficiently, we shall reduce the amplitude of the external vibrations, and on reaching a certain frequency depending upon the size of the sphere, they will entirely vanish. That is, at a certain frequency, a simple harmonic impressed force, acting uniformly within a spherical space in a dielectric, produces no external effect whatever. The vibrations are stationary or standing, and are purely internal, or confined to the sphere itself. By general experience of statical problems this would seem to be impossible. But we must expect to find strange manners when we go into strange lands.

This strange behaviour arises from the general electromagnetic property that the true source of disturbances due to impressed force (electric or magnetic) is the curl thereof. The impressed force itself, on the other hand, is associated with the supply of energy. (*See* § 87.) In the present case the surface of the sphere of impressed force is the seat of its curl, its intensity varying as the cosine of the latitude, if the polar axis be parallel to the impressed force; and consequently, as already described for the case of setting up the steady state, the electromagnetic waves proceed both ways from the surface at which they originate. The waves going inward contract, then cross at the centre and expand again. We have, therefore, two trains of outward waves, and the external effect may be imagined to be due to their superimposition. That they may sometimes assist and sometimes partly cancel one another, according to relative phase, may be readily conceived, and the theory indicates that at any one of an infinite series of definite frequencies there is complete cancellation externally. This refers to the simple-harmonic state. There is always an external effect initially, at any frequency—namely, the initial electromagnetic shell of depth equal to the diameter of the sphere of impressed force, because it consists entirely of the beginning part of the first outward wave, the second one only reaching to its rear; but on the inner side of this shell there may be no disturbance, except within the sphere of impressed force. The waste of energy is not continuous at the critical frequencies, but consists merely of the energy in the initial shell. The impressed force works continuously, but as much positively as negatively within a period, so that it is inactive on the whole.

Effects of the above described kind are principally remarkable from the contrast they present towards more familiar

effects, especially those of a statical kind. The behaviour of
electrification on a sphere may be taken as an illustrative contrast. We know that if we have a spherical surface in a uniform dielectric covered with electrification, the displacement is
partly internal and partly external. Only when the electrification is uniformly spread does the displacement vanish
internally, and become wholly external. This was formerly explained by the inverse-square law. Every element
of electrification was supposed to exert force equably in
all directions round itself, internally as well as externally,
and a mathematical consequence of this is the complete
cancellation of the internal force when the electrification
is uniformly spread on the surface of a sphere. The same
reasoning was applied when the sphere was not of the same
nature as the external medium—when it was a conductor, to
wit. Now the internal force is also zero in this case. Moreover, it is zero whatever be the nature of the internal material
(without electrical sources). But it is certainly untrue that
the electric force due to a point-source of displacement is the
same in all directions around it when the medium is not homogeneous. So the reason for the absence of internal force is
entirely wrong, although it seems right in one case, viz., that
of homogeneity. The true, and sufficient, and comprehensive
reason for the equilibrium, and the absence of internal force is
the satisfaction of the second circuital law when the external
electric force is perpendicular to the surface. This is independent
of the nature of the internal matter, and is obviously to be preferred to a hypothesis that is only valid sometimes. Varying the
internal material will make the law of spreading of force be
different in any number of ways, and, if we like, different
for every element of electrification, by arranging various
kinds of heterogeneity. But independently of this, we might
vary the law of force in many ways in special cases.
Thus, in the case of a uniform spread of electrification
on a sphere, we may imagine the displacement from any
particle of electrification to radiate in any symmetrical manner
that will give zero displacement inside and a uniformly
radial displacement outside. The simplest case after the usual
uniform spreading in all directions, is a planar spreading. Let
every element of electrification on the surface send out its

displacement in the tangential plane only, though equally in all directions in that plane. This will give the correct static result. It is not altogether a fantastic example, for we can make an electromagnetic problem of it by letting the electrified surface expand at the speed of light. Then the assumed law of spread is the actual law, for every element of electrification is the core of a plane electromagnetic wave. The resultant magnetic force due to all the waves, however, is zero, and the resultant electric force is as in the static problem. (*See* §§ 61 and 164.)

Now when we have vibrating electromagnetic sources on a spherical surface producing electromagnetic waves simple-harmonically, we also in general have both internal and external effects, for there is an internal as well as an external train of waves. But whilst they cannot cancel internally, they may do so externally. Here is one contrast with the static problem, and along with it is another, viz., that it is the nature of the external medium that is now indifferent (with a reservation), whereas in the static problem we have independence of the nature of the internal medium. The reservation hinted at is connected with the initial uncancelled electromagnetic shell. The external medium should allow it to escape, or absorb it somehow, so that it will not interfere with the effects under consideration. To regard the zero external effect as being actually due to the coexistence of two trains of waves which cancel one another is merely a mathematical artifice, however, because neither train of waves exists. If they did exist we should have to vary their nature to suit the constitution of the external medium, just as in the problem of static equilibrium we require different laws of force to suit the nature of the internal medium. The comparison of the static with the kinetic problems shows a complete reversal of relations as regards internal and external.

What we can do with a single surface electromagnetic source we can repeat with others inside it. We see, therefore, that the whole sphere of impressed force may be filled up with vibratory sources of the most vigorous nature without producing any (except initial) external effect, if their periods be properly chosen in relation to their situation, which problem admits of multiple solutions. This is suggestive as regards the stores of energy bound up with matter.

Relations between E and H in a Pure Wave. Effect of Self-Induction. Fatuity of Mr. Preece's "KR law."

§ 180. In the above very little has been said about the distribution of electric and magnetic force from part to part of an electromagnetic wave. This varies greatly in different kinds of waves, and cannot be explained without the formulæ. But there is one very important property of pure electromagnetic waves (which also holds good, more or less approximately, in general) which may be described at present. As already mentioned, the electric and magnetic energies are equal in a pure electromagnetic wave. Now the density of the electric energy is $\frac{1}{2}cE^2$, if E is the electric force (intensity), and that of the magnetic energy is $\frac{1}{2}\mu H^2$. If we equate these, we obtain a relation between E and H. Thus

$$\tfrac{1}{2}cE^2 = \tfrac{1}{2}\mu H^2. \quad \ldots \quad \ldots \quad (1)$$

Also $\qquad\qquad \mu cv^2 = 1. \quad \ldots \quad \ldots \quad (2)$

Therefore $\qquad\qquad E = \pm \mu v H. \quad \ldots \quad \ldots \quad (3)$

The positive or negative sign depends on which way the wave is going. Disregarding this, the electric and magnetic forces have a constant ratio. They are therefore in the same phase, or keep time together in all their variations without lag or lead. Along with this, they have the property of being perpendicular to one another (that is, **E** and **H** are perpendicular, not E and H), and their plane is in the wave-front, or the direction of motion of the wave is perpendicular to **E** and to **H**. It is the direction of the flux of energy. These properties hold good in all parts of a pure electromagnetic wave, although the magnitudes and directions of the electric and magnetic forces may vary greatly from one part of the wave to another.

The above is also the state of things that obtains, more or less perfectly or imperfectly, in long-distance telephony over copper circuits of low resistance, and by lowering the resistance per mile we may approximate as nearly as we please to the state of pure electromagnetic waves. It is the self-induction that brings about this state of things, showing such a contrast to the more familiar relations between the electric and magnetic forces on circuits in general. Like a kind of fly-wheel, the self-induction imparts inertia and stability, and keeps the waves going. It is the long-distance telephoner's best friend who was, not many years since, spurned with contempt from the door. Some people thought there was a very absurd fuss made about self-induction, and that it was made a sort of fetish of ; so, knowing no better, they poured much cold water on the idol. But, whatever opinions we may hold regarding their competence as judges, there can be no question about the stern logic of facts. Self-induction came to stay, and stayed it has, and will stay, having great staying power. Whatever should we think of engineers who declined to take into account the inertia of their machinery? There was also some considerable fuss made about a supposed law of the squares, or KR law as it was or is called, according to which you could not telephone further than KR = such or such a number, because the speed of the current varied as the square of the length of the line, or else inversely. But in spite of the repeated attempts made to bolster up the KR law, the critical number has kept on steadily rising ever since. I see now that it has gone up to 32,000.* But it need not stop there. Make your circuits longer, and it will go up a lot more.

As regards the ether, it is useless to sneer at it at this time of day. What substitute for it are we to have? Its principal fault is that it is mysterious. That is because we know so little about it. Then we should find out more. That cannot be done by ignoring it. The properties of air, so far as they are known, had to be found out before they became known.

Wave-Fronts; their Initiation and Progress.

§. 181. Still keeping to a simple dielectric, we may always, by consideration of the fact that the speed of propagation is v, find the form of the wave-front due to any collection of point-sources, and trace the changes of shape and position it under-

---

* See *The Electrician*, December 30, 1892, p. 251, for data, in article by Mr. Jos. Wetzler. The 32,000 is for the New York-Chicago circuit of 1,000 miles at 4·12 ohms per mile, or 2·06 ohms per mile of wire. Comparison of Mr. Wetzler's with Mr. Preece's figures is interesting. Good telephony is got by Mr. P. at 10,000, and by Mr. W. at 45,000 ; excellent by Mr. P. at 5,000, and by Mr. W. at 31,000 ; and so on. But there is still less of a KR law in the American than in the English cases.

goes as it progresses.   For obviously, if a point P be at a distance *vt* from the nearest source, no disturbance can have
reached P from it if it started into action *after* the moment
$t = 0$; and generally, the effect at P at a given moment arising
from a particular source depends upon its state at the moment
r/v earlier (if r be the distance from the source to P), and upon
the previous state of the source, causing residual or cumulative action at P.   We therefore know the limiting distance of
action of the sources for every one of their momentary states.
The wave-front belonging to a set of disconnected point-sources
consists initially of disconnected spheres.   But as they expand,
they merge into one another to form a continuous extreme wave-
front consisting of portions of spheres. When the point-sources
are spread continuously over a surface to form a surface-source, we
have a continuous wave-front from the first moment ; or rather,
two wave-fronts, one on each side of the surface.   There is an
exception, to be mentioned later, when the sources send waves
one way only from the surface.   This is not the already
described case of the cancelling of two trains of waves at particular frequencies, but is a unilateral action obtained by a
special arrangement of surface-sources.   Passing over this,
observe that when we have once got a wave-front we may ignore
the sources which produced it, and make the wave-front itself
tell us what its subsequent history will be.   For, to trace the
course of the wave-front from one moment to the next, we have
merely to move any element of the surface in the direction of
the normal to that element through a small distance *a* to
obtain the new position at the moment a/v later.   This being
done for the whole surface, gives us the new position of the
wave-front, and by continuing this process we may follow the
wave-front in its progress as long as we please. Thus the wave-
front coming from the surface of a sphere is a sphere ; from a
round cylinder, if infinitely long, also a round cylinder ; but if
of finite length, then a cylinder with rounded ends ; from a cube,
initially a cube with rounded corners, but becoming more and
more spherical as it expands ; and so on.   It is easily seen that
at a sufficiently great distance from a finite collection of sources
of any kind, the wave-front tends to become spherical, or the
complex source tends to become equivalent, at a great distance,
to a point-source of some complex kind.

Effect of a Non-Conducting Obstacle on Waves. Also of a Heterogeneous Medium.

§ 182. Now consider the effect of an obstacle brought into our medium, say a non-conducting dielectric mass of different inductivity or permittivity. The manner of propagation in it is similar in kind to that in the external medium, but it varies in detail, and the speed will usually be different. Then, when we have a wave coming from (say) a point-source in the first medium, the presence of the obstacle at first makes no difference. It has no immediate action, and, until the wave reaches it, might as well not be there. But immediately the wave does reach it a change occurs. The interface of the two media becomes the seat of sources of fresh disturbances, or they may be considered fictitious sources, in contrast with the original, for they do not bring in any fresh energy. Thence arises a new effect, viz., reflection. Not the whole, but only a part of the wave disturbance enters the new medium. The rest is thrown back, and forms a reflected wave. In the same way as we may trace the course of primary waves from a source in a uniform medium may we trace the course of the secondary waves set up by the obstacle. The disturbance outside it is then due to the superposition of the primary and secondary waves, and, if there be just one obstacle, this state of things continues. Of course the secondary wave-front may itself be complex, because the disturbances going into the second medium and transmitted therein according to its nature, may reach the interface again at other parts, and there suffer reflection and transmission anew. This complication is done away with by making the obstacle infinitely big, with a plane boundary. Then we have just one wave in the obstacle and two in the medium containing the source, viz., the primary and the first (and only) reflected wave.

But if there be a second obstacle, not only will it, like the first, give rise to a secondary wave when the primary meets it, but each obstacle will act similarly towards the secondary wave from the other, whence arises a pair of tertiary waves, and so on. Thus, with only two obstacles, we shall have a succession of infinitely numerous waves crossing and recrossing one another, arising out of the action of a point-source, with great complications. But, however complex in detail, we have the important fact that the mere knowledge of the speed of propagation allows us to lay down the whole course of the waves generated, and the position of the wave-fronts belonging to a given epoch at the source. The mathematics of the full treatment may be altogether beyond human power in a reasonable time; nevertheless, we can always predict with confidence that the results must have such and such general properties relating to the course of the waves, besides the properties involved in the persistence of energy.

In the above we were concerned with discontinuous heterogeneity, or an abrupt change in the value of one or both of the constants c and μ at an interface. When the medium is continuously heterogeneous, or the "constants" change in value continuously from place to place, then the speed v is made a function of position. The process of partial reflection and partial transmission which occurred at the interface now takes place in general wherever the value of c or μ changes. These changes being continuous, so are the results, so that we do not have distinctly separable trains of waves, but rather a continuous distortion. We can, however, follow the course of a wave expanding from a point by communicating to the wave-front the speed proper to its position, such speed being, as above, definitively known. A spherical wave may remain spherical, only varying in its speed; or more likely, it may change its form as it expands. It may become ellipsoidal, for example, and of course must do so if the speed in different parts of the medium vary suitably.

It is possible for the inductivity and permittivity to change, either abruptly or continuously, without any reflection, or casting behind of the disturbance as a wave progresses, although its speed varies. In such a case it will be found that the ratio μ/c is constant. The change in μ, or the ratio of induction to magnetic force, then fully compensates the change in c, the ratio of displacement to electric force, either of which changes by itself would cause reflection.

In connection with a heterogeneous medium (including abrupt changes of nature), we may notice its behaviour as regards steady states, in contrast with that of a thoroughly homogeneous medium. In the latter case, as we have already

described, the introduction of a steady point-source causes
the steady state to begin immediately at and spread around it,
after which no change occurs. That is, the medium being
homogeneous, there is no reflex action. But when it is heterogeneous the case is quite different. The steady state of displacement due to impressed voltage depends upon the permittivity in all parts of the medium, and only comes about as
the final result of the infinitely numerous reactions between
different parts, due to the reflections that take place, which
modify the final distribution of displacement as well as (usually)
the total amount. There is a similar contrast in the mechanism
of the mathematics involved in the calculation of the steady
state. When the medium is homogeneous we may find it
through the potential of fictitious matter at the source alone.
When heterogeneous, we require to have fictitious matter all
over the medium, or rather in all parts where it changes its
nature, and, owing to the infinitely numerous reactions, we
should, in a general or complete solution, require to perform
not one space-integration, but an infinite series of successive
space-integrations, before we could arrive at the true potential
function, allowing for the variations of permittivity everywhere. Thus, by electromagnetic considerations we obtain
some insight of the true nature of transcendental static problems, involving potential functions and assumed instantaneous
action at a distance.

Effect of Eolotropy.   Optical Wave-Surfaces.
Electromagnetic versus Elastic Solid Theories.

§ 183. When the medium is not isotropic as regards either
the displacement or the induction, or as regards both, we have
very remarkable effects, known in the science of optics as
double-refraction. There can be two distinct wave-speeds for a
given position of the wave-front, and these speeds change when
the direction of the normal to the wave-front changes. Hence
double refraction, or the separation of a wave entering an
eolotropic medium into two waves travelling independently of
one another at different speeds. If the medium is only electrically eolotropic, although the displacement and induction
are still in a wave-front (of either wave), and are perpendicular
to one another, the electric force is inclined to the wave-front,

so that the ray, which indicates the direction of the flux of energy, is inclined to the normal to the wave-front. Similarly, with magnetic eolotropy alone, it is the magnetic force that is inclined to the wave-front. When the medium is both electrically and magnetically eolotropic, both the electric and the magnetic force are inclined to the wave front, whilst the displacement and induction, though still in the wave-front, are no longer perpendicular to one another.

By imagining plane electromagnetic sheets to traverse the medium in all possible directions about a point, and comparing their positions with respect to the point at equal times after crossing it, we arrive at the conception of the wave-surface. This is obviously a sphere in an isotropic medium. But when it is eolotropic, the two speeds, and their variation according to the direction of motion of the waves, make the wave-surface become a very singular double surface. With electric eolotropy alone, which is the practical case, it is Fresnel's wave-surface. It is also another Fresnel surface with magnetic eolotropy alone. But when the medium is eolotropic as regards both the induction and the displacement, the wave-surface is of a more general and symmetrical character, including the former two as extreme examples. It is still a double surface, however, except in one case. We have already mentioned that in an isotropic medium there is a peculiar behaviour when the ratio μ/c is constant, although μ and c vary. We might anticipate some peculiarity in the wave-surface when μ/c is constant. This constancy now means that the directional properties of μ are exactly paralleled by those of c. That is, the principal axes of μ and those of c are coincident, whilst the value of the ratio of the permittivity to the inductivity is the same for the three axes. The result is to reduce the double wave-surface to a single surface, which is an ellipsoid.

Assuming light to consist of vibrations of an " elastic solid " medium, single or ordinary refraction is explained by an alteration in the density or the rigidity of the ether. Not only is the theory quite hypothetical in considering the kinetic energy to be the energy of vibrational motions of displacement, but the alteration of density or rigidity assumed is a further hypothesis on the top of the main one, for there is no evidence that there is such a change. Moreover, the explanation will not work properly. On the other hand, the electromagnetic theory says that light consists of electromagnetic vibrations in the ether. This, too, is a hypothesis. But the auxiliary part, that refraction is caused by change of permittivity from one medium to another, is not a hypothesis, but a fact. More-over, the theory works.

Similarly, double refraction in elastic solid theories of light is explained by eolotropy as regards elasticity, or by something similar relating to the density. This is also hypothetical, and not without its troubles. But, on the other hand, Maxwell declared that double refraction occurs because the doubly refracting medium is electrically eolotropic. Now this is a fact too, and the theory is a clear one.

These remarks will serve to illustrate what I mean by the far greater intrinsic probability of the electromagnetic theory, apart from the experimental work of late years. It is much less hypothetical than elastic solid theories. We know nothing about the density or the rigidity of the ether, or how they vary in different bodies, or if they vary. But we do know a good deal about electric permittivity and magnetic inductivity. In-stead of dealing with possibilities, we are dealing with actual facts.

The old objection that a mechanical theory of light was surely to be preferred to an abstract electromagnetic theory was very misleading. The electromagnetic theory is mechan-ical, without, however, a precise specification of the mechanism. An elastic solid theory is merely a special mechanical theory. It cannot satisfy the electromagnetic requirements, but this failure, though immensely important in itself, is not the point here. Even if it did satisfy them, it would probably be less true than the electromagnetic theory, which, being abstract, does not assert so much. There may be many "mechanical" solutions of an abstract theory. Elastic solid theories are a great deal too precise in saying what light consists of, and mechanical speculations in general should be received with much caution, and regarded rather as illustrations or analogies than expressions of fact. We do not know enough yet about the ether for dogmatising.

One can imagine that a clear-headed man might be able to work his way through all the theories of light yet propounded, assimilate what was useful and true, eliminate what was useless

or false, and finally construct a purified theory, not professing
to explain what light is, but still connecting together in a perfectly unobjectionable manner, free from hypotheses, all the
principal facts, and of so compliant a nature as to readily adapt
itself to future discoveries, and finally settle down to a special
form expressing *the* theory of light. It is hardly likely that
the clear-headed man will be found for the purpose, and perhaps his theory would not be very popular. It would be so
very abstract. What is far more likely is that the electromagnetic theory (itself abstract and possessing many of the
desired qualifications), which has already begun to find its
way into optical treatises at the end, will gradually work its
way right through them to the beginning, and in doing so,
oust out the most of the old-fashioned hypotheses. The
result will be to have optical theory expressed throughout in
electromagnetic language. To do it properly, it is hardly
necessary to say that the preposterous 4π, which the B.A.
Committee seem to want to perpetuate, should be ignored from
the beginning.

A Perfect Conductor is a Perfect Obstructor, but does not absorb the Energy of Electromagnetic Waves.

§ 184. So far, in varying the nature of the medium, we have
not introduced any property causing its local waste, such as
the existence of conductivity (finite) necessitates. In all the
varied journeys of electromagnetic waves in a (theoretical) non-
conducting dielectric with non-conducting obstacles there is no
local waste of energy, and the work done by impressed forces is
entirely accounted for by the electric and magnetic energy.
Assuming a certain amount of work done up to a certain
moment, and none later, that amount expresses the energy of
the electromagnetic field, and, however it may vary in distribution, the total remains the same, just as if it were a quantity
of matter moving about, having continuity of existence in time
and space. The useless complication introduced by the circuital indeterminateness of the energy flux may be ignored.
The only way, then, to get rid of the energy is to absorb it at
the sources by working against impressed forces. Putting
that on one side, there is the waste of energy by dissipation
in space, which cannot be stopped. The energy is still in exis-

tence, and is still in the electric and magnetic forms, and must always be at a finite distance from its source and an infinite distance from infinity; but since it gets out of practical reach, and is constantly going out further, it is virtually wasted. Thus, ultimately, the energy of all disturbances generated in a boundless non-conducting dielectric goes out of range and is wasted.

To prevent this, we may interpose a reflecting barrier. Imagine a screen to be introduced, at any finite distance, enveloping the sources, of such a nature as to be incapable of either transmitting or absorbing the energy of waves impinging upon it. Then clearly the dissipation of the energy is stopped, and it all keeps within the bounded region. The waves from the source will be reflected from the boundary, and their subsequent history will be an endless series of crossings and recrossings. The only way to destroy the induction and displacement is to employ artful demons (or impressed forces), so situated and so timing themselves as to absorb the energy of waves passing them, instead of generating more disturbances. In the absence of this demoniacal possession of the region, the energy will remain within it in the electromagnetic form and be in constant motion.

But any opening in the screen, establishing a connection through ether between the inner and outer regions, will at once put a stop to this local persistence of energy. For energy will pass through the opening, and once through, cannot get back again (though a part may), but will escape to infinity. A mere pinhole will be sufficient, if time enough be given, to allow all the energy to pass through it into the external region, and there go out unimpeded to an infinite distance, in the absence of a fresh complete screen to keep it within bounds.

A screen to perform the above-described functions may be made of that very useful scientific substance, the perfect conductor, which is possibly existent in fact at the limiting zero of absolute temperature, the latest evidence being the recent experiments of Profs. Dewar and Fleming, measuring the resistances of metals at very low temperatures. If a perfect conductor, it is also a perfect obstructor, or is perfectly opaque to electromagnetic waves, without, however, absorbing their energy superficially.

Conductors at Low Temperatures.

§ 185. For this reason, a perfectly conducting mass exposed to electromagnetic radiation in ether cannot be heated thereby, for heating would imply the absorption of energy. If, then, the law (whose existence has long been suspected) that metals tend to become perfect conductors when their temperature is sufficiently lowered, were absolutely true, it would follow that a metal, if once brought to zero temperature, would remain there, provided its only source of heat-energy be electromagnetic vibrations. At the same time it is conceivable, and, indeed, inevitable, that it should receive energy from them to some extent in another way, viz., by the electromagnetic stress producing motion in bulk, even though the establishment of irregular molecular motions be prevented.

Similarly, we should expect that a metal which obeys the law approximately would show very small absorptive power for radiation at very low temperatures. This refers to the reception of the energy of true radiation in ether—that is, in vacuo. A body may become heated in other ways, by the impacts of air particles, for example, if air be present. Prof. Dewar's late experiments are suggestive in this respect, but it is too soon to draw conclusions from partially-published experiments.

But even as regards strongly absorptive bodies, the perfectly black body of the thermal philosophers, for example, we may expect a similar diminution of absorptive and emissive power with fall of temperature. Stefan's law of radiation asserts that the emissivity varies as the fourth power of the absolute temperature. Since, however, in deriving this law from electromagnetic principles, as has been recently done by B. Galitzine,* we seemingly require to invoke the aid of reversible cycles and the second law of thermodynamics, whose range of application is sometimes open to question, we may well be excused from an overhasty acceptance of the law as the expression of absolute fact. The second law of thermodynamics itself needs to be established from electromagnetic principles, assisted by the laws of averages, so that we may come to see more clearly the validity of its application, and obtain more distinct notions of the inner meaning of temperature.

---

* *Phil. Mag.*, February, 1893.

Equilibrium of Radiation.   The Mean Flux of Energy.

§ 186.   As before said, a perfectly conducting screen enclosing a dielectric region supporting electromagnetic disturbances,
keeps in their energy, which remains in the electric and magnetic forms, and if there be no sources of energy present, the
total energy remains constant.   Some interesting questions
arise regarding the subsequent history of the electromagnetic
disturbances, when left to themselves, subject to the obstruct-
ing and reflecting action of the screen.

In certain cases the initial state will continue absolutely
unchanged; for example, when it is the steady state due to
electrification, or associated therewith.   Similarly as regards
an initially steady state of magnetic force—that, for instance,
associated with a linear current (without resistance) in the
enclosed region.   Again, in other cases, although the subsequent history of an initial state may be one of constant
change, yet there will be regular recurrence of a series of
states ; as when a periodic state of vibration persists without
any tendency to degrade.   It is sufficient to mention the very
rudimentary case of a plane wave running to and fro between
parallel plane reflecting boundaries, without the slightest tendency to change the type of the vibrations.   We see from
these examples, which may be multiplied, that there is no
necessary tendency for the initial state, even when vibratory,
to break up and fritter down into irregular vibrations.   Nevertheless, there does appear to be a general tendency to this
effect, when the initial states are not so artfully selected as to
prevent it happening.   Even when we start with some quite
simple type of electromagnetic disturbance, the general effect
of the repeated reflections from the boundary (especially when
of irregular form) and the crossing of waves is to convert the
initial simplicity into a highly complex and irregular state of
vibration throughout the whole region.   This cannot happen
universally, as we have seen, and therefore a general proof of
conversion from any initial state to irregularity cannot be
given; but there can be little doubt as to the usual possibility of the phenomenon.   Especially will this be the case
if the initial state be itself of an irregular type, such as that
due to ordinary radiation from matter, when it is tolerably

clear that the irregularity will persist, and become more complete.

Let us, then, assume that we have got the medium within our screen into this state in its extreme form, without enquiry into the intermediate stages. Then the very irregularity gives rise to a regularity of a new kind, the regularity of averages. The total energy, which is a constant quantity, will be half electric and half magnetic, and will be uniformly spread throughout the enclosure, so that the energy density (or energy per unit volume) is constant. As regards the displacement and the induction, they take all directions in turn at any one spot, quite irregularly, but so that their time-averages show no directional preference. Similarly, the flux of energy, which is a definite quantity at a given moment at a fixed spot, is constantly changing in amount and direction. But in virtue of the constancy of the mean density and the preservation of the normal state by constant exchanges of energy, there is a definite mean energy flux to be obtained by averaging results. This mean flux expresses the flux of energy per second across a unit area anywhere situated within the enclosure.

To estimate its amount, let the mean density of the energy be U. This is to include both kinds. Now fix attention upon a unit area, A, fixed in position anywhere within the enclosure, and consider the flux of energy through it under different circumstances. First of all, if the energy all moved the same way at the same speed v (that of propagation), as in simple plane progressive waves, and the direction of its motion were perpendicular to the fixed unit area A, then the energy passing through it would belong to a ray (or bundle of rays) of unit section, and the energy flux would be Uv simply. This is the maximum. But this is impossible, because energy would accumulate on one side of A at the expense of the other. The next approximation, to prevent the accumulation, is to let half the energy go one way and half the other; still, however, in the same line. This brings us down to $\frac{1}{2}$Uv. To go further, we must take all possible directions of motion into account. The original ray conveying Uv must assume all directions in turn, and the mean value of the flux through A must be reckoned. Now, if the ray of unit section be turned round so as to make an angle θ with the normal to A, the effective section of the ray is reduced from 1 to $\cos \theta$; that is, $\cos \theta$ is the fractional part of the ray which sends its energy through A, through which the flux per second is therefore $Uv \cos \theta$, due to the ray at inclination θ. The true flux through the area A is therefore the mean value of $Uv \cos \theta$ for all directions in space assumed by the ray. Now the mean value of $\cos \theta$ for a complete sphere is zero, and therefore the mean flux through A is zero. This is right, as it asserts that as much goes through one way as the other. To obtain the amount going either way we must average over a hemisphere only. The mean value of $\cos \theta$ is then $\frac{1}{2}$. But we are only concerned with half the total energy, or $\frac{1}{2}U$, when we are confined to one hemisphere. Consequently we have

$$W = \tfrac{1}{4}Uv, \quad . \quad . \quad . \quad . \quad . \quad . \quad (4)$$

to express the flux of energy W per second each way through any unit area in the enclosure.

Another way of getting this result, which is, however, essentially the same, is to divide the original ray of unit section along which the flux is Uv into a very great number n of equal rays of unit section, each conveying $1/n$ part of the same, and placed at such inclinations to the normal to A that no direction in space is favoured. This amounts to dividing the surface of a sphere whose centre is that of the area A into n equal parts, the centre of every one of which defines the position of one of the n rays. Any ray now sends $(Uv/n) \cos \theta$ through A per second. Now sum this up over the whole hemisphere and the result is W. In the limit, when n is infinitely great, we have

$$W = \int_0^{2\pi} \int_0^{\frac{\pi}{2}} \frac{Uv \cos \theta}{4\pi} \sin \theta \, d\phi d\theta = \tfrac{1}{4}Uv, \quad . \quad . \quad (5)$$

as before. The 4π divisor in the integral is not the unspeakable 4π of the B.A. units. It is the area of the sphere of unit radius, whilst the other factor $\sin \theta \, d\theta d\phi$ is the area of an element of the sphere.

As this is an important fundamental result in radiation, it is desirable to establish it as generally and simply, and with as much definiteness of meaning as possible. Bartoli made it come to $\frac{1}{2}Uv$ (apart from electromagnetic considerations, which are, however, only accidentally involved). On the other hand, Galitzine makes it $\frac{1}{4}Uv$ by considering a special case, viz., that of a cylindrical enclosure with flat ends, one of which is a radiant source (perfectly black) at constant temperature, whilst the other end and the round surface are perfectly reflecting. It would appear, however, from the above method, that the result is general, and is independent of sources of heat, and of the emissivity and temperature. Since we made no use of the screen after introducing it to keep in the radiation, it may be dispensed with, provided the stationary condition be still maintained. Thus, if a portion of the screen be made "perfectly black," maintained at constant temperature, the quantity W, which represents the amount of heat falling upon it per unit area per second, is completely absorbed by it. But this is perfectly compensated by the emission from the black surface of an equal amount of heat. So $\frac{1}{4}Uv$ measures the total emissivity under the circumstances.

The Mean Pressure of Radiation.

§ 187. Another important fundamental quantity is the mean pressure of radiation. In a simple ray the electric and magnetic stresses unite to form a pressure U along the ray, with no pressure or tension in lines perpendicular to the ray, that is, in the plane of the wave, as described in § 86. But when the radiation is balanced as in the last paragraph, there is no directional preference, and the pressure is all ways in turn, and therefore, on the average, simulates a hydrostatic pressure. Its value may be readily estimated in a manner similar to that employed above concerning the mean flux of energy. When we make the energy U go all one way in a ray of unit section through the area A situated anywhere, the pressure in the ray is U, and this is the pressure on A if the ray is perpendicular to A. But when the ray is inclined to the normal to A at an angle θ, only the fraction $\cos \theta$ of the ray is concerned in the action upon A. Furthermore, the line of pressure is inclined to the normal to A at an angle θ, so that the effective normal pressure is still further reduced by the factor $\cos \theta$ a second time. The pressure on A is therefore only $U \cos^2 \theta$. This must now be averaged for all directions in space that we may give to the ray, without preference. Now the mean value of

$\cos^2\theta$ for the complete sphere is $\frac{1}{3}$. So the pressure, say p, is given by

$$p = \tfrac{1}{3}U. \quad . \quad . \quad . \quad . \quad . \quad (6)$$

Notice particularly that we take the mean for the complete sphere, not merely for the hemisphere as in the former calculation. The reason is that whether a ray goes through A from right to left or from left to right, the pressure is the same : so both ways have to be reckoned.

Or, we may divide the original ray in which the pressure is U into a great number n of rays also of unit section, in each of which the pressure is U/n. Let the axes of these rays be defined by the middle points of n small equal areas into which we may divide the surface of a sphere, and sum up the normal pressures on A. We obtain, in the limit,

$$p = \int_0^{2\pi} \int_0^{\pi} \frac{U \cos^2\theta}{4\pi} \sin\theta \, d\phi d\theta = \tfrac{1}{3}U. \quad . \quad . \quad (7)$$

This result was given by Boltzmann some years ago, and Galitzine confirms it for the special case of his straight cylinder with a radiant surface at one end. By the above method we see that the result is general, resulting from the uniformity of radiation in all directions, as the previous formula for the mean flux of energy did.

Emissivity and Temperature.

§ 188. This pressure p is not only the mean pressure throughout the enclosure, but also the pressure on its envelope, which exerts an equal back pressure. If it move, then work is done by or against the enclosed radiation through the agency of its pressure, and the enclosure loses or gains energy to an equal extent. Observe that the idea of temperature does not enter explicitly when the boundary of the enclosure is of ideal perfectly reflecting material. But when it, or a part of it, is made absorptive and emissive, the physics of the matter becomes far more difficult and to some extent dubious. The notion of temperature comes in, and with it the second law of thermodynamics. Assuming its full applicability, Stefan's law follows easily enough, as Galitzine has shown. Let the enclosure be a cylinder of unit section with two pistons A and B. Let A be fixed, and be a perfectly black body, whilst B is movable to and from A and is a perfect reflector, as is also the round cylinder.

Now start with B close to A; the volume of the enclosure is then nil. Draw B very slowly away from A through the distance h, keeping A at constant temperature t all the time, and then stop it. During this operation the source A keeps the enclosure filled with energy to density U, and pressure p, corresponding to the temperature t at which A is maintained. The pressure p does the external work ph. Besides that, there is energy Uh in the enclosure at the end of the operation. Their sum is therefore the heat lost by A (excess of heat emitted by A into the enclosure over that returned to A). Say

$$H = (U + p)h. \quad . \quad . \quad . \quad . \quad . \quad (8)$$

Now we know p in terms of U, so that we have, by (6),

$$H = \tfrac{4}{3}Uh. \quad . \quad . \quad . \quad . \quad . \quad (9)$$

Now, B being fixed, let the cylinder cool down to zero temperature. The whole of the energy Uh in the enclosure goes out through A. Lastly, push B back to A without working, and then raise A to temperature t. A cycle is then completed. Applying the second law, we have

$$\frac{H}{t} = \int_0^t \frac{1}{t}\frac{dH}{dt}dt, \quad . \quad . \quad . \quad . \quad (10)$$

where H is as before, and dH is an element of the heat lost in the cooling process. On the left side put $(U + p)h$ or $\tfrac{4}{3}Uh$ for H, because external work was done in the first operation. On the right side leave out the p term, because during the cooling B was fixed. So we get

$$\frac{\tfrac{4}{3}U}{t} = \int_0^t \frac{1}{t}\frac{dU}{dt}dt, \quad . \quad . \quad . \quad (11)$$

by omitting the factor h. Differentiate with respect to t, and we get

$$\frac{dU}{dt} = \frac{4U}{t}, \quad . \quad . \quad . \quad . \quad (12)$$

from which we conclude that U, and, therefore, also p and W, vary as the fourth power of the temperature, the result above mentioned as applied to the emissivity W.

We tacitly assume that the ether is able to escape freely from the cylinder through its envelope, or else that it is freely compressible, without resistance. This difficulty in connexion with the ether is a very old one.

Internal Obstruction and Superficial Conduction.

§ 189. The properties of a perfect conductor are derived from those of common conductors by examining what would happen if the resistivity were continuously reduced, and ultimately became zero. In this way we find that a perfect conductor is a perfect obstructor, for one thing, which idea is singularly at variance with popular notions regarding conductors. But it is also a perfect conductor literally, though in a different sense to that commonly understood. Ohm's law has played so important a part in the development of electrical knowledge, especially on the practical side, that it is really not at all a matter of wonder that some practicians should have been so reluctant to take in the idea of a conductor as an obstructor. Scientific men who can follow the reasoning by which the functions of conductors follow from known facts have no difficulty in pursuing the consequences far beyond experimental observation. Again, younger men, with fewer prejudices to surmount, do not find much trouble with superficial conduction and internal obstruction. But the old established practitioner with prejudices, who could not see the reason, was put into a position of some difficulty—resembling chancery. If you have got anything new, in substance or in method, and want to propagate it rapidly, you need not expect anything but hindrance from the old practitioner—especially if he sat at the feet of Faraday. Beetles could do that. Besides, the old practitioner is apt to measure the value of science by the number of dollars he thinks it is likely to bring into his pocket, and if he does not see the dollars, he is very disinclined to disturb his ancient prejudices. But only give him plenty of rope, and when the new views have become fashionably current, he may find it worth his while to adopt them, though, perhaps, in a somewhat sneaking manner, not unmixed with bluster, and make believe he knew all about it when he was a little boy! He sees a prospect of dollars in the distance, that is the

z

reason. The perfect obstruction having failed, try the perfect conduction.

You should make your converts out of the rising generation and the coming men. Thus, passing to another matter, Prof. Tait says he cannot understand my vectors, though he can understand much harder things. But men who have no quaternionic prejudices can understand them, and do. Younger men are born into the world with more advanced ideas, on the average. There cannot be a doubt about it. If you had taught the Calculus to the ancient Britons you would not have found a man to take it in amongst the whole lot, Druids and all. Consider too, what a trouble scientific men used to have with the principle of the persistence of energy. They could not see it. But everybody sees it now. The important thing is to begin early, and train up the young stick as you want it to grow. Now with Quaternions it is different. You may put off till to-morrow what you cannot do to-day, for fear you commence the study too soon. Of course, I refer to the Hamilton-Tait system, where you have to do violence to reason by making believe that a vector is a quaternion, and that its square is negative.

According to Ohm's law alone, a perfect conductor should be one which carried an infinite current under a finite voltage, and the current would flow all through it because it does so ordinarily. But what is left out of consideration here is the manner in which the assumed steady state is established. If we take this into account, we find that there is no steady state when the resistance is zero, for the variable period is infinitely prolonged, and Ohm's law is therefore out of it, so far as the usual application goes. In a circuit of no resistance containing a finite steady impressed voltage E, the current would mount up infinitely and never stop mounting up. On the other hand, if we insert a resistance R in the former circuit of no resistance, there will be a settling to a steady state, for the current in the circuit will tend to the value E/R, in full obedience to Ohm's law. The current is the same all round the circuit, although a part thereof has no resistance. We conclude that that portion has also no voltage.

But this is only a part of the story. Although we harmonise with Ohm's law, we overlook the most interesting part. The

smaller the resistance the greater the time taken for the
current to get into the conductor from its boundary, where it
is initiated. In the limit, with no resistance, it never gets in
at all. Where, then, is the current? For, as we have said,
it mounts up to a finite value if there be a finite resistance
inserted along with the perfect conductor, and mounts up
infinitely if there be no resistance.

We recognise the existence of electric current in a wire by
the magnetic force round it, and in fact measure the current
by its magnetic force. Therefore, according to this, there is
the same total current in the wire, if the magnetic force outside it remains the same. If, then, the magnetic force stops
completely at the surface of the wire, whose interior is entirely
free from magnetic force, the measure of the current is just
the same. The uniformly distributed current of the steady
state appropriate to finite conductivity becomes a mere surface
current when the conductivity is infinite. In one case we
have a finite volume-density of current, and in the other a
finite surface-density. When the current inside the wire is
zero so is the electric force, in accordance with Ohm's law
again. The electric and magnetic phenomena are entirely in
the dielectric outside the wire, the entrance of any similar
manifestations into it being perfectly obstructed by the absence
of resistance. For this purpose the thinnest skin would serve
equally well. In the usual sense that an electric current is a
phenomenon of matter, it has become quite an abstraction, for
there is no matter concerned in it. It is shut out completely.
In the circuit of finite resistance, a portion of which is a wire
of no resistance, supporting a steady current, there is no
difference whatever in the external magnetic force outside the
resisting and non-resisting parts, though in one case there is
entrance of the magnetic force and waste of energy, whilst in
the other there is no entrance and no waste. These conclusions do not rest upon Maxwell's theory of dielectrics, but
upon the second circuital law of electromagnetism applied to
conductors. But it is only by means of Maxwell's theory that
we can come to a proper understanding and explanation of the
functions of conductors.

The sense in which a perfect conductor is a perfect conductor in reality as well as in name is that it allows electro-

z 2

magnetic waves to slip along its surface in a perfectly free manner, without waste of energy. Though perfectly obstructive internally, it is perfectly conductive superficially. It merely guides the waves, and in this less technical sense of conduction the idea of a perfect conductor acquires fresh life.

The Effect of a Perfect Conductor on External Disturbances. Reflection and Conduction of Waves.

§ 190. The conditions at the interface of a perfect conductor and a dielectric are that the electric force in the dielectric has no tangential component and the magnetic induction no normal component. Or

$$\text{VNE} = 0, \qquad\qquad \text{NH} = 0,$$

if N be the unit normal from the conductor. Thus, when there is electric force at the boundary it is entirely normal, with electrification to match; and if there is magnetic force it is entirely tangential, with electric current to match. Both electrification and current are superficial. The displacement measures the surface density σ of the one, and the magnetic force that of the other, say c, thus

$$\sigma = \text{ND}, \qquad\qquad \text{c} = \text{VNH},$$

in rational units, without any useless and arbitrary 4π constant, such as is required in the B.A. system of units, of amazing irrationality. If, then, we have electromagnetic disturbances given in a dielectric containing a perfect conductor, the latter first of all is free from disturbance, and next causes such reflected waves as to annihilate the tangentiality of the electric force and the normality of the magnetic force.

As regards steady states, the influence of a perfect conductor on induction due to foreign sources is to exclude it in the same manner as if the inductivity were made zero; that is, the induction goes round it tangentially instead of entering it. This is usually ascribed to an electric current-sheet induced upon its surface, whose internal magnetic force is the negative of that due to the external field. This is right mathematically, but is deceptive and delusive physically. There is no internal force, neither that of the external field nor that of the superficial current. The current sheet itself merely means the abrupt stoppage of the magnetic field, and cannot really be supposed to be the source of magnetic force in a body which cannot permit its entrance. The previously mentioned case of a perfectly conducting wire inserted in a circuit of finite resistance supporting a steady current, will serve to bring out this point strongly. The supposed induced superficial current is now actually the main current in the circuit itself.

It is different with the steady state due to external electric sources. The displacement is just as much shut out from the perfect conductor (which may also be a dielectric) as was the magnetic induction, but in a strikingly different manner, terminating upon it perpendicularly, as if it entered it in the manner that would happen were the conductor nonconducting, but of exceedingly great permittivity, so that it drew in the tubes of displacement.

Although a perfect magnetic conductor is, in the absence of knowledge even of a finite degree of magnetic conductivity, a very far-fetched idea, yet it is useful in electromagnetic theory to contrast with the perfect electric conductor. A perfect magnetic conductor behaves towards displacement just as a perfect electric conductor does towards induction; that is, the displacement goes round it tangentially. It also behaves towards induction as a perfect electric conductor does towards displacement; that is, the induction meets it perpendicularly, as if it possessed exceedingly great inductivity, without magnetic conductivity. This magnetic conductor is also perfectly obstructive internally, and is a perfect reflector, though not quite in the same way as electric conductors. The tangential magnetic force and the normal electric force are zero.

As regards waves, there are two extreme ways in which a perfect conductor behaves—that is, extreme forms of the general behaviour. It may wholly conduct them, or it may wholly reflect them. In the latter case we may illustrate by imagining a thin plane electromagnetic sheet, consisting of crossed electric and magnetic forces in the ratio given by $E = \mu v H$, moving at the speed of light, to strike a perfect conductor flush—that is, all over at the same time, by reason of parallelism of the sheet and conducting surface. The incident sheet is at once turned into another plane sheet, which runs away from the conductor as fast as it came. If the conductivity be

of the electric kind the reflected sheet differs from the incident
in having its displacement reversed, but in no other respect.
This is perfect reflection with reversal of **E**. During the act
of reflection, whilst the incident and reflected sheets partly coincide, **E** is zero and **H** is doubled. Both are tangential ; but
there can be no tangential **E**, so the reflector destroys **E** and
initiates the reflected sheet, in which **H** is the same as in the
incident sheet, whilst **E** is reversed.

On the other hand, when the conductivity is of the magnetic
kind, the reflected wave sheet differs from the incident only in
having its induction reversed. The displacement persists,
being doubled during the act of reflection, whilst the induction
is then annulled.

The other extreme occurs when a plane electromagnetic sheet
hangs on to a conductor perpendicularly. It then slips along
the conductor at the speed of light, with perfect slip. This
may occur with a plane reflector, but, of course, the most
striking and useful and practical case is that of a straight
cylinder—a wire, in fact, though it need not be round, but may
have any form of section. The wave then runs along the wire
at constant speed v, and without change of type, at least so
long as the wire continues straight and of unchanged section.
If the section vary regularly, so that the wire is a cone, then it
is a spherical wave that is propagated along it without change
of type. This case includes an infinitely fine wire, when we
may have either spherical or plane waves. Other interesting
cases may be made up by varying the angle of the cone, or
using a double cone, or a cone and a plane, &c.

Now, in the first main case of perfect reflection (flush) the
incident and reflected sheets are wholly separated from one
another, except just at the reflecting surface, where there is
a momentary coincidence. On the other hand, when a plane
wave runs along a wire, or, say, more conveniently here, along
a plane, we only see one wave. It is the case of reflection at
grazing incidence, and may be considered the limiting case of
permanent union of the incident and reflected waves. Between
these two cases we have the general case of incidence and reflection at any angle. There are two plane waves (sheets, most
conveniently for reasoning and description) one going to and the
other coming from the plane reflector, where they join together,

making equal angles with it. In the overlapping region, close
to the plane, the displacement due to the union of the two
waves is normal to the surface (that is, with an electric conductor) which is electrified, and the electrification runs along
the surface at a speed depending upon the angle of incidence,
being v at grazing incidence (of rays) and $v/\cos \theta$ at incidence
angle θ, varying, therefore, between v and infinity. It may be,
perhaps, rather a novel idea to some readers that electrification
can run through space at any speed greater than that of light,
but the matter is made simple enough by considering the rate
of incidence upon the reflecting surface of different parts of the
plane sheet. In the case of nearly flush incidence of a sheet,
its different parts strike the surface nearly simultaneously, so
that there is an immensely great speed of motion of the electrification along the surface. The electrification is the same in
amount always, and is continuously existent, so we are somewhat justified in speaking of the electrification moving; but
we may equally well regard it as a case of continuous generation of electrification at one end and of annihilation at the
other end of the part of the conducting surface which is
momentarily charged, the generation and annihilation being
performed by the different parts of the incident sheet and the
reflected sheet as they reach and leave the surface. Details of
these simple cases, leading to a plainer understanding, will
come later, when these general notions are got over. In the
limiting case of normal incidence of rays, when the incident
sheet strikes the surface flush, the electrification is non-existent.
It goes out of existence just as its speed becomes infinite.

The above describes one extreme kind of reflection of a plane
sheet at any angle, and is what occurs when H in the incident,
and, therefore, also in the reflected wave, is tangential to the
reflector, whilst E is in the plane of incidence. But when it
is H in the incident wave that is in the plane of incidence,
and E is tangential, we have quite another kind of composition
in the overlapping part near the reflector. There is no E within
it at all, and also no electrification; whilst the H within
it is parallel to the reflector, and simply joins together the H's
in the parts of the waves which do not overlap, the H in one
wave being directed towards the surface in the plane of incidence, and in the other away from the surface. In both waves, of

course, H is in the plane of the wave. In other respects we have a similarity in propagation. It is now a surface current, instead of electrification, that runs along the surface at speed $v/\cos\theta$.

If the reflector be a magnetic conductor, we have two very similar main cases, in one of which magnetification (the analogue of electrification), and in the other a magnetic current, runs, or appears to run, along the surface.

The Effect of Conducting Matter in Diverting External Induction.

§ 191. The theory of the effect of a finite degree ot conductivity in the medium on electromagnetic waves is far more difficult and complicated than that of the effect of infinite conductivity. Nevertheless, we may gain a general idea of the nature of the effect by means of the substitution of simple problems for the real ones that present themselves, and also in this way obtain a knowledge of very important properties concerned. There are two ways in which we may regard the question of conductivity. First, starting from the theory of perfect conductors surrounded by perfect dielectric non-conductors, we may examine the effect of introducing a slight amount of resistivity, to be then increased more and more until at last we come to conductors of high resistivity, or infinite, when we have dielectrics merely. The other way is to start with electromagnetic waves in a perfect dielectric, and examine the effect produced upon them of introducing first a small amount of conductivity, then more and more, until we come to perfect conductors again. Both ways are instructive—none the less because they give very different views of the same matter.

In the first place, it may be readily conceived that if a conductor have only a slight amount of resistivity it may behave approximately the same as if it had none, and may obstruct waves nearly perfectly internally, and likewise reflect and conduct them superficially nearly perfectly. This is true, but the element of time has to be taken into account, as it becomes of great importance when there is some resistivity, however little, instead of quite none. Suppose, for example, we have an initially steady magnetic field, and bring a conductor into it very quickly from a distance, and keep it there. If this be done quickly enough, the first effect is nearly the same as if the conductor were perfect. That is, the induction of the field will be driven out of the space it previously occupied, which is now occupied by the conductor, and pass round it tangentially, as if the conductor were of zero inductivity. It will, therefore, be mechanically acted upon by the stress in the field in a manner resembling the action upon a perfect diamagnetic body—that is, there will be a force tending to drive it from stronger to weaker parts of the field. In another form, we may say that external force must be applied to the conductor to bring it into the field and to hold it there. But this state of things will not continue. The magnetic force, which is at first only skin deep, will penetrate into the interior in time, according to a law resembling that of the diffusion of heat. Given time enough, it will assume the same distribution as if there were no conductivity, although our assumption is that there is nearly no resistivity— that is, in the ultimate state tended to, it is merely the inductivity that settles how the magnetic induction will distribute itself. Initially, there is a skin current, its total being measured by the difference in the magnetic force just outside and a little way inside the conductor, in the bulk of which there is practically no magnetic force or electric current. As in the case of a perfect conductor, Ohm's law is fully obeyed. There is no internal current, because there is no internal electric or magnetic force. They simply have not had time to get in, on account of the obstructive action of the high conductivity. It is as useless and misleading to say that one electric force is cancelled by another, as it is to ascribe the absence of magnetic force to the counteracting magnetic force of the skin current. As the magnetic force spreads into the conductor, so does the electric current, which is the curl of the former. In the end, when the magnetic force has got steady, the current ceases. There may now still be moving force on the conductor, but not of the same kind as before, it being simply the ordinary paramagnetic attraction or diamagnetic repulsion according as the value of the inductivity of the body exceeds or is less than that of the external medium.

It will be very much the same thing if we start with the conductor at rest in a neutral state in a neutral field, and then establish a steady magnetic field by some external cause. There will still be the same internal obstruction offered by the conductor, and consequent delay in the assumption of the steady state throughout it. In both cases, too, there is a necessary waste of energy involved in the process, according to Joule's law of the generation of heat by the existence of electric current in conducting matter. Thus two things happen when the degree of conductivity is not infinite. First, the reflection is imperfect at the boundary of the conductor, a portion of an incident disturbance being transmitted into it. Next, in the act of transmission and the attenuation involved, there is a loss of energy from the electromagnetic field. We shall see later more precisely the nature of the attenuating process. At present we may note that if the external field be not steady or do not tend to a steady state, so that the conductor is exposed to fluctuating forces, then the internal part of the conductor need never acquire any sensible magnetic force. Thus, if the external field be the sum of a balanced alternating, and of a field which would be steady in the absence of the conductor, only the latter part will penetrate fully into it. The former alternating part will penetrate imperfectly, the more so the greater the conductivity, and, as before said, not at all when the conductivity is perfect. The other field is then also excluded. With rapid alternations the region of sensible penetration is only skin-deep, consisting of layers of opposite kinds (as regards direction of the magnetic force, which is nearly tangential), with, however, so very rapid an attenuation of intensity in going inward that practically only one wave need be considered (except for short waves, like light). Of course, the heat generation is now confined to the skin. What goes in further does so by ordinary heat diffusion.

The time-constant of retardation of a conductor varies as the conductivity, as the inductivity, and as the square of the linear dimensions. This refers to the intervals of time required to establish a definite proportion of the steady state under the action of steady forces—in bodies of different size, conductivity, and inductivity, but geometrically similar. Here the two properties, conductivity and inductivity, act conjointly, so that, for example, iron is far more obstructive than copper, although its conductivity is much inferior. It is different with the heat-generation. There the inductivity and conductivity act in opposite senses, for, with the same electric force, the waste varies as the conductivity, or, with the same current-density, as the resistivity. It results that in cases of skin-conduction of rapidly alternating currents, the resistance per unit area of surface varies directly as the square root of the product of the resistivity (not conductivity), inductivity, and frequency. Thus, whilst we may increase the resistance by increasing the frequency, with a given material, and also by increasing the inductivity, we decrease it by increasing the conductivity, in spite of the fact that the internal obstruction varies as the conductivity and inductivity conjointly. The point to be attended to here is that mere internal obstruction is no necessary bar to effective skin conduction, although, of course, in a given case the resistance is greater than if the conduction were more widespread. It depends upon how it is brought about, whether by conductivity or inductivity. This is how it comes about that with the complete internal obstruction of a perfect conductor, with the effective skin reduced to nothing, there is still no resistance, and the slip of electromagnetic waves along them is perfect. But it is different when we obtain the internal obstruction by increasing the inductivity, preserving the conductivity constant. Perfect internal obstruction then means infinite resistance, and no proper slipping of waves at all. If the obstruction be not complete, it will be accompanied by very rapid attenuation of waves running along the surface when the obstruction arises from high inductivity, and by relatively very slight attenuation when it arises from high conductivity.

The repulsive force which was referred to in the case of a perfect conductor brought into a magnetic field, or when a magnetic field is created outside the perfect conductor, arises from its obstructive action, combined with the fact that it is only the lateral pressure of the magnetic stress that acts on the conductor, owing to the tangentiality of the magnetic force. This repulsive force is naturally also operative, though in a less marked form, when the conductivity is not perfect. In fact, it is operative to some extent whenever there is a sufficiently rapid alternation of the field for the conductance to cause a sensible departure from the undisturbed state of the magnetic force, and is, therefore, strongly operative with ordinary metallic conductors with quite moderate frequency of vibration. The

remarkable experiments of Prof. Elihu Thomson on this "electromagnetic repulsion" will be remembered. The phenomenon
has nothing specially to do with electromagnetic waves. It is
magnetic repulsion, rather than electromagnetic, using the
word "magnetic" in its general sense, apart from the special
fact of magnetisation when the inductivity of the conductor
is not the same as that of the ether.*

When a conductor is brought into an electric instead of a
magnetic field, the case is somewhat different. There is in
any case merely skin conduction, for there is an actual destruction of the flux displacement by conductivity. The final result
therefore, is that the ultimate permanent state in the conductor is a state of perfect neutrality, just as if the conductivity were perfect. Electric conductivity destroys displacement, but it cannot destroy induction. Similarly, magnetic
conductivity would destroy induction, but would be unable to
destroy displacement. Thus, a magnetic conductor brought
into an electric field would, in time, permit its full penetration, but if brought into a magnetic field the final result would
be a state of internal neutrality, however low the conductivity.
If, however, it be very low, then, whether it be of the electric
or the magnetic kind, there will be an initial nearly complete
penetration (owing to the removal of the obstruction), followed
by subsidence to zero of the flux appropriate to the conductivity, electric or magnetic respectively. The persistence of
magnetic induction, in spite of the presence of electric con-

---

* This reference to Elihu Thomson's experiments must not be understood as a full explanation, which is sometimes complex. The idea in the
text has been of a lump of metal. When made a disc or a linear circuit
we have special peculiarities, and the theory may be perhaps best done
in terms of inductances and resistances. The principle concerned of
the temporary diversion of magnetic induction by conducting matter
remains in force, however, whether the matter be in a lump or in a closed
line. In the latter case the tendency of the conductance is to keep the
total induction through it constant. Consider first an infinitely conducting disc which completely diverts induction ; and next, a ring made
by removing nearly all the disc except the outer part. Induction now
goes through the circuit, of course, when brought into a magnetic field,
but its total is zero, by reason of the infinite conductance and the current
"induced" in the ring. As is well known, Maxwell considered perfectly
conducting linear molecular circuits in applying his views to Weber's
theory of diamagnetism.

ductivity, is a very important and significant fact, of which I shall give a simple proof in a later Section, in amplification of my proof of 1887-8. The present method of passing from perfect to finite conductivity is unsuitable for the purpose, because in good conductors the dielectric permittivity is altogether swamped, and is therefore ignored; whilst, on the other hand, in very bad conductors the permittivity is a factor of the greatest importance. Now we can pass continuously from a non-conducting dielectric to a conducting one, up to perfect conduction, but we cannot pass the other way without having the permittivity in view all the time, which makes the matter difficult.

Parenthetical Remarks on Induction, Magnetisation, Inductivity and Susceptibility.

§ 192. As people's memories are very short, and there is some discussion on the subject, I may repeat here that the so-frequently-used word inductivity is not intended for use as a mere synonym for permeability. The latter is the ratio of the inductivity of a medium to that of ether, and is therefore a mere numeric. On the other hand, inductivity has a wider meaning, namely, such that $\frac{1}{2}\mu H^2$ is the density of the magnetic energy, irrespective of dimensions. We can only make it a numeric by assumption. Even then, it has only a fictitious identity with permeability—a forced numerical identity. Similar remarks apply to some other quantities, but they are particularly necessary in the case of inductivity, on account of the obscure and misleading manner in which the connections between induction, magnetic force, and magnetisation were formerly commonly presented (and still are sometimes), together with the misleading connection between the susceptibility and the permeability. We should write

$$p = 1 + \kappa,$$

if p is the permeability and κ the susceptibility, instead of

$$p_i = 1 + 4\pi\kappa_i,$$

where the suffix letter refers to the common irrational reckoning. The 4π is, as usual, simply nonsense, unworthy of scientific men near the end of the nineteenth century. Now introduce

Maxwell's ether theory, and make $p = \mu/\mu_0$, where μ_0 is the inductivity of ether, and μ that of some other substance (with the usual reservations), then

$$\mu = \mu_0(1 + \kappa)$$

takes the place of the common

$$\mu_i = 1 + 4\pi\kappa_i,$$

which is misleading in two respects, first as regards the obstreperous 4π, and next in making μ and κ be quantities of the same kind. But κ is always and essentially a numeric, whilst μ is not. We see that the use of inductivity rather than permeability is necessary in electromagnetic theory, as a matter of logical common sense as well as for the purpose of scientific clearness. But this need not interfere with the use of permeability in its above-described sense of a ratio. If, on the other hand, one of the two words should be abolished, there can, I think, be little doubt as to which should go.

In the case of purely elastic magnetisation (without intrinsic) we have

$$\mathbf{B} = \mu\mathbf{F} = \mu_0(1 + \kappa)\,\mathbf{F},$$

where $\mathbf{B}$ is the induction and $\mathbf{F}$ the magnetic force. Here $\mu_0\mathbf{F}$ is what the induction would be in ether, so that the additional part $\mu_0\kappa\mathbf{F}$ expresses the effect of the matter present, which becomes magnetised. The ratio κ, therefore, naturally expresses the susceptibility for magnetisation of the matter. Perhaps, in passing, it might be thought that $\mu_0\kappa$ should express the susceptibility. But this will not do, because magnetisation and induction are similar. The induced magnetisation is $\mu_0\kappa\mathbf{F}$. In strictness it should not be called the intensity of magnetisation, but rather the density, if we properly carry out Maxwell's principles about forces and fluxes, or intensities and densities. $\mathbf{B}$ is a flux, therefore so is $\mu_0\kappa\mathbf{F}$, to be measured per unit area. Now, the common form is

$$\mathbf{B}_i = \mu\mathbf{F}_i = (1 + 4\pi\kappa_i)\mathbf{F}_i,$$

or
$$\mathbf{B}_i = \mathbf{F}_i + 4\pi\mathbf{I}_i,$$

if $\mathbf{I}_i = \kappa_i\mathbf{F}_i$. Here we have, apart from the 4π absurdity, an irrationality of a different kind, viz., that induction and magnetisation are made identical in kind with magnetic force, since we have the difference of two flux densities expressed by an "intensity," which is referred to unit length. Now, this may matter very little in practical calculations, but it is more than mischievous in theory. Suppose, for example, we are working with a kind of mathematics that takes explicitly into account the two ways of measuring vector magnitudes, with reference to length and to area respectively, according as they are regarded as " flux " densities or "force " intensities. Then, if we do not recognise and take account of the radical difference between B_i and F_i in the last equation, we may expect to be led to singular and unaccountable anomalies. This is, I think, what has happened in Mr. MacAulay's recent paper on the theory of electromagnetism. The remedy is easy. There should be no special limitations imposed upon the quantities concerned such as occur when permeability and inductivity are made the same.

It is also highly desirable, for the same purpose of obtaining scientific clearness and freedom from distressing anomalies, that the distinction between "induced" and intrinsic magnetisation should be clearly recognised and admitted in the formula. In the above there has been no intrinsic magnetisation. Let this now be I_0. An equivalent form is μh_0, where h_0 is the corresponding intrinsic magnetic force. This I_0 is of the same nature as B. The complete induction becomes

$$B = \mu(F + h_0) = \mu H,$$

where H is the complete force of the flux B. This is the best way of exhibiting the induction. If H be split at all, let it be into the part involving the intrinsic force h_0 and the rest. Or,

$$B = I_0 + \mu F.$$

The other separation, namely, of μF into the ether induction $\mu_0 F$ and the additional part due to matter, is less useful. If done, then

$$B = I_0 + I + \mu_0 F.$$

If we now amalgamate I_0 and I, to make, say, I_1, the total magnetisation, intrinsic and induced, we have

$$B = I_1 + \mu_0 F,$$

which, translated into irrational units, makes

$$B_i = 4\pi I_{1i} + \mu_0 F_i;$$

and lastly, omitting the μ_0 by assuming it to be unity, we obtain the common

$$\mathbf{B}_i = \mathbf{F}_i + 4\pi \mathbf{I}_{1i},$$

containing three faults, the arbitrary 4π, the equalising of an intensity and a flux, and the unnatural union of physically distinct magnetisations.

"Different men have different opinions—some like apples, some like inions." But can anyone possibly really like the roundabout and misleading way of presenting the magnetic flux relations which I have above criticised? There is no excuse for it, except that it was employed by great men when they were engaged in *making* magnetic theory, before they had assimilated its consequences thoroughly. When the rough work of construction is over, then it is desirable to go over it again, and put it in a better and more practical form. We should copy the virtues of great men, if we can, but not their faults.

Men who are engaged in practical work can hardly be expected to fully appreciate the importance of these things, because their applications are of so highly specialised a nature, in the details of which they may become wholly absorbed. They may even go so far as to say that the paper theory of magnetic induction is not of the least moment, because they are concerned with iron, and although there may be a certain small range of application of the theory even in iron, yet they are scarcely concerned with it, and, therefore, it is of no consequence. There could not be a greater mistake. A complete theory of magnetic induction, including hysteresis, must necessarily be so constructed as to harmonise with the limited theory that has already been elaborated, which is understandable when exhibited in a purified form, freed from 4π's and other anomalies. First of all, we have the ether, in which $\mathbf{B} = \mu_0 \mathbf{F}$ or $\mathbf{B} = \mu_0 \mathbf{H}$, because of the absence of intrinsic magnetisation. Next we come to elastically magnetised bodies in which the relation between flux and force is linear. Then $\mathbf{B} = \mu \mathbf{H}$, where μ differs from μ_0, being either greater or smaller, and is either a constant scalar, or else (with eolotropy) a linear operator. If there is no intrinsic magnetisation, $\mathbf{F}$ and $\mathbf{H}$ are still the same, and the curl of either is the current density. But should there be intrinsic magnetisation, then $\mathbf{H} = \mathbf{h}_0 + \mathbf{F}$, whilst $\mathbf{B} = \mu \mathbf{H}$ still, and it is the curl of $\mathbf{F}$ that is the current density. Or, $\mathbf{B} = \mathbf{I}_0 + \mu\mathbf{F}$. The next step is to make $\mu\mathbf{F}$ be not a linear, but some other function of $\mathbf{F}$, to be experimentally determined, if it be possible to express $\mathbf{B} - \mathbf{I}_0$, which is the free induction, as a function of $\mathbf{F}$. Of course, it can be done approximately. Then comes the difficult question of hysteresis. This involves the variation of $\mathbf{I}_0$ with $\mathbf{F}$, with consequent waste of energy. If this little matter be satisfactorily determined, we may expect to have a sound mathematical theory of magnetic induction in an extended form which shall properly harmonise with the rational form of the elementary theory. The divergence of $\mathbf{B}$ is zero all through. The experimental justification of this generalisation is the fact that no unipolar magnets have yet been discovered.

Effect of a Thin Plane Conducting Sheet on a Wave. Persistence of Induction and Loss of Displacement.

§ 193. Coming now to the effects produced on electromag‑netic waves by a small amount of conductivity, to be after‑wards increased, we shall adopt a particular device for simplify‑ing the treatment. Imagine, first, the dielectric medium to possess a uniformly‑distributed small conductivity. Evidently, the action of the conductivity on a wave is a continuous and cumulative one. Next, localise the conductance in parallel sheets—that is, substitute for the uniform conductivity a great number of parallel plane conducting sheets, between which the medium is non‑conducting. If we increase their number suffi‑ciently, their action on a wave whose plane is parallel to that of the sheets will approximate, in the gross, to the effect of the uniform conductivity which the conductance of the sheets replaces. We have, therefore, to examine the influence of a single very thin conducting sheet upon a wave. This is not difficult.

Imagine, then, a simple plane electromagnetic sheet to be running through the ether at the speed of light. This is the natural state of things ; and, in the absence of conductivity or other disturbing causes, there will be no change. Now insert a plane conducting sheet in the path of the wave. It should be so thin that the retarding effect of diffusion within it is quite insensible. Let the wave strike it flush. The theory shows that it is immediately split into two similar waves, one of which is transmitted beyond the plate, whilst the other is reflected. The transmitted wave differs from the incident in no respect except strength. It is attenuated in a certain ratio depending upon the conductance of the sheet, being greatly attenuated when the conductance is large, and slightly attenuated when it is small. The formulæ are reserved. This transmitted wave moves on just as the incident wave did, and nothing further happens to it. The reflected wave, on the other hand, having its direction of motion opposite to that of the incident and transmitted waves, travels back the way it came, and nothing further happens to it.

The direction of the magnetic force in the three waves is the same. This is one general property, irrespective of the amount of conductance. But a much more striking one connects the intensity of the magnetic force in the three waves. The sum of the intensities in the reflected and transmitted waves equals the intensity in the original incident wave. That is, the conductance, with dissipation of energy, has had no effect whatever on the total induction. It has merely redistributed it, by splitting it into two parts, which then separate from one another. The "number of tubes" in the reflected wave may be made to bear any ratio we please to the number in the transmitted wave by altering the conductance of the plate; but their sum is always exactly the number of tubes in the incident wave. This property exemplifies, in a manner which may be readily understood, the persistence of induction, in spite of conduction and waste of energy.

But as regards the displacement, the case is quite different. From the fact that the reflected wave runs back, whilst its magnetic force preserves its original direction, we see that the electric force must be reversed. On the other hand, it is unchanged in the transmitted wave. If, then, their sum equalled the electric force in the incident wave, it would imply that the transmitted wave was of greater amplitude than the incident. But it is smaller, invariably. So there is a loss of displacement. Thus, if H in the incident becomes $(1-n)H$ in the transmitted wave, where n is some proper fraction, it becomes nH in the reflected wave. At the same time E in the incident becomes $(1-n)E$ in the transmitted, and $-nE$ in the reflected wave. The loss of E is, therefore, $2n$E, or the loss of displacement is $2n$D. By the loss we mean the excess of the displacement in the incident over the sum of the displacements in the two resulting waves, transmitted and reflected. This loss occurs at the very moment the incident wave coincides with the plate. The plate itself may be regarded as a dielectric homogeneous with the ether outside, or perhaps of different permittivity, but with the conducting and dissipating property superposed. When thin enough, the permittance of the plate is of insensible influence, and may be disregarded. But strictly, a conducting dielectric is a dielectric which cannot support displacement without wasting it, so that a continuous supply of fresh displacement is needed to keep it up. The rate of waste of energy is proportional to the electric stress.

But it should be carefully noted that the loss of energy and the loss of displacement are entirely distinct things, which are not proportional ; and that the loss of displacement itself may sometimes require to be understood in a somewhat artificial sense. For the loss may be greater than what there is to lose. It must then be understood vectorially. This occurs when n is greater than $\frac{1}{2}$. When $n = \frac{1}{2}$, the reflected and transmitted waves are equally strong, and only differ in the direction of the displacement. The loss of displacement is, therefore, complete. The loss of energy in the plate is simultaneously at its maximum, being equal to one-half of the energy of the original wave. If we reduce the conductance of the plate, we increase the transmitted wave, and reduce the waste of energy in it and the loss of displacement. The amount of the latter still remains positive, therefore, assuming it to be positive in the incident wave. The extreme is reached when the plate has no conductance. Then the incident wave goes right through without any splitting and reflection, and therefore without attenuation, and there is no waste of energy. On the other hand, if we increase the conductance above the critical value making $n = \frac{1}{2}$, we reduce the transmitted wave and increase the reflected, whilst we simultaneously reduce the waste of energy in the plate and increase the loss of displacement. The extreme is reached when the plate is a perfect conductor. There is then no transmitted wave and no loss of energy, whilst the reflected wave is of full size, but with the displacement reversed as compared with the incident, so that the loss of displacement, in the sense described, is the greatest possible, viz., 2D.

It will be seen from these details that whilst the absorption or dissolving of induction by conductance is a myth, the idea of an absorption of displacement is not without its inconveniences when the conductance is great, and that this becomes extreme when there is really no loss of energy in the plate, when, in fact, the incident wave does nothing in it, but is wholly rejected with its displacement reversed. It seems, then, more natural to consider the waste of energy from the field caused by the plate as loss. This takes place equally from the electric and magnetic energies, since they are necessarily equal in every one of the three waves. But in the application made later, the plate is to have very slight conductance (in the limit an infinitely small amount), so that the total displacement cannot change sign, but merely suffers a slight loss. Then the idea of loss of displacement by conductance becomes useful again.

The loss of energy takes place as the incident wave is traversing the plate. Its successive layers each cause a minute attenuation of the wave passing, and this applies equally to the induction and displacement, so that the transmitted wave emerges from the plate a pure electromagnetic wave, a reduced copy of the incident. The successive layers, too, each cast back a minute portion of the wave traversing them, with unchanged sign of the induction, but with displacement reversed; and these rejected fluxes make up the reflected wave. There are evidently residual effects due to the internal reflections of minute portions of the main reflected wave, but these residuals tend to vanish when the plate is thin enough.

If the plate be a magnetic instead of an electric conductor, the theory is quite similar. The transmitted wave is an attenuated continuation of the incident. The reflected wave is also a copy of the incident, also reduced. But it is now the induction that suffers loss, because its direction in the reflected wave is opposite to that in the incident and transmitted. On the other hand, the displacement now fully persists, being merely split into two parts by the plate.

Notice that if the plate be both an electric and a magnetic conductor, its attenuating effect from these two causes on the transmitted wave will be additive, so that it will emerge a pure wave with extra attenuation. But as regards the reflected wave, we have a peculiar result. The action of the magnetic conductance is to reverse the induction whilst keeping the displacement straight; whilst that of the electric conductance is to reverse the displacement and keep the induction straight. The result is that the reflected wave is reduced in magnitude by the addition of magnetic conductance to previously existent electric conductance. With a proper proportioning of the two conductances, the reflected wave may be brought nearly to evanescence from a plate of finite conductance. In the limit the compensation is perfect, and the incident wave goes right through without reflection, though it suffers extra attenuation. This is the explanation of the distortionless propagation of waves in a dielectric medium possessing duplex conductivity, electric and magnetic. Whilst there is no reflection in transit, there is a continuous loss both of displacement and of induction.

The Persistence of Induction in Plane Strata, and in general. Also in Cores and in Linear Circuits.

§ 194. Now return to the case of electric conductivity alone, and, as described, let it be locally condensed into the conductance of any number of parallel plates. We know that the effect of any one of them on a thin electromagnetic sheet is to split it, as previously described. If we like, therefore, we can follow each of the resulting waves, and observe how they are, in their turn, split by the first plates they meet, giving rise to four waves, to be a little later split into eight, and so on. This process may seem cumbrous, but it is also an instructive one.

Thus, consider what happens to the total induction. We know that it persists in amount and direction when a single split occurs. Now the same property applies to every successive split a wave suffers in our dielectric medium containing parallel conducting plates. So the total induction remains constant. It is redistributed and spreads out both ways, but without the least loss. There is a small loss of energy at every split, but this does not affect the total induction. This applies when we start from a single pure electromagnetic sheet moving either way. It therefore applies when the initial state consists of any number of such sheets, of any strengths, forming a perfectly arbitrary initial distribution of induction and displacement in parallel plane layers. There is still persistence of the induction. Finally, the same applies when we split up the conducting plates themselves into plates of smaller conductance, and spread them out at uniform distances. The ultimate limit of this process is reached when the conductance is quite uniformly spread, so that we have a perfectly homogeneous medium under consideration. It is, fundamentally, a dielectric propagating disturbances at speed v; but it is, in addition, a conductor as well, and distorts the waves and dissipates their energy. The speed v is $(\mu c)^{-\frac{1}{2}}$, with the proper values of μ and c. The conductivity does not interfere with this property of propagation at finite speed. But observe that if we choose to ignore the displacement, then the corresponding speed is infinitely great. We conclude from the above that plane sheets of induction in electric conductors always preserve the total induction constant in amount, irrespective of the amount of elastic displacement, or whether there is any at all. That is, induction cannot be destroyed by conductance.

If, then, it suffers destruction, this must be due to some other cause. It may be merely a cancellation by the union of oppositely-directed inductions. This may be termed a vectorial cancellation. It may occur, of course, with plane strata of induction. Thus if, in an infinitely large conductor, the total induction be initially zero, which does not require the induction density to be zero, the final effect will be a complete annihilation of the induction by mutual cancellations. Should, however, the total induction be not zero, it will persist. The induction density will tend to zero, but that will be merely on account of its attenuation by spreading, not because there is any destruction by the conductance or resistance when either of them is finite. To prevent the attenuation to zero we may interpose infinitely conducting barriers, one on each side, in planes parallel to the sheets of induction. Then the final result will be that the induction will spread itself out uniformly between the barriers and maintain a finite density.

To illustrate this property in a somewhat less abstract manner, consider a large ring, say of copper, though iron will do equally well except as regards some complications connected

with its magnetisation. Let it be inductised by an enveloping
coil-current so that the induction goes along the core in a
complete circuit. When it is steadily set up, if we remove
the coil-current (and the coil too, preferably for our present
purpose) the induction in the core will, in time, all come out of
it. But if we clap an infinitely-conducting skin upon the core,
it will not come out. Then we have a certain flux of induction
locked up, as it were, in a conducting material, which has no
effect upon it. It can neither be destroyed by the conductance
of the core nor can it get through the perfectly-obstructive
skin. If the skin is clapped on after the induction has
partially escaped (which escape begins on the outside, before
the interior is sensibly affected), there is a redistribution of
induction, which continues until a new state of equilibrium is
reached. During this process there is electric current in the
core and some waste of energy. But there is no waste of the
induction. The final induction is the mean value of the
original induction across the section of the core.

In further illustration, let the core be hollow and be inductised circularly—instead of along its length—by means of two
currents on its boundaries, inner and outer, oppositely directed,
following the length of the core. When this is done, remove
the currents and clap on perfectly conducting skins internally
and externally. There will be a similar persistence of the
induction, although its tubes now go round the inner boundary
circularly. There may be an initial settling down, but the
outer skin will not let the induction expand outwardly, and the
inner skin will not let it contract inwardly. If the latter could
happen we might have cancellation. To get this effect remove
the inner skin. Then, whether we fill up the hollow with
finitely conducting matter or leave it nonconducting, we allow
the induction to spread internally and permit cancellation. The
induction will now wholly disappear, in spite of the external skin.
That is to say, there will be a continuous passage of the induction out of the initially inductised region, accompanied by electric current therein, which will continue until the whole of the
magnetic energy is wasted as heat in the core.

The same property is exemplified, though in a less easily understandable manner, with a single closed line or circuit of infinite
conductance. If it embrace a certain amount of induction it

will always do so, in the absence of impressed force to alter the amount. The induction is locked in and cannot pass through the infinitely-conducting circuit to dissipate itself. If the conductance of the circuit be finite, then it can get through. The time-constant varies as the conductance. The disappearance of the induction is manifested by the waste of energy in the circuit, the electric current in which is supported by the voltage of the decreasing induction through it. But the current is there all the same (measured magnetically) when the conductance is infinite. The induction is steady, and there is no voltage in the circuit. But none is needed.

On the other hand, if there is initially no induction through the circuit, there will continue to be none when a magnetic field is created in its neighbourhood. But although the tubes of induction cannot cut through the infinitely conducting circuit so as to make the induction through it be a finite quantity, yet they do pass through a surface bounded by the circuit, as much positively as negatively. The resulting induction distribution is to be got by superimposing the external induction and that due to a current in the circuit of such strength as to make the total induction through it be zero. The property is a general one, for if the circuit be moved about in a magnetic field, there is always, in virtue of its impermeability to the magnetic flux, zero total induction through it if its conductance be infinite ; whilst if it be finite but great, there is an approximation to this result so long as the motion is kept up, or the external field be kept varying. At the same time, the least amount of resistance in the circuit will be sufficient, if time enough be given, to allow the external induction, when due to a steady cause, to get past it to the full extent, when of course the current in the circuit will cease.

In a similar manner, displacement can be locked up by a circuit of perfect magnetic conductance. There is also persistence of displacement in spite of a finite degree of magnetic conductivity in a continuous medium, unless it be electrically conducting as well.

The Laws of Attenuation of Total Displacement and Total Induction by Electric and Magnetic Conductance.

§ 195. Next consider the effect of a conducting medium upon the total displacement. We know that the latter decreases with the time, and the law of decrease may be readily found from the theory of a single conducting plate. We found that when its conductance exceeded a certain value, the loss of displacement exceeded the original. But, in regarding the action of a homogeneous conductor upon a wave as the limit (in the gross) of that of an assemblage of parallel plates in which the conductance is localised (which process may not seem unassailably accurate beforehand, but which is justified by the results), it is easy to see that we have merely to deal with plates of such very low conductance that the loss at each is extremely small, so that the above-mentioned difficulty does not enter. Thus, let the loss at one plate be such as to reduce the initial displacement D in a wave to mD, where m is a fraction nearly equal to unity. Here mD is the sum of the displacements in the transmitted and reflected waves, the latter being very small and of the opposite sign to the initial D. As these waves separate, they reach other plates and are split anew. If these plates have each the same conductance as the first, the total mD is further attenuated by them to m^2D when the two waves become four. Next, when these four waves are split into eight by the next plates that are reached, the total displacement becomes m^3D ; and so on. These successive displacement totals decrease according to the law of a geometrical series. It follows that, in the limit, we shall have the total displacement represented by an exponential function of the time, say by

$$D = D_0 \epsilon^{-nt}, \quad \ldots \ldots \quad (1)$$

where D_0 is the initial value, and D what it becomes at time t. To find the value of n, we have merely to examine the form of the fraction m, observe how it depends on the conductance of one plate, and proceed to the limit by making the number of plates infinite, whilst their conductances are infinitely small. The result is that the constant n has the value k/c, where k is the conductivity and c the permittivity of the homogeneous conducting medium.

In the irrational units of the B.A. Committee this quantity is represented by $4\pi k/c$, which is, of course, nonsense, like the quaternionic doctrine about the square of a vector. They are both going to go. The above reasoning applies to any initial distribution of displacement in plane layers, instead of merely one elementary sheet. Therefore, equation (1) shows that the total displacement subsides according to the time-factor $\epsilon^{-kt/c}$. Now, this represents Maxwell's law of subsidence of displacement in a conducting condenser (apart from "absorption" and hysteresis), or of the static distribution of displacement associated with electrification in a conducting medium. We see that the law has a far more general meaning. The initial displacement need not be static, but may be accompanied by magnetic induction, and may consequently move about in the most varied manner, whilst its total amount decreases according to the static law. A homogeneous medium is presupposed, and modifications may be introduced by the action of boundaries.

Passing next to the analogous case of a magnetic conductor, in which the total displacement remains constant whilst the total induction subsides, it is unnecessary to repeat the argument, but is sufficient to point out the law according to which the subsidence occurs. If B_0 be the initial total induction, and B what it becomes at time t, we shall have

$$B = B_0 \epsilon^{-gt/\mu}, \quad . \quad . \quad . \quad . \quad . \quad (2)$$

where g is the magnetic conductivity and μ the inductivity. The time-constant c/k of the former case has become μ/g.

Returning to the former case, it should be noted that when the initial distribution is of the static nature, unaccompanied by magnetic force, it retains this property during the subsidence. For, since the displacement subsides everywhere according to the same time-factor, its distribution does not alter relatively, or it remains similar to itself. Since, then, there is no magnetic force, there is also no true electric current. There is also no flux of energy. That is, the electric energy is converted into heat on the spot.

A considerable extension may be given to this property. If there be a conducting dielectric in which the permittivity varies from place to place, containing a static distribution of displacement, then, if the conductivity vary similarly from place to place, so that the time-constant c/k is the same everywhere, the displacement will subside everywhere alike, without magnetic force or flux of energy, and with purely local dissipation of the electric energy. For the solution is represented by

$$\mathbf{E} = \mathbf{E}_0 \epsilon^{-kt/c}, \quad \mathbf{H} = 0,$$

where $\mathbf{E}_0$ is the initial electric force of the static kind, having no curl, and $\mathbf{E}$ that at time t. Both the fundamental circuital laws are satisfied, the first because the true current is the sum of the conduction and displacement currents, and the second because $\mathbf{E}_0$ has no curl and k/c is constant. If it were not constant then, obviously, the property considered would not be true; there would be different rates of subsidence at different places, and the distribution of displacement would change, along with magnetic force, electric current and transfer of energy.

The corresponding property in a magnetic conductor requires the constancy of the time-constant μ/g. Then, whether μ and g are themselves constant or variable from place to place, a static distribution of induction subsides everywhere alike, and without the generation of electric force.

Returning again to plane strata of displacement in an electrically-conducting homogeneous dielectric, it may be inquired how the property (1) of the subsidence of the total displacement will be affected by the simultaneous existence of magnetic conductivity. This will undoubtedly affect the phenomena in detail, but will have no effect on the property in question. Similarly, the law (2) of the subsidence of total induction will not be affected by the presence of electric conductivity. That is, in general, when there are both conductivities present, and both the fluxes displacement and induction present, the total displacement subsides according to one law and the total induction according to the other, without interference. These properties have their parallels in the theory of telegraph circuits, as we shall see later.

It should be remembered that we are dealing always with matter in the gross, and not with molecules at all ; or, equivalently, we assume a homogeneous constitution of the elements of volume. Thus, when displacement subsides in an electric conductor without generating magnetic force, the possibility and necessity of which are clearly indicated by the two circuital laws, it may be that if we go in between the molecules there is magnetic force. It is, in fact, difficult to conceive how displacement in a heterogeneous medium of molecular consti- tution could be done away with without the generation of magnetic force, considering that the energy of the displace- ment is converted into heat energy.

This matter, however, does not belong to the skeleton theory of electromagnetism, but is rather to be considered as a side- matter involving physical hypotheses to account for the influ- ence of matter upon the electromagnetic laws.

The Laws of Attenuation at the Front of a Wave, due to Electric and Magnetic Conductance.

§ 196. Besides the above simple laws relating to the subsi dence of the total fluxes (sometimes true for the elementary parts) there are equally simple laws relating to the subsidence of the fluxes at the front of a wave advancing into previously undisturbed parts of the medium, which sometimes admit of extension to the body of the wave. To understand this it may be mentioned first, that the front of a wave in a non-conduct- ing dielectric is always pure ; that is, the electric and magnetic fluxes are in the wave-plane, and are perpendicular and in constant ratio. The body of the wave need not be of this pure type, owing to the change of form of the wave-front and other causes, but the property of purity always characterises the wave-front. This may be disguised in the case of a thin electromagnetic shell, when it is regarded as the front, for the shell itself may be complex. Then the mere front of the shell may be the only quite pure part. But taking cases free from this complication, we should next note that the introduction of conductivity into the medium makes no difference in the form of the wave-front or its position at a given stage of its progress, provided, of course, that the two quantities upon which the speed of propagation depends—the inductivity and permittivity—are not altered. Now, as has been already explained in connection with the theory of a thin conducting plate, as the wave advances through a continuously conducting medium its successive layers are being continuously subjected to a reflecting process, a minute portion of every layer being thrown back, whilst the bulk is transmitted. In the body of a wave, therefore, there is a mixed-up state of things. At the very front, on the other hand, there is no such mixture, for the disturbance consists wholly of what has been transmitted of the front layer. We may, therefore, fully expect that the law of its attenuation in transit is of a simple nature.

To find it, locally condense the conductance into that of any number of equal conducting plates. Let any one of these plates attenuate a wave traversing it from E to mE. If initially pure it emerges a pure wave, and passes on to the next plate, where it suffers a second attenuation—viz., to m^2E, and again emerges pure. At the third plate it becomes m^3E, and so on. The reflected portions we wholly ignore at present. The limit of this process, when the plates are infinitely closely packed and of infinitely small conductance, so as to become a homogeneous dielectric possessing finite conductivity, is that the time-factor of attenuation takes the exponential form. The result is

$$E = E_0 \, \epsilon^{-kt/2c}, \quad . \quad . \quad . \quad . \quad . \quad (3)$$

E_0 being the initial, and E the value at time t. The time-constant $2c/k$ is just double that of the subsidence of total displacement. Whilst, for example, the total displacement in a plane wave attenuates to, say, $\frac{1}{100}$ of its initial value, the disturbance at the wave front has only attenuated to $\frac{1}{10}$ of its original value.

The property (3) applies to the magnetic as well as to the electric force and flux. It does not apply merely to plane waves, but to any waves, because the superficial layer only is involved, and any elementary portion thereof may be regarded as plane. So it comes about that the exponential factor given in (3) makes its appearance in all investigations of waves in electrical conductors when the permittivity is not ignored. It is a more fundamental formula than the previous one with the time-constant c/k, which is the final result of the complex process of mixture of reflected waves, or is equivalent thereto.

The corresponding property in a magnetic conductor is that the disturbance at the front of a wave is attenuated in time t according to the time-constant $2\mu/g$. Thus,

$$E = E_0 \epsilon^{-gt/2\mu}. \quad . \quad . \quad . \quad . \quad . \quad (4)$$

Here the time-constant is twice that of the subsidence of the total induction. Like the former formulæ, these modified ones, (3) and (4), have their representatives in the theory of a telegraph circuit, in spite of the absence of magnetic conductance. It is replaced by something that produces approximately the same result.

In a conductor possessing duplex conductivity, electric and magnetic, their attenuative actions at the wave-front are independently cumulative, or additive. The attenuation is expressed by

$$E = \mu v H = E_0 \epsilon^{-(k/2c + g/2\mu)t}. \quad . \quad . \quad . \quad (5)$$

It is really the attenuative actions of a single conducting plate that are additive. This applies separately to every successive thin conducting layer through which the front of the wave runs, with the result (5), where the time-factor is the product of the two former time-factors of (3) and (4).

In the theory of coils and condensers, not only do we meet with the time-constants L/R and S/K, the ratios of inductance to resistance and of permittance to conductance, but also with the double values. Their ultimate origin may be traced in the theory of the effect of a thin conducting plate upon a wave.

The exponential time-factors concerned in (3) and (4), and the more complex one in (5), also make their appearance in connection with the disturbance in the body of a wave, though in a less simple manner. This will be returned to.

The Simple Propagation of Waves in a Distortionless Conducting Medium.

§ 197. Coming now to the influence of conductivity on a wave elsewhere than at its extreme front, where we have recognised that the influence is simply attenuative, the easiest way of treating the matter is not to pass from the known to the unknown, but to reverse the process and pick out the cases which theory indicates are most readily understandable. This is to be done by a process of generalisation. The theory of a conductor with duplex conductivity is, in a certain case, far simpler than that of a real electric conductor. We have already mentioned that the reflective actions of two plates, one an electric, the other a magnetic conductor, are of opposite natures.   The first reverses the displacement, and the second reverses the induction when throwing back a portion of the wave.   The joint action of the two plates when coexistent and coincident, or the action of a single plate with duplex conductance, results in a complete disappearance of the reflected wave when the conductances are in proper ratio and the plate is infinitely thin.   We then have transmission with attenuation but without reflection.   This occurs, in a homogeneous medium, when $k/c = g/\mu$.   Reflex action being abolished, we are reduced to a kind of propagation of unique simplicity.

To see the full meaning of this, start from any initial distributions of induction and displacement in a non-conducting dielectric.   Imagine that we have obtained the full solution showing the subsequent history of the disturbances.   Now, if we introduce only one kind of conductivity, say electric, we shall, with the same initial state, have a profoundly different subsequent history.   Again, with magnetic conductivity alone, we shall have a course of events different from both the previous.   But if we add on magnetic conductivity to previously existent electric conductivity, we shall partly counteract the distorting influence of the latter.   This counteraction becomes complete when the value of the magnetic conductivity is raised so high as to produce equality of the time-constants of attenuation due to the two conductivities separately.   Further increase of the magnetic conductivity will overdo the correction and bring on distortion again, though of a different kind.

Similarly, the distortion due to magnetic conductivity alone is diminished by introducing electric conductivity, and becomes completely abolished when there is enough of the latter to equalise the time-constants.   Further increase brings on the distortion again, which is now of the electric kind.

When the state of balance occurs, and the distortion is wholly removed, the course of events following any initial state is precisely the same as in a non-conducting medium, but with a continuous attenuation expressed by equation (5) above specialised to suit the equality of the time-constants.   That is, the time-factor of attenuation is now $\epsilon^{-kt/c}$.   This removal of distortion applies to every kind of wave.

This distortionless state in conducting media furnishes a sort of central basis for investigating the more recondite effects accompanying distortion. Nevertheless, its consideration would possess only a theoretical value, on account of the non-existence of the second kind of conductivity involved, were it not for the remarkable practical imitation of the distortionless state of things which is presented in the theory of telephone and other circuits under certain circumstances. If we abolish the fictitious magnetic conductivity throughout the medium traversed by the waves, we should, to have distortionless transmission, also abolish the electric conductivity. This is only to be attained by using wires of no resistance to guide the waves through a non-conducting medium. But they have resistance, of greater or lesser importance according to circumstances. Of what nature, then, is the distortion of waves produced by the resistance of a wire along which they run? The answer is, that it is approximately of the k d due to magnetic conductivity in the medium generally. On the other hand, the different kind of distortion due to electric conductivity in the medium generally remains in action, being the effect of the leakage-conductance of the insulating medium surrounding the wire, or the average effect of other kinds of leakage at distinct and separate spots along the circuit. Thus we obtain an approximate reproduction of the theory of magnetic conductivity acting to neutralise the distorting effect of electric conductivity. The time-constants μ/g and c/k become L/R and S/K in a telegraph circuit, L being inductance, R resistance, S permittance, and K leakage-conductance. Their equalisation produces the distortionless circuit, which may turn up again later on. In the meantime I may remark that if the reader wishes to understand these things, he must give up any ancient prejudices he may be enamoured of about a "KR law" and the consequent impossibility of telephoning when "KR" is over 10,000. When pointing out, in 1887, the true nature of the telephonic problem and the absurdity of the "KR law" applied thereto generally, I predicted the possibility of telephoning with "KR" several times as great. It has since been done. In America, of course. A short time since, in noticing the KR = 32,000 reached by the New York-Chicago circuit, I further predicted that it would go up a lot more. It did very shortly after. The record is now about 50,000 (Boston-Chicago) for practical work, I believe. It means a good deal more for possible work. But there is no need to stop at 50,000. That can be largely exceeded in an enterprising country.

The Transformation by Conductance of an Elastic Wave to a Wave of Diffusion. Generation of Tails. Distinct Effects of Electric and Magnetic Conductance.

§ 198. We are now prepared to somewhat understand the nature of the changes suffered by electromagnetic waves in transit through a conducting medium. It being the distortion due to the conductance alone that is in question, we eliminate that due to other causes by choosing plane waves for examination, since these do not suffer any distortion in a homogeneous dielectric when it is non-conducting. Imagine, then, a simple electromagnetic plane wave-sheet of small depth to be running through a dielectric at the natural speed conditioned by its inductivity and permittivity. At any stage of its progress, let the medium become slightly electrically conducting all over, not merely in advance of the wave but behind it as well, for a reason that will presently appear. What happens to the wave now that the fresh influence is in operation?

A part of the answer we can give at once, by the previous. The wave-sheet will move on just as before, but will attenuate as it goes, according to the time-factor $\epsilon^{-kt/2c}$. Since we suppose the conductivity to be slight, it follows that a great distance may be traversed before there is notable attenuation. We also know that the total induction remains constant. The rest of it—that is, what is not in the sheet at any moment —is therefore left behind. The rejecting process commences the moment the conductivity is introduced, and continues to act until the plane wave is attenuated to nothing. The rejected portions travel backwards. But they are themselves subject to the same laws as the main plane wave, and so get mixed up. The result is that at time t after the introduction of the conductivity, the whole region of disturbance extends over the distance $2vt$, half to the right and half to the left of the initial position. At the advancing right end we have a strong condensed disturbance, viz., the original wave attenuated, and behind it a weak diffused one. We can therefore, without misunderstanding refer to them as the head and the tail, without any body to complicate matters. Now the nature of the tail is quite different as regards the displacement and the induction. It is therefore convenient to regard one of them alone in the first place, and, of course, we select the induction, on account of the simple property of persistence that it possesses. We can distinguish three or four different stages in its development.

The first stage is when the attenuation of the head is not great, say, whilst the head decreases from 1 to 0·75. Whilst this occurs, the total induction in the tail rises from 0 to 0·25. The tail is long and thin, and tapers to a point at its extreme end, or tip, at distance $2vt$ behind the head, and is thickest where it joins on to the head.

The second stage roughly belongs to the period during which the head further attenuates from 0·75 to 0·5 or 0·4. The total induction in the tail then increases from 0·25 to 0·5 or 0·6. During this stage we find that the tail, which has, of course, greatly increased in length, does not go on increasing in thickness at the place where it is developed, but stops increasing and shows a maximum at or near that place.

The third stage occurs during the further attenuation of the head to, say, 0·1, whilst the total induction in the tail increases to 0·9. The maximum thickness of the tail is now a long way from the head, and at the end of the stage is nearer to the middle than to the head. Of course, since the head itself is now so small, the additions made to the tail must also become smaller.

The fourth stage is when the head practically disappears and all the induction is in the tail. The maximum thickness is now nearly in the middle—on the right side, however—and the tail is nearly symmetrical with respect to its middle, where there is a swelling, beyond which the tail tapers off both ways to its two tips.*

The final state is the consummation of the previous, and is one of perfect symmetry with respect to the middle of the tail, which is situated exactly where the plane wave was when the

---

* As the division into distinct stages is somewhat arbitrary, this description of the transition from an elastic to a diffusion wave should be understood to be only roughly approximate. It is made up, not from the formula, but by a numerical process of mixture.

spreading began.   The spreading now takes place according to the pure diffusion law, as of heat by conduction.

Now as regards the displacement in the head and tail in the different stages.   In the first and second stages the displacement is wholly negative in the tail, assuming it to be positive in the head, where, it should be remembered, it attenuates in the same manner as the induction which accompanies it.   Thus, when the head has fallen to 0.9, the total displacement in the head and tail has fallen to $(0.9)^2$ or 0.81, so that the total negative displacement in the tail is of amount 0.09, which is not much less than the coincident induction.   And when the head has attenuated to 0.8, and the total displacement to $(0.8)^2$ or 0.64, the negative displacement in the tail amounts to 0.16. But, unlike the induction, the displacement increases in the tail from the head up to not far from the tip, where, of course, it falls to zero.   There is no tip at the forward end.   But as the tail stretches out further to the left, and has fresh additions made to it on the right side, the decrease of the density of displacement in passing towards the head continues, until somewhere about the end of the second stage, it becomes zero next the head.   This node is approximately at the place where the induction has its maximum.   When the head has fallen to 0.4, we have the total displacement attenuated to 0.16, so that the negative displacement in the tail amounts to 0.24.

In the third stage the displacement is negative from the tip up to somewhere near and beyond the maximum of induction, and increasingly positive in the remainder, up to the head. That is, the region of positive displacement now extends itself from the head a good way into the tail.   At the same time the place of maximum negative displacement moves forward.

During the fourth stage, the place of maximum negative displacement shifts itself to nearly the middle of the region between the tip and the node, beyond which the positive displacement has a nearly similar distribution, with a maximum. But this positive distribution is only three-parts formed, as the head is still of some importance.   The fifth stage completes the formation of the tail, with the displacement negative in one half and positive in the other, and nearly symmetrical with respect to the middle.   Finally, we come to a state of perfect symmetry, with one maximum of induction and two (or a maximum and a minimum) of displacement. This may be readily understood by considering that in the expression for the true electric current, $k\mathbf{E} + c\mathbf{E}$, the second term finally becomes a small fraction of the first, or the current is practically the conduction-current only. Then $\mathbf{E}$, and therefore the displacement, is proportional to the space-variation of the induction. Owing to the tail being made up of a complicated mixture of infinitesimal electromagnetic waves going both ways, we lose sight of the fundamental property of elastic wave-propagation in a dielectric, the resultant effect being propagation by diffusion now, or very nearly so. The approximation to this result is closest in the middle of the tail. At the tips, on the other hand, we still have elastic waves, but, of course, of insensible strength.

Now, suppose that the initial state is one of induction only, though still in a plane sheet. It may be regarded as the coincidence of two plane electromagnetic sheets of half the strength, with similar inductions and opposite displacements. The subsequent history may, therefore, be deduced from the preceding. The initial induction splits into two halves, of which one moves to the right at speed v, and the other to the left. The history of the first wave as distorted by the conductance has been given. That of the second wave is the same, if we allow for the changed direction of motion and sense of the displacement. So we have only to superpose the two systems to show how an initial distribution of induction in a plane sheet splits and spreads, and the accompanying electric displacement. We have two equal heads, separating from one another at speed $2v$, whilst the two tails unite to make a stouter kind of tail (referring to the induction) joining the two heads. This tail is always thickest in the middle. In fact, the distributions of induction and displacement are symmetrical with respect to the initial position from the first moment. In the final state that is tended to, when the tails have vanished, the induction is distributed in the same way as in the former case, in spite of the remarkable difference in the initial phenomena.

If, again, the initial state is one of displacement alone in a plane sheet, this generates two oppositely-travelling electromagnetic waves in which the displacements are similar and the inductions are opposite. The result is therefore to be got by combining the previous solution for a wave of positive induction and displacement moving to the right, with a similar wave of negative induction and positive displacement moving to the left.

Whatever be the magnitude of the conductivity, if finite, the same phenomena occur during the conversion of elastic waves to diffusion waves, and they may be represented by the same diagrams, if suitably altered in the relative scale of ordinates and abscissæ. But by increasing the conductivity from the small value which makes the above-described process take place over a considerable interval of time, we make it occur in a small interval only; and when we come to what are usually considered good conductors, viz., metals, then the interval of time is so small that we may say practically that the heads vanish almost at once, and before the propagation has proceeded any notable distance. The rest of the story is the spreading by diffusion or mixture. Thus we have an important practical distinction between the very good and the very bad conductor, as regards the manner of propagation of induction. In the former the electric displacement is of no account hardly, except perhaps for very short waves, and the practical theory is the theory of diffusion. At the other extreme are perfect non-conductors, in which the propagation is entirely by elastic waves. Between the two we have, in bad conductors, a mixture of the two kinds, though with a continuous transition from one kind to the other. The mathematical treatment of elastic waves is the easiest. The next in order of difficulty are the waves of pure diffusion, by the ordinary Fourier mathematics. The most troublesome are the intermediate forms of changing type, and the full analytical results are best got by a generalised calculus. But the general nature of the results may be obtained approximately by easy numerical calculations, with diagrammatical assistance. This is explainable without difficulty, and may be entered upon later on, when we come to detailed problems.

In the above the conductivity has been electric only, and we have seen that there is a great difference between its effects on the electric and on the magnetic flux, which arises from the positive reflection and persistence of the latter, and the negative reflection and subsidence of the former flux. The corres-

ponding effects in a magnetic conductor may be readily
deduced by the proper transformations, and exhibit analogous
differences. Thus, starting with a pure plane wave sheet
moving to the right, it is now the displacement that is conserved and divided between the head and tail, being initially all
in the head, and finally all in the tail. On the other hand, the
induction is negative in the tail at first and until the attenuation of the head is considerable. After this, the region of positive induction extends from the head into the tail. As time
goes on, and the head disappears, we have negative induction
in one half and positive in the other (forward) half, whilst the
displacement is positive all along it, with its maximum nearly
in the middle. In the final state of diffusion we have symmetry
with respect to the initial position of the wave, and the induction is proportional to the space-variation of the displacement.
The second term of the magnetic current $g\mathbf{H} + \mu\dot{\mathbf{H}}$ is then a
small fraction of the first term.

Now, in the medium with duplex conductivity we must take
the distortionless condition for the standard state. When this
obtains there is no tailing, although the head (displacement and
induction together) attenuates according to the time-factor $\epsilon^{-kt/c}$.
But if the electric conductivity be in excess (that is, greater than
is required to make $k/c = g/\mu$) there is tailing of the kind described as due to electric conductivity, with the induction positive and the displacement negative in the tail at first. And if
the magnetic conductivity is in excess, the tailing is of the
other kind, with positive displacement and negative induction
at first. But of course the details are not the same, because
both the fluxes now attenuate to zero in time, in the tail as well
as in the head.

Application to Waves along Straight Wires.

§ 199. We must now endeavour to give a general idea of
the tailing of waves when they run along conducting wires, in
co-ordination with the previous. In the first place, observe
that although the lines of electric and magnetic force in a pure
electromagnetic sheet in ether must be always perpendicular to
one another, yet they need not be straight lines, except in their
elementary parts. On the contrary, we may have an infinite

variety of distributions of the electric and magnetic fluxes in curved lines in sheets which shall behave as electromagnetic waves. Keeping to plane sheets, we may take any distribution of displacement in a sheet that we please, and make it an electromagnetic wave by introducing the appropriate distribution of induction, also in the sheet. If left to itself it will move through the ether one way or the other at speed v, according to the directions of the fluxes. It should be noted, however, that the fluxes should have circuital distributions. For if not, and there is electrification or its magnetic analogue, they, too, should be moved through the ether so as to exactly keep up with the wave. If the electrification does not move thus, we have a changed state of things. But there is a way out of this difficulty. Let the displacement, when it is discontinuous, terminate upon infinitely-conducting lines, or surfaces of cylinders, according to circumstances. The interference produced by holding back the electrification will then be done away with. It is sufficient for explicitness to take a single definite case.

Let the displacement in a plane electromagnetic sheet terminate perpendicularly upon a pair of perfectly-conducting cylindrical tubes placed parallel to one another. One is for the positive and the other for the negative electrification. The induction will then go between and round the tubes. One tube may enclose the other, but, preferably, we shall suppose this is not the case, so that our arrangement resembles a pair of parallel wires. Now the wave will run through the ether in the normal manner, without distortion or change of type, and will carry the electrification (on the conductors) along with it.

Similarly if the tubes be magnetic conductors, if the induction terminates perpendicularly upon them, whilst the displacement goes between and round them. So far, then, the theory of the propagation of the wave is unaltered.

But this property may be greatly extended. The medium outside the tubes through which the wave is moving may be made electrically conducting. The theory is then identical with that of § 198, as regards the distortion produced and the gradual destruction of the wave, ending in the process of diffusion. This will be true whether the tubes are perfect electric or magnetic conductors, provided we have the displacement terminating upon them in the first case and the induction in the second.

Again, if the medium is magnetically conductive, the theory still holds good, with the same reservations as regards terminating the fluxes. But now the distortion is of the other kind, with persistence of the displacement and subsidence of the induction.

Finally, the theory is still true when the medium is a duplex conductor, whilst the tubes are either perfect electric or magnetic conductors, according to choice. The distortion is now of the electric or of the magnetic kind, according as k/c is greater or less than g/μ.

But not one of these cases is quite what we want to represent propagation along real wires. The nearest approach is the case of finite electric conductivity in the medium combined with perfect electric conductance of the tubes. If we make their conductance be imperfect, we then come close to the real problem. Now when we do this, the theory, when done precisely, becomes excessively difficult, for two reasons. First, the waves in the ether outside the tubes are no longer plane; and next, the penetration of the disturbances into the substance of the tubes in time by means of cylindrical waves has to be allowed for. The latter is, of course, more necessary when solid wires are in question. A practical working theory is seemingly impossible of attainment by strictly adhering to the actual conditions. But one is possible by taking advantage of the fact that the waves in the ether are very nearly plane under ordinary circumstances. Of this we may assure ourselves by considering that the tangential component of the electric force at the surface of a wire (upon which the penetration depends) is usually a very small fraction of the normal component of the same outside it. The practical course, then, is to treat the waves in the dielectric as if they were quite plane. This does not prevent our allowing for the distorting effect of the resistance of the wires. It has the effect of making our solutions approximate instead of complete. But the important thing is to have a theory that, whilst sufficiently accurate, is practically workable, and harmonises with more rudimentary theories. This is precisely what we do get, as done by me in

1886. The theory is brought to such a form that we may employ it in several ways, according as circumstances allow us to ignore this or that influence. We may, for example, treat the wires as mere resistances (constant), and this is quite sufficient in a variety of applications. Or, using the same equations with more general meanings attached to the symbols, we may find the effects due to the imperfect penetration of the magnetic induction and electric current into the wires when subjected to varying forces at their boundaries, either simple harmonically or otherwise. Furthermore, the theory is in such a form that it admits readily and without change of the introduction of terminal or intermediate conditions of the kind that occur in practice, whose effects are brought in according to the usual equations of voltage and current in arrangements of apparatus.

What we are immediately concerned with here, however, is the connection between this theory and the general theory of waves in a medium of duplex conductivity. When the tubes are so thin that penetration is practically instantaneous (for the wave-length concerned), a constant ratio is fixed (assisted by Ohm's law) between the intensities of magnetic and of tangential electric force at the boundary of the tube, where it meets the dielectric. And when we incorporate this result in the second circuital law applied to any section of the circuit formed by the parallel tubes, we find that the result is to turn it to the form expressing the existence of magnetic conductivity in the outer medium, without resistance in the wires. That is, the resistance of the wires has the same effect in distorting and dissipating the waves outside as the fictitious magnetic conductivity in the medium generally. This is true in the first and most important approximation to the complete theory. It is a point that is not altogether easy to understand, because the magnetic conductivity is fictitious, whereas the resistance is real. This, however, may be noted, that the theory of propagation of plane waves in a medium of duplex conductivity bounded by perfect conductors for slipping purposes, professes to be a precise theory, whereas the other, although concerning the same problem in most respects, professes to be only approximate so far as the influence of the wires is concerned.

Transformation of Variables from Electric and Magnetic Force to Voltage and Gaussage.

§ 200. In the consideration of the transmission of waves along wires, especially, of course, in practical calculations, it is more convenient to employ the line-integrals of the electric and magnetic forces as variables rather than these forces themselves. This transformation of variables involves other necessary transformations, and the connection between the old quantities, suitable for waves in general, and the new, should be thoroughly understood, if something more than a superficial knowledge of the subject be desired.

Thus, commencing with a circuit consisting of a pair of parallel straight conducting tubes, of no resistance in the first place, take for reference any plane which crosses the tubes perpendicularly. It is the plane of the wave at the place, and the lines of electric and magnetic force lie in it, the former starting from one (the positive) and ending upon the other (the negative) tube, the latter passing between and round them. These lines always cross one another perpendicularly, but their distribution in the plane may be varied by changing the size and form of section of the tubes by the plane and the distance of separation. Now there is no axial magnetic induction—that is, induction parallel to the tubes. Nor is there any tangential component of electric force on the surface of the tubes. It follows, by the second circuital law, that the line-integral of the electric force from one tube to the other is the same by any path in the reference plane. It is not the same by any path with the same terminations if we depart from the reference plane, but that is not yet in question. Call this constant line-integral V. It is the transverse voltage. In another form, we may say that the circuitation of the electric force in the reference plane is zero.

Next consider the magnetic force. Its circuitation is also zero in the reference plane, provided the one tube or the other is not embraced, because there is no axial electric current in the dielectric. But when a tube is embraced the circuitation is finite. Call this quantity C. It is "the current" in the tubes, or, at any rate, is the measure thereof, and is positive for one tube and negative for the other. These two quantities,

V and C, the transverse voltage and the current, are the new variables. They are as definite as **E** and **H** themselves in a given part of the reference plane under the circumstances.

Next, as regards the fluxes, displacement and induction. Introduce a second reference plane parallel to the first and at unit distance in advance of it along the tubes. Anywhere between the planes we have $\mathbf{D} = c\mathbf{E}$, where **E** and **D** vary according to position. Now, as we substitute the complete voltage for **E**, so we should substitute the complete displacement for **D**. The complete displacement is the whole amount leaving the positive and ending upon the negative tube, between the reference planes. That is, it is the " charge " per unit length of the tubes, say Q. As **D** is a constant multiple of **E**, so is Q a constant multiple of V. Say $Q = SV$. This S is the permittance of the dielectric, per unit length axially. It is proportional to c, the permittivity, but of course involves the geometrical data (brought in by the tubes) as well.

Similarly, we have $\mathbf{B} = \mu\mathbf{H}$ everywhere between the reference planes, **H** and **B** varying from place to place. But if we substitute the complete circuitation of **H**, we should also substitute the complete integral of **B**. This means the total flux of induction passing between and round the tubes, between the reference planes. Call it P. It is the magnetic momentum per unit length axially. As **B** is a constant multiple of **H**, so is P a constant multiple of C. Thus $P = LC$, where L is the inductance per unit length axially. This L varies as the inductivity μ, and involves the geometrical data.

In virtue of the relation $\mu c v^2 = 1$, the inductance and permittance are reciprocally related. Thus, $LSv^2 = 1$. Since v is constant, it might appear that the inductance was merely the reciprocal of the permittance, or the elastance, with a constant multiplier, only to be changed when the dielectric is changed. But there is much more in it than this. Different physical ideas and effects are conditioned by inductance and permittance. Inductance and inductivity involve inertia, whilst permittance and permittivity involve compliancy or elastic yielding.

Observe that P and Q are analogous, being the total magnetic and electric fluxes, whilst V and C are also analogous, the voltage and gaussage respectively. So the ratios L and S are

also strictly analogous, in spite of their quantitative recipro-
cality.   They are both made up similarly to conductance, only
in the case of permittance it is reckoned across the dielectric
from tube to tube, and in the case of inductance round and between the tubes—that is, the magnetic circuit is closed, and
the electric circuit unclosed.   There are other ways of setting
out the connections, but the above way brings out the analogies
between the electric and magnetic sides in a complete manner.
The quantity C may be quite differently regarded when the
magnetic field penetrates into the substance of a tube (then to
be of finite conductance), viz., as the total flux of conduction
current, or sometimes of displacement current as well, in a
tube. But the above method is independent of the penetration,
holding good whether there is penetration or not, saving small
corrections due to the waves not being quite plane.   The magnetic view of C, as the gaussage, is also the nearest to experimental electrical knowledge.   The other view is darker,
because we cannot really go inside metals and observe what
is going on there, or the forces in action.   They must be
inferential in a greater degree than the external actions.

The electric energy per unit volume being $\frac{1}{2}ED$ or $\frac{1}{2}cE^2$,
when this is summed up throughout the whole slice of the
medium contained between the reference planes at unit dis-
tance apart, we obtain the amount $\frac{1}{2}VQ$ or $\frac{1}{2}SV^2$, which is
therefore the electric energy per unit of length axially.

Similarly, the magnetic energy density $\frac{1}{2}HB$ or $\frac{1}{2}\mu H^2$, when
summed up throughout the slice, amounts to $\frac{1}{2}CP$ or $\frac{1}{2}LC^2$.
The reader should, in all these transformations, compare the
transformed expressions with the original, and note the proper
correspondences.   It is all in rational units, of course, to avoid
that unmitigated nuisance, the 4π factor of the present B.A.
units.

The density of the flux of energy, which is the vector product of E and H in general, is of amount simply EH when
the forces are perpendicular, as at present.   When this is
summed up over the reference plane, the result is VC, the product of the voltage and current.   The flux of energy is parallel
to the guiding tubes, and EH is the amount for a tube in
the dielectric of unit section, whilst VC is the total flux of
energy.

The relation $E = \mu v H$, or, which means the same, $H = cvE$, which obtains in a pure electromagnetic wave, becomes converted to $V = LvC$, or $C = SvV$. We see that Lv is of the dimensions of electric resistance, and Sv of electric conductance. Similarly, because V and C are the line-integrals of E and H, we see that μv is of the dimensions of resistance, and cv of conductance. The activity product VC is what engineers have a good deal to do with, now that "electrical energy," or energy which has been conveyed by electromagnetic means, is a marketable commodity. However mysterious energy (and its flux) may be in some of its theoretical aspects, there must be something in it, because it is convertible into dollars, the ultimate official measure of value.*

Transformation of the Circuital Equations to the Forms involving Voltage and Gaussage.

§ 201. Now, still under the limitation that the guiding tubes have no resistance, and that the dielectric has no conductance, consider the special forms assumed by the circuital laws. These are, in general,

$$\text{curl } H = c\dot{E}, \quad \ldots \ldots \ldots (1)$$

$$-\text{curl } E = \mu\dot{H}. \quad \ldots \ldots \ldots (2)$$

Now, in applying these to our plane waves, we may observe at the beginning that E and H and their time-variations are in the reference plane, and have no axial components. So the only variations concerned in the operator "curl" are such as occur axially, or parallel to the guides.† Let x be distance along them, then the circuital laws become

$$-\frac{dH}{dx} = c\dot{E}, \quad \ldots \ldots \ldots (3)$$

$$-\frac{dE}{dx} = \mu\dot{H}, \quad \ldots \ldots \ldots (4)$$

---

* *See* the Presidential Address to the Institution of Electrical Engineers, January 26, 1893.

† That is, curl reduces from $\nabla\nabla$, where ∇ is $i\nabla_1 + j\nabla_2 + k\nabla_3$ in general, to $i i \nabla_1$ simply. It is now more convenient to use the tensors E and H, as in (3) and (4), ignoring their vectorial relations.

which apply anywhere in the reference plane. Transforming to V and C, we obtain

$$-\frac{dC}{dx} = SV, \quad \text{...... (5)}$$

$$-\frac{dV}{dx} = L\dot{C}. \quad \text{...... (6)}$$

By (3) and (4) we see that the space-variation of H is the electric-current density, and the space-variation of E the magnetic-current density in the dielectric. By (5) and (6) we express the same truths for the totals concerned.

We may see the meaning of (5) thus. Draw a closed line in the reference plane on and embracing the positive guiding tube. The circuitation of H there is the total surface current C. Let the closed line be shifted to the next reference plane at unit distance forward. Then $-dC/dx$ is the amount by which C decreases during the shift, and (5) tells us that it is equivalently represented by the time-rate of increase of the charge on the tube between the reference planes, or the rate of increase of the total displacement outward. This process may be applied to any closed line in the reference plane, provided it embraces the tube. When it is shifted bodily forward through unit distance to the next reference plane, the amount by which the circuitation of H, that is, C, decreases from the first to the second position measures the displacement current through the strip of the cylinder swept out by the closed line. This is the time-rate of increase of the charge per unit length of the tubes, and is the total transverse current from one tube to the other.

The meaning of (6) requires a somewhat different, though analogous, elucidation. Join the tubes by any line in the first reference plane. The line-integral of E along it is V, the transverse voltage. Shift the line bodily along the tubes to the next reference plane. The transverse voltage becomes $V + dV/dx$, still reckoned from the positive tube to the negative. But if we join the two starting points on the positive tube together, and likewise the ending points on the negative tube, we make a complete circuit, having two transverse sides and two axial sides. Now reckon up the voltage in this circuit. The axial portions contribute nothing, because the electric force is perpendicular to them. The voltage required is, therefore, simply the difference in the values of the transverse voltage in the other two sides, or dV/dx. By the second circuital law, it is also measured by the rate of decrease of induction through the circuit, that is, by $-L\dot{C}$. Whence follows equation (6).

The reader who is acquainted with the (at present) more "classical" method of treating the electromagnetic field in terms of the vector and scalar potentials cannot fail to be impressed by the difference of procedure and of ideas involved. In the present method we are, from first to last, in contact with those quantities which are believed to have physical significance (instead of with mathematical functions of an essentially indeterminate nature), and also with the laws connecting them in their simplest form. Notice that V is not the difference of potential in general. It sometimes degenerates to difference of potential, viz., in a perfectly steady state. But when the state changes, we cannot express matters in terms of an electric potential.

Now, still keeping the guiding tubes perfectly conducting, let the medium in which they are immersed be slightly conducting. The electric-current density, which was $c\dot{E}$ before, now becomes $kE + c\dot{E}$, where the additional kE is the conduction-current density. Along with this there is waste of energy at the rate kE^2 per unit volume. This waste of energy may also be regarded as a storage of energy, viz., as heat in the medium. But as it is not recoverable by the same means (reversed) which stored it, it is virtually wasted, and we have no further concern with it.

We have next to consider the total waste in the slice of the medium between the two reference planes, in terms of the transverse voltage. It sums up to KV^2, where K is the transverse conductance of the medium per unit length axially. This is proportional to the conductivity k, and involves geometrical data in the same way as the transverse permittance, as may be readily seen without symbolical proof, on considering that the conduction-currrent lines and the displacement lines are similarly distributed, whilst both are controlled by the transverse voltage. In fact, the conduction-current density kE sums up to KV, the complete transverse conduction-current, in the same way as the displacement density $c\mathbf{E}$ sums up to the total transverse displacement SV, in terms of the permittance and voltage.

The first circuital law (1) becomes changed to

$$\operatorname{curl} \mathbf{H} = k\mathbf{E} + c\dot{\mathbf{E}}, \quad \ldots \ldots \quad (7)$$

by the addition of the conduction current. And its modified form (3) for our plane waves becomes

$$-\frac{d\mathbf{H}}{dx} = k\mathbf{E} + c\dot{\mathbf{E}}; \quad \ldots \ldots \quad (8)$$

whilst in terms of the new variables, voltage and gaussage, we have this extended form of the equation (5),

$$-\frac{d\mathbf{C}}{dx} = \mathbf{KV} + \mathbf{S}\dot{\mathbf{V}}. \quad \ldots \ldots \quad (9)$$

Next, let the medium become magnetically conductive as well, no other change being made. The conduction-current density is $g\mathbf{H}$, and its distribution resembles that of the induction itself, whilst its amount is controlled by the quantity C, the gaussage. The total magnetic-conduction current may therefore, be represented by RC, where R is the magnetic conductance, which varies as the magnetic conductivity g, and involves the geometrical data in the same manner as the inductance does.

Similarly, the rate of waste of energy due to the magnetic conductivity is $g\mathbf{H}^2$ per unit volume, and the total rate of waste in the slice of the medium between the two reference planes amounts to $\mathbf{RC}^2$ correspondingly.

Finally, the effect of the magnetic conductivity is to turn the second circuital law from the elementary form (2) to

$$-\operatorname{curl} \mathbf{E} = g\mathbf{H} + \mu\dot{\mathbf{H}}, \quad \ldots \ldots \quad (10)$$

and, correspondingly, the special form (4) for our plane waves becomes turned to

$$-\frac{d\mathbf{E}}{dx} = g\mathbf{H} + \mu\dot{\mathbf{H}}. \quad , \ldots \ldots \quad (11)$$

This, in terms of V and C, is equivalent to

$$-\frac{dV}{dx} = RC + L\dot{C}, \quad \ldots \ldots \quad (12)$$

which is the proper companion to the equation (9) expressing the first circuital law. These equations (9) and (12) are the practical working equations.

We have already remarked that the structure of the electric permittance and conductance are similar. From this it follows that the time-constant c/k is identically represented by S/K. Similarly, the time-constant μ/g is identical with L/R.

The density of the flux of energy is, with the two conductivities, still represented by the vector product of **E** and **H**, and therefore the total flux is still the activity product VC. To corroborate this statement, and at the same time show the dynamical consistency of the system, consider that if the quantity VC is really the flux of energy across a reference plane, the excess of its value at one plane over that at a second plane further on at the same moment represents the rate of storage of energy between the planes. Therefore, the rate of decrease of VC with x is the rate of storage between two reference planes at unit distance apart. Now,

$$-\frac{d}{dx}(VC) = -V\frac{dC}{dx} - C\frac{dV}{dx}. \quad \ldots \quad (13)$$

On the right side use the circuital equations (9) and (12), and it becomes

$$V(KV + S\dot{V}) + C(RC + L\dot{C}), \quad \ldots \quad (14)$$

or, which is the same,

$$KV^2 + RC^2 + \frac{d}{dt}(\tfrac{1}{2}SV^2 + \tfrac{1}{2}LC^2). \quad \ldots \quad (15)$$

By the previous, the first term is the total electric waste, the second is the total magnetic waste, the third is the increase of total electric energy, and the fourth is the increase of total magnetic energy, per unit of time, between the two reference planes. This proves our proposition, with a reservation to be understood concerning the circuital indeterminateness of the flux of energy. There is, therefore, no indistinctness anywhere, nor inconsistency. We have now to show in what manner the above is affected by the resistance, &c., of the guides.

The Second Circuital Equation for Wires in Terms of V and C when Penetration is Instantaneous.

§ 202. Let the parallel conducting tubes be of finite resistance. As a result, the external disturbance penetrates into them, and a waste of energy follows. As a further result, the external disturbance becomes modified. This occurs in a two-fold manner. By cumulative action, the nature of a wave sent along the leads may be profoundly modified as it progresses. Besides this, there is what we may term a local modification, whereby the wave at any place is no longer strictly a plane wave. It is merely with this effect that we are now concerned. The departure from planarity requires a considerably modified and much more difficult theory, no longer expressible in terms of V and C simply, in order to take it into account. But in the construction of a practical theory, we take advantage of the fact that the departure from planarity is slight, to which we have already referred towards the end of § 199. A line of electric force does not now start quite perpendicularly from the positive lead and end similarly on the negative lead. It has a slight inclination to the perpendicular, and therefore curves out of the reference plane to a small extent. In the dielectric itself, this peculiarity is of little importance. But the slight amount of tangentiality of the electric force at the surface of the leads, which is conditioned thereby, is of controlling importance as regards the leads themselves, and eventually through them, to the waves in the dielectric, by attenuating and altering the shape of waves (considered axially), as just referred to. We should now consider what form is assumed by the second circuital law in terms of V and C, when we admit that there is tangential electric force on the surface of the leads, but on the assumption that the minor effects in the dielectric itself, which are associated with the presence of the tangential force, are of insensible influence.

We may still regard V as the transverse voltage in the reference plane. Whether we go straight across from the positive to the negative lead, keeping in the reference plane, or leave it slightly in order to precisely follow a line of force in estimating the voltage, is of no moment, because the results differ so slightly. The quantity C is subject to a similar slight

difference, according to the way it is reckoned. We may, perhaps, most conveniently reckon it as the circuitation of the
magnetic force round either lead upon its surface and in the
reference plane, because this way gives the exact value of the
conduction current in the lead. But the circuitation in other
paths in the reference plane will not give precisely the same
value now that the waves are not quite plane. This difference
we also ignore in the practical theory. In truly plane waves
the electric current in the dielectric is wholly transverse. There
is now really an axial component as well, but being a minute
fraction of the transverse current, it is ignored. In short, we
have to make believe that the waves are planar when considering their propagation through the dielectric, whilst at the same
time we take into account the departure from planarity in
considering the influence of the leads and what occurs in them.
It is unfortunate to have to refer to small corrections, as it
confuses the statement of the vitally important matters. Let
us, then, set them aside now.

Construct the second circuital equation in the manner followed in § 201, in elucidating the meaning of equation (6).
Consider a rectangle consisting of two transverse sides in
reference planes at unit distance apart, beginning upon the
positive and ending upon the negative lead, and of two axial
sides of unit length upon the leads themselves. Reckon up the
voltage in this rectangle. The transverse sides give V and
$V + dV/dx$. The axial sides give E_1 and E_2 say, if E_1 is the
tangential component in the direction of increasing x of the
electric force at the boundary of the positive tube, and E_2 the
same on the negative tube, but reckoned the other way. The
complete voltage is then

$$E_1 + \left(V + \frac{dV}{dx} \right) + E_2 - V,$$

or
$$\frac{dV}{dx} + E_1 + E_2.$$

It is also equal to $- L\dot{C}$, as in § 101. (There is no magnetic
conductance of the dielectric now.) So

$$-\frac{dV}{dx} = E_1 + E_2 + L\dot{C}, \quad . \quad . \quad . \quad . \quad (16)$$

is the form assumed by the second circuital law.

Observe that we have made no specification of the nature of the leads, so that the equation possesses a high degree of generality. As regards the first circuital law, that is unchanged, being expressed by (5) when the dielectric is non-conductive, and by (9) when conductive, since the minor changes alluded to as regards V and C are not significant.

Being general, the equation (16) needs to be specialised before it can be worked. If our electromagnetic variables are to continue to be V and C, we require to express E_1 and E_2 as functions of C. Fortunately this can be done, sometimes very simply, and at other times in a more complicated way. To take the simplest case, let the leads be tubes (or sheets) of so small depth that penetration is practically instantaneous as waves pass along them. Then E_1 is not only the boundary tangential electric force, but is also the axial electric force throughout the substance of the positive tube. Similarly as regards E_2 for the negative tube. It is true that, in virtue of the transverse electric current from tube to tube, there is also transverse current in the tubes, and, therefore, transverse electric force, but this is to be ignored, because it is a small fraction of the axial. The current density in the positive tube is therefore axial, of strength k_1E_1, if k_1 is the conductivity of the material; and the total current in the tube is K_1E_1, where K_1 is the axial conductance per unit length, or, which is the same, E_1/R_1, if R_1 is the resistance per unit length. But this quantity is also the previously investigated quantity C, the circuitation of the magnetic force on the boundary of the tube. So we have the elementary relations

$$E_1 = R_1C_1, \qquad E_2 = R_2C_2, \quad . \quad . \quad . \quad (17)$$

which are, be it observed, essentially the connections of the electric and magnetic forces at the boundaries, though brought to a particularly simple form by the instantaneous penetration. The second circuital equation (16) therefore takes the form

$$-\frac{dV}{dx} = (R_1 + R_2)C + L\dot{C}. \quad . \quad . \quad . \quad (18)$$

Or, finally, if R is the resistance per unit length of the two leads,

$$-\frac{dV}{dx} = RC + L\dot{C}, \quad . \quad . \quad . \quad . \quad (19)$$

which is the practical equation for most purposes. Observe, comparing it with (12), that there is an identity of form, but with a changed meaning of the symbol R. In the exactly stated problem of plane waves running through a medium of duplex conductivity bounded by perfect conductors, the quantity R is the external magnetic conductance per unit length axially. In the present approximately stated problem of very nearly plane waves running through an electrically conducting medium bounded by resisting wires, the quantity R is the resistance of the wires, per unit length axially. What we have to do, therefore, in order to turn the real problem into one relating to strictly plane waves admitting of rigorous treatment, is to abolish the resistance of the leads and substitute equivalent magnetic conductance in the dielectric medium outside. The two materially different properties are nearly equivalent in their effects. Equations (13), (14), (15), still hold good, only RC^2 is now to be the rate of waste in the leads, instead of in the medium generally due to the (now suppressed) magnetic conductance. The flux of energy is now not quite parallel to the leads everywhere, but has a slight slant towards them. But the usual formula gives the waste correctly. The product E_1H_1 of the tangential electric force E_1 and the magnetic force, say H_1, at the boundary of the positive lead, is the rate of supply of energy to the lead per unit surface. Therefore, by circuitation, the product E_1C is the rate of supply per unit length axially. This is the same as the previous R_1C^2.

The equivalence of effect of magnetic conductance externally and of electric resistance in the leads, is undoubtedly a somewhat mysterious matter, principally because it is hard to see (apart from the mathematics) why it should be so. As regards the above reasoning, however, it is essentially simple, and is a direct application of fundamental electrical principles. It may, therefore, cause no misgivings, except the doubt that may present itself whether the ignored small effects in the dielectric due to departure from planarity of the waves are really ignorable. As a matter of fact, they are not always of insensible effect, and plenty of problems can be made up and worked out concerning tubes and wires which do not admit of the comparatively simple treatment permissible when V and C are the variables, and the results are exceedingly curious and interesting, though quite unlike those at present in question. But these are not practical problems, and have little bearing upon the question of the free propagation of waves along long parallel straight wires. They do not start as full grown plane waves from the source of disturbance, but they very soon fit themselves on to the wires properly, and then follow the laws of plane waves pretty closely.

Although thin tubes for leads were specially mentioned in connection with equation (19), yet any sort of leads will do provided the penetration be sufficiently rapid to be instantaneous " within the meaning of the Act." It is obviously true for steady currents, when the inertial term disappears and R is the steady resistance. And if the variations of current be not sufficiently rapid to cause a sensible departure from uniformity of distribution of current in the conducting wires, then, of course, (19) may still be safely used. At the same time, it should be mentioned that the inductance L should, under the circumstances, be increased by a (usually) small amount due to induction in the conductors themselves, as will be presently noted more closely. Furthermore, remember that no allowance has been made for the influence of parallel conductors, should there be any. The earth does not count if the leads be alike and equidistant from the ground, as its influence can be embodied in the values of L and S, the inductance and permittance.

The Second Circuital Equation when Penetration is Not Instantaneous. Resistance Operators, and their Definite Meaning.

§ 203. The next question is what to do when penetration is not instantaneous within the meaning of the Act. We should first go back to (16), which remains valid, and inquire whether the tangential electric forces cannot be expressed in terms of the current (or conversely) in some other way than by a linear relation. Suppose we say

$$E_1 = R''_1 C, \qquad E_2 = R''_2 C. \quad . \quad . \quad (20)$$

Is it possible to give a definite meaning to the symbol R''? Is there a definite connection between C, regarded as a function of the time, and E_1 or E_2 also regarded in this way?

Imagine a wire to be free from current, and, therefore, electrically neutral. Now expose its boundary to tangential electric force, beginning at a certain moment, varying in some particular way with the time later, and then ceasing. The result is that C, the total current in the wire, will run through a particular sequence of values, and then finally cease. If we begin again with the applied force, and make it run through the same values in the same manner, we shall again obtain the same values of C at corresponding moments. So far, then, the connection is a definite one. Moreover, if we make the applied force run through the same values increased in a certain ratio, the same for all, then the current will have its previous values increased in the same ratio. But if we change the nature of the applied force as a function of the time (irrespective of size) we shall find that C is not merely changed as a function of the time, but also as a function of the applied force. That is, the mere value of one does not necessitate any particular simultaneous value of the other. So, if we keep to the usual sense meant when algebraists say $u = f(x)$, or u is a function of x, we cannot say that C is a function of E. It is, nevertheless, true that the march of C is strictly connected with that of E, so that when the latter is given, the former is obligatory. To deny this would be equivalent to the denial of there being definite controlling laws in operation. The full connection between E and C, however, involves not merely their values, but also the values of their first, second, third, &c., differential coefficients up to any order. That is, the symbol R'', taken by itself, is a function of the differentiator d/dt. To illustrate by a simple example, suppose

$$R'' = R + Lp + (Sp)^{-1}, \quad \cdot \quad \cdot \quad \cdot \quad \cdot \quad (21)$$

where R, L, S, are constants, and p stands for the differentiator. This means that

$$E = \{R + Lp + (Sp)^{-1}\}C, \quad \cdot \quad \cdot \quad \cdot \quad \cdot \quad (22)$$

or
$$E = RC + L\dot{C} + S\!\int\!Cdt, \quad \cdot \quad \cdot \quad \cdot \quad \cdot \quad (23)$$

in the common notation of integrals. Now imagine the march of C to be given. This implies that the march of $\dot{C}$ is also known, and likewise that of $\int Cdt$. Consequently the march of

E is explicitly known. It is not obvious that when the march
of E is given, that of C is known, by the same operator implicitly. That is, by (22) we have

$$C = \frac{E}{R + Lp + (Sp)^{-1}}, \quad \cdots \quad (24)$$

and given E as a function of the time, C is known as a function
of the time, or if not known, can be found without ambiguity.
It is not obvious, because we do not immediately see how the
operation indicated in (24) is to be carried out, whereas, in the
case of (22), it is visible by inspection. Nevertheless, the fact
that the march of E, physically considered, conditions that of
C, makes the above equation (24) not only definite, but complete. In the usual treatment of the theory of differential
equations, there is no such definiteness. Arbitrary constants
are brought in to any extent, to be afterwards got rid of. Now
this is all very well in the general theory of differential
equations, where arbitrary constants form a part of the theory
itself, but for the practical purpose of representing and obtaining
solutions of physical problems, the use of such arbitrary and
roundabout methods (which are too often followed, especially
by elementary writers) leads to a large amount of unnecessary
work, tending to obscure the subject, without helping one on.
It would not, perhaps, be going too far to say that such a
misuse or inefficient application of analysis often makes rig-
marole.

When we say that $E = R''C$, where R'' is the resistance
operator (so called because it reduces to the resistance in
steady states), we assert a definite connection between E and
C, so that when C is fully given as a function of the time, and
the operations contained in R'' are performed upon it, the function E results, and similarly, when it is E that is given, then
the inverse operator $(R'')^{-1}$ (or the conductance operator) acting upon it will produce C. There may be an infinite number
of differentiators in R'', as p, p^2, p^3, and so on, where p^n means
d^n/dt^n. But there is not a single arbitrary constant involved
in $E = R''C$, nor, indeed, anything arbitrary.

Returning to the wires, the form of the resistance operator,
as a function of p, depends upon the electrical and geometrical

data. It has been determined for round wires and round tubes, and plane sheets. We may take

$$-\frac{dV}{dx} = (R''_1 + R''_2)\, C + LC, \quad . \quad . \quad . \quad (25)$$

as the form of the second circuital law, and make the determination of the operators a matter of separate calculation.

As already mentioned, when C is steady, R''_1 degenerates to R_1 and R''_2 to R_2, the steady resistances (per unit length) of the wires, tubes, rods, or cylinders of any shape that may be employed. At the same time the inertial term disappears. Also, when C varies, it is sometimes sufficient to take into account only the first approximation to the form of the operators. This is, for a solid round wire,

$$R''_1 = R_1 + \tfrac{1}{2}\mu_1 p, \quad . \quad . \quad . \quad . \quad . \quad (26)$$

and similarly for R''_2. This $\tfrac{1}{2}\mu_1$ is the value of the steady inductance of the wire, μ_1 being its inductivity. When not solid, or not round, some other expression is required. Notice that this brings (25) to the elementary form (19), because the inductance of the wires may be included in L itself, which then becomes the complete steady inductance (per unit length axially), including that due to the dielectric and that due to the two wires. This usually means only a small increase in the value of L, unless the wires be of iron.

Simply Periodic Waves Easily Treated in Case of Imperfect Penetration.

§ 204. But besides the above simplification, there is an exceedingly important general case in which a similar reduction takes place. This occurs when the sources of disturbance vary simple periodically with the time. Then the electric and magnetic fluxes everywhere vary ultimately according to the same law with the same period, provided the relations of the fluxes to the forces are linear ; that is, when the conductivity, permittivity, and inductivity, are constants at any one place. Now, when this comes to pass, both E and C in the equation $E = R''C$ vary simple periodically with the time. But when the sine or cosine is differentiated twice, the result is the same function negatived, and with a factor introduced. Thus, if the frequency

is $n/2\pi$, so that $\sin(nt+\theta)$ may be considered to be the time-
factor in the expression for either E or C, we have the property
$d^2/dt^2 = -n^2$. Put, therefore $p^2 = -n^2$ in the expressions for the
resistance operators, and we reduce them to

$$R''_1 = R'_1 + L'_1 p, \qquad\qquad R''_2 = R'_2 + L'_2 p, \quad . \quad . \quad (27)$$

where R' and L' are functions of n^2. They are, therefore, constants at a given frequency, and this is a very valuable property,
as it is clear at once that we reduce the equation (25) to the
simple form

$$-\frac{dV}{dx} = (R'_1 + R'_2)\, C + (L + L'_1 + L'_2)\, pC, \quad . \quad (28)$$

or, more briefly and clearly,

$$-\frac{dV}{dx} = R'C + L'\dot{C}, \quad . \quad . \quad . \quad . \quad (29)$$

which is the same as (19), valid when the penetration is instantaneous, but with different values of the constants involved.
The steady resistance R is replaced by R', the effective resistance (of both tubes per unit length) at the given frequency,
and L the steady inductance (inclusive of the parts due to the
wires) by L', the effective inductance.

Owing to the reduction of the second circuital equation to
the primitive form, we are enabled to express the propagation
of simply periodic waves along wires by the same formulæ,
whether there be or be not imperfect penetration. We simply
employ changed values of the constants, resistance and inductance, which may be independently calculated, or left to the
imagination, should the calculation be impracticable. This is,
when it can be effected, the best way of making extensions of
theory. Do the work in such a way that harmony is produced
with the more rudimentary results, and so that they will work
together well, and the new appear as natural extensions of the
old. An appearance of far greater originality may, indeed, be
produced by ignoring the form of the elementary results, but
the results would be cumbrous, hard to understand, and unpractical.

The effect of increasing the frequency from zero to a high
degree is to first lessen the penetration, and end in mere skin

penetration. Consequently, the quantity L′, the effective inductance, goes from the full steady value $L + L'_1 + L'_2$, and finishes at L simply, the inductance of the dielectric. But the difference need not be great. In the case of suspended copper telephone wires the inductance of the dielectric is far larger than the rest, so that there is no important variation in the value of the inductance possible, nor would there be were the frequency increased up to that of Hertzian vibrations. But as regards the effective resistance, the case is different. As the distribution of current in a wire changes from that appropriate to the steady state, the resistance increases, and the increase is not always a negligible matter. If long-distance telephony were carried on along iron wires it would be a very important effect. But the Americans, who were the introducers of long-distance telephony, soon found that iron would not do, and that copper would. The reason of the failure is mainly the largely increased resistance of iron. In copper, on the other hand, it is an insensible effect at the lower limit of telephonic frequency of current waves, with the size of wire employed, and is not very important at a frequency three or four times as great. On the other hand, in the numerous experiments with very rapid vibrations of recent years, due to Hertz, Lodge, Tesla, and many others, the increased resistance due to imperfect penetration becomes a very important matter, and is one of the controlling factors that should be constantly borne in mind.

Long Waves and Short Waves. Identity of Speed of Free and Guided Waves.

§ 205. By replacing fictitious magnetic conductance of the medium outside the pair of leads by real electric resistance of the leads themselves, we have obtained the same results as regards the propagation of waves, subject to certain reservations referring to the practical applicability of the theory. It is worth while noticing, in passing, a certain peculiarity showing roughly when we may expect the theory to be admissible, and when it should fail. We know that in the transmission of waves along perfectly conductive leads the waves are continuously distorted if the medium be electrically conductive. Also, that by introducing magnetic conductivity into the medium we may reduce this distortion, and ultimately abolish it when the magnetic conductivity reaches a certain value. Now observe here that the correcting influence of the magnetic conductivity is exerted precisely where it is required, namely, in the body of the wave itself. The wave is acted upon in every part by two counteracting distorting influences, so that the correction is performed exactly; in other words, we obtain an exact theory of distortionless transmission.

But, on the other hand, when we employ the resistance of the leads to perform the same functions, the correcting influence is not exerted uniformly throughout the body of the wave, but outside the wave altogether; at its lateral boundaries, in fact. Nevertheless, for reasons before stated, we still treat the waves as if they preserved their planarity under the influence of the resisting leads. Whilst, therefore, we fully recognise and employ the finite speed of propagation of disturbances axially, or parallel to the leads, we virtually assume that the correcting influence of the leads is transmitted laterally outwards instantaneously. We may, therefore, perceive that the wave-length is a matter of importance in determining the applicability of the practical theory. It is of no moment whatever in the exact theory employing magnetic conductance; but, when we remove this property from the medium generally, and (virtually) concentrate it at the leads, we should at the same time keep the wave-length a considerable multiple of the distance between the leads if the practical theory is to be applicable. To see this, it is sufficient to imagine the case of waves whose length is only a small fraction of the distance between the leads, when it is clear that their correcting influence could no longer be assumed to be exerted laterally and instantaneously, as if the small portion of the leads between two close reference planes belonged to and was associated solely with the slice of the medium between the same planes.

We have referred in the above to the action of the resistance of the leads as a correcting one, neutralising the distortion due to another cause. But the same reasoning is applicable when the action is not of this nature. Thus, when the external medium is non-conductive, and the leads are non-resistive, we have perfect transmission. Making the leads resistive therefore now brings on distortion. We may now say that in order that this distortion should occur in the same way as under the influence of magnetic conductivity in the external medium, the wave-length should not be too small, as specified above. The practical theory is therefore the theory of long waves, and in the interpretation of the word " long," some judgment may be exercised as regards the leads and other matters, because there is no hard and fast distinction, and what may be a long wave under some circumstances may be a short one under others.

To illustrate this point, imagine that we have got an ingenious instrument (not yet made), for continuously recording the electromagnetic state of a non-conducting medium, say the air at a certain place (just as we have thermometric and barometric recorders), and that this instrument is so immensely quick in its action as to take cognisance of changes happening in very short intervals of time, say one thousand-millionth of a second. Now in applying this instrument to register the state of air traversed by electromagnetic waves, it is clear that the size of the waves must be considered in relation to the size of the instrument. If the instrument were one decimetre across, then waves of one kilometre in length could as well be regarded as of infinite length. But if only one metre in length, though it could still be used, there would be no longer the same accuracy of application. And if of only a centimetre in length, then it is plain that several waves would be acting at once on the apparatus in different parts, and the resultant effect recorded would not represent the history of the waves by any means.

Now as regards waves sent along parallel leads, it is obvious that light waves are totally out of the question, being immensely too short. On the other hand, telephonic waves are to be treated as long waves—very long, in fact—though they are short compared with telegraphic waves. But it is somewhat curious that the electromagnetic waves investigated by Hertz are sometimes so short as to come within the scope of the above reservational remarks, or of others of a similar nature. Whether the generation of waves by an oscillator be considered, or their effect on a resonator, or their transmission along leads, their shortness in relation to the apparatus employed may sometimes vitiate the results of approximate theories, and render caution necessary. For example, the plane-wave theory indicates that the attenuation factor for waves running along parallel leads is $\epsilon^{-Rx/2Lv}$ in the distance x, or $\epsilon^{-Rt/2L}$ in the equivalent time of transit t. This is when R and L are constants. If not, and the waves be simply periodic, then we may use the same formula with the effective values of R and L at the frequency concerned. But if this be true for long waves, we cannot expect it to continue true on shortening the waves to the transverse distance of the leads, more especially if we are ignorant of the precise type of the waves. At the best, we should not expect more than results of a similar kind. But not much has been done yet in the quantitative examination of Hertzian waves, for sufficiently obvious reasons.

In one respect, however, a formerly very strange anomaly has been cleared up satisfactorily. When Hertz opened people's eyes and made them see the reality of Maxwell's ether as a medium propagating electromagnetic disturbances at the speed of light, by showing their transmission across a room and reflection by a metallic screen, the full acceptation of Maxwell's theory was considerably hindered for a time by his finding that the speed of waves sent along wires was much less than that of free waves. The discrepancy was a large one, and gave support apparently to the old view regarding the function of wires, which made the wires the primary seat of transmission, and effects outside secondary, due to the wires. And it came to pass that people, whilst admitting the truth of Maxwell's theory, yet made a distinction between waves in free space and " in wires." This was thoroughly out of harmony with Maxwell's theory, which makes out that the wires, though of great importance as guides, are nevertheless only secondary. On the other hand, it should be mentioned that Lodge found no such large departure from the speed of light in his experiments. But the matter has been explained by the discovery that an erroneous estimate was made of the permittance of the oscillator in the experiments which apparently showed that the speed was largely reduced. When corrected, there is not left any notable difference between the speed of a free wave and of one guided by wires. Of course, there is no reason why reference to waves " in wires " should not be dropped, unless the laterally-propagated cylindrical waves are meant. These are secondary, and have no essential connection with the propagation of the primary waves through the external dielectric, although modifying their nature.

The Guidance of Waves.   Usually Two Guides.   One sufficient, though with Loss.   Possibility of Guidance within a Single Tube.

§ 206. When waves are left to themselves in ether without the presence of conductors, they expand and dissipate themselves. Even if they are initially so constituted as to converge to centres or axes, they will subsequently expand and dissipate. To prevent this we require conducting guides or leads. Now this usually involves dissipation in the leads ; but the point at present under notice is the property of guidance only. We can stop the expansion in a great measure, and cause a wave to travel along wherever we wish it to go. Practically there are two leads, as a pair of parallel wires ; or if but one wire be used, there is the earth, or something equivalent, to make another. But it is still much the same, as regards guidance, when there is but one wire, if we choose to imagine the case of a single infinitely-long straight wire alone by itself in ether. If we make it the core of a plane electromagnetic sheet, this sheet will run along the wire just as well and in the same way as if there were a second guide. But the energy of such a sheet, even though of finite depth, and containing electric and magnetic forces of finite intensity, would be infinite. The quantity L, the inductance per unit length of guide, is infinite under the circumstances. We could not, therefore, set up such a wave from a finite local source. If we cause an impressed voltage to act axially for a very short interval of time across any section of the guide, say in a reference plane, the result is an approximately spherical wave. (To be perfectly spherical, the wire should be infinitely fine. The case is then that of a spherical wave-sheet with conical boundaries, already referred to, with the angle of the cone made infinitely small.) Its centre is at the origin of the wave, and as it expands, the portions of the wave-sheet nearest the wire become approximately parallel plane waves, one going to the right, the other to the left along the wire. But not being pure plane waves they are weakened as they progress, by the continuous expansion of the spherical wave of which they form a part.   Nevertheless, we have the propagation of nearly plane waves of finite energy, or of a perfectly plane wave-sheet of infinite energy, along a single guide.

Now, this takes place outside the conducting guide, and the question presents itself whether we cannot transmit an electromagnetic wave along the interior of a tube, in a manner resembling a beam of light ?   We can certainly do so if we have a second guide within the first tubular one, for this does not differ substantially from the case of two parallel wires, each outside the other.   But it does not seem possible to do without the inner conductor, for when it is taken away we have nothing left upon which the tubes of displacement can terminate internally, and along which they can run.   A theoretical expedient is to carry the electrification forward at the proper speed.   But we want the process to be automatic, so to speak, hence convection will not do.   Again, if we make the displacement start from one portion of an electrically conducting tube and terminate upon the rest, we must insulate the two portions from one another, and then there will be a division of the charges between the interior and exterior, so that the result will be an external as well as an internal wave, or rather, one wave occupying both regions.

It would appear that the only way of completely solving the problem of the automatic transmission of plane waves within a single tube is a theoretical one, employing magnetic as well as electric conductance.   To see this, imagine any kind of purely plane wave being transmitted in the normal manner through the ether, and fix attention upon a tube of the flux of energy, or a beam, using optical language.   This beam cuts perpendicularly through the reference planes, in which the lines of electric and magnetic force lie, which again cross one another perpendicularly.   The shape of the section of the beam by a reference plane may be arbitrary.   But let it be quadrilateral, and so that it is bounded by magnetic lines on two opposite sides, and by electric lines on the other two.   Now let the two sides of the tube upon which the displacement terminates perpendicularly, be electrically conductive thin sheets, and the other two sides, upon which the induction terminates perpendicularly, be magnetically conductive sheets.   If the conduction be perfect, we shall not interfere with the transmission of the beam within the tube.   That is, we have solved the problem stated. We may also notice that the external portion of the wave is not interfered with by the tube.   But the interposition of the tube in the manner described renders the external and internal waves quite independent of one another, and either of them may be suppressed.

Similarly, if the interior region be made finitely conductive, electrically or magnetically, or both together, we can still investigate the transmission of waves along it in the same way as previously described for complete plane waves in a homogeneous medium made conductive, and the same applies to the external region, independently of the internal.   Going further, we may do away with the diffused conductances, and concentrate equivalent resistances in the plates bounding the tubes, in the same way as we replace magnetic conductance of the external medium by equivalent electric resistance of the wires in passing from the exact plane-wave theory to the practical theory of wires in terms of V and C.   That is, the two plates on which the displacement ends may be made electrically resistive, the resistance taking the place of the magnetic conductance in the interior ; whilst the other two plates should be made magnetically resistive, if the interior electric conductance is also to be abolished.   We may then express the propagation of waves in the tube in a manner resembling the practical theory of wires, though it will no longer be an exact theory. Nor will the interior and exterior regions be quite independent of one another now that the tube is only finitely conductive.

Interpretation of Intermediate or Terminal Conditions in the Exact Theory.

§ 207.   Leaving these somewhat abstruse considerations, return to the practical theory concerning wires and its connection with the exact theory involving magnetic conductance.   In the working out of the practical theory (which is not yet, however, the theory of official representatives of practice, though they are decidedly getting on) we have often to consider the effects due to intermediate insertions of resistance in the circuit of the leads, or of shunts across them, and other modifications.

D D

The question now is, what do these insertions represent in the exact theory ?

Consider first a pair of parallel leads of no resistance. This will admit of the application of the exact theory, provided there be either no leakage or the external medium be uniformly conductive electrically. Now we know that when the leads have resistance, we make the transformation to the exact theory by substituting uniform magnetic conductivity of the external medium of the proper amount. Plainly, then, if we insert an electrical resistance in a lump in the circuit of the leads at any place, we should, in the exact theory, transfer it equivalently (changed to magnetic conductance) to the whole of the corresponding reference plane. That is to say, we must make the reference plane (outside the leads) uniformly conductive magnetically, so that the total magnetic conductance of the equivalent plate inserted equals the electric resistance which it replaces. We may then investigate the action of the plate on the waves traversing it in the manner described in a previous paragraph (§ 193 and after).

Similarly, the effect of an electrically conducting bridge across the circuit is equivalent to that of an electrically conducting plate at the reference plane. Their conductances are here not only equal, but of the same kind. We merely rearrange the leakage conductance so that it shall act uniformly, to suit plane waves.

Thus, a terminal short-circuit should be replaced by a terminal perfectly conductive plate, reflecting H positively and E negatively, or with reversal of sign ; or, which is the same, C positively and V negatively. This maintains E and V permanently zero at the short-circuit, unless there be impressed force there. And a terminal disconnection may be represented, in the exact theory, by a terminal perfectly magnetically conductive plate, reflecting E positively and H negatively, or V positively and C negatively, thus maintaining C permanently zero at the disconnection.

A somewhat different kind of transformation is required in connection with impressed forces. Suppose, for instance, we insert an impressed voltage in the circuit of the leads at a given reference plane. How is it to be equivalently transferred to the whole plane, or to that part of it outside the leads, so that

it may generate plane waves of the precise type belonging to
the given leads ?   This is by no means so easy to follow up as
the previous plate substitutions ; but the following may sufficiently describe the essence of the transformation.   First, when
the generation of disturbances is concerned, it is not impressed
electric force e, but its curl, that is effective.   We must therefore find the curl of the given impressed force.   If the latter
is half in each wire, acting opposite ways, and uniform across
their sections by the reference plane, then the curl of e,
say f, is situated on the boundaries of the wires, and encloses
them.   Having ascertained the integral amount of f, it must
be transferred to the whole reference plane outside the wires.
Then the question arises how to distribute it properly.   To
answer this, it may be mentioned that when a distribution of
f starts into action, the first effect is to generate induction
following the lines of f.   This will be made clear later.   If,
therefore, we distribute the lines of f over the reference plane
in such a way as to exactly copy the natural distribution of
the magnetic lines in plane waves having the given leads for
cores, we shall obtain what we want, viz., the sources so
distributed as to generate plane waves of the required type on
the spot.

By means of the above and similar devices, most telephonic
and telegraphic problems may be converted to problems in an
exact plane-wave theory.   But we must always be careful to
distinguish between a theory and the application thereof.   The
advantage of a precise theory is its definiteness.   If it be dynamically sound, we may elaborate it as far as we please, and be
always in contact with a possible state of things.   But in
making applications it is another matter.   It requires the
exercise of judgment and knowledge of things as they are, to
be able to decide whether this or that influence is negligible or
paramount.

The Spreading of Charge and Current in a long Circuit, and their Attenuation.

§ 208.   The detailed application of the principles already
discussed to the data which occur in practice, or which may
occur in later practice, would lead us into technical complications which would be out of place in the present stage of

development of this work.   Only a few general considerations
can be taken up here in connection with the theory of plane
waves.   We have seen that there are four distinct quantities
which fundamentally control the propagation of "signals" or
disturbances along a circuit, symbolised by R, K, L, and S,
the resistance, external conductance, inductance, and permittance; whilst there are two distinct variables, the transverse
voltage V and the circuital gaussage C, or the difference of
potential (in an extended sense) of the leads in a reference
plane, and the current in the leads.   The words voltage and
gaussage were proposed in § 27 to represent electromotive force
and magnetomotive force, and have been employed experimentally to see how they would work.   They work very well
for theoretical purposes, but I observe that they are not
generally or universally approved of, perhaps on account of
the termination "age," or because of the commencements volt
and gauss being unitary names.   It is open, of course, to anyone to find names which shall be more suitable or meet with
general approval.

But I stick to inductance.   Of all the words which I have
proposed, that one seems to me to be, more than any other
except impedance and conductance, the right word in the right
place, and I am bound to think it possible that some of those
who prefer the old "coefficient of self-induction," do not fully
appreciate the vital significance of inductance in a theory of
intermediate action of a medium.   The idea of the direct
mutual action of current-elements upon one another is played
out.   And as regards the name of the practical unit of inductance, I think the best name is mac, in honour of the man who
knew something about self-induction, of course.   Many other
names have been proposed, but none so good as mac.   Some
critic has made etymological objections.   But what has
etymology got to do with it?   The proper place for etymology
is the grammar book.   I always hated grammar.   The teaching of grammar to children is a barbarous practice, and should
be abolished.   They should be taught to speak correctly by
example, not by unutterably dull and stupid and inefficient
rules.   The science of grammar should come last, as a study
for learned men who are inclined to verbal finnicking.   Our
savage forefathers knew no grammar.   But they made far

better words than the learned grammarians.   Nothing is more admirable than the simplicity of the old style of short words, as in the A sad lad, A bad dog, of the spelling book.   If you transform these to A lugubrious juvenile, A vicious canine, where is the improvement?

Now, as we have already explained the characteristics of distortion of plane waves by conductance in the medium generally, of two kinds, acting differently on the electric and magnetic fluxes, the corresponding relations in a long circuit of leads need be only briefly mentioned.   Consider an infinitely long circuit of parallel and equal resisting leads under uniform external conditions, and governable by the relations laid down. Let it be, at a given moment, wholly free from charge or current except at one place, where there is merely a charge, say Q_0.   Or, say $Q_0 = SV_0$ on unit length only at the place.   Left to itself, this will spread.   But it will do so in widely diverse manners under different circumstances, although there are some common characteristics.   The speed of propagation of disturbances is $v = (LS)^{-\frac{1}{2}}$, independent of the values of R and K, the dissipation constants.   Thus, at time t after the spreading begins, we are bound to find the charge, or what is left of it, within the region of length $2vt$, being vt on each side of the origin.   Beyond this region there is no disturbance.

Next as regards the amount of the charge.   It cannot increase, but it may diminish.   If the insulation be perfect, which means that $K = 0$, the charge remains constant.   It is unaffected by the resistance of the leads, although energy is wasted in them.   It only decreases by the external conductance, or by equivalent leakage.   Then we have

$$Q = Q_0 \epsilon^{-Kt/S}, \quad \cdots \quad \cdots \quad (1)$$

to express the charge Q at time t, decreasing according to the well-known exponential law.

As regards the manner of spreading, either with or without the subsidence, this is represented in the exact theory by the spreading and subsidence of displacement in a medium which is magnetically as well as electrically conductive, although the former property does not affect the total amount of displacement.   We see, therefore, that the distortionless case, which occurs when the time constants L/R and S/K are equal, forms a natural division between two distinct kinds of resultant distortion. In this neutral case, the charge Q_0 splits into two halves which at once separate from one another at relative speed $2v$. At time t later they are each attenuated to

$$\tfrac{1}{2}Q = \tfrac{1}{2}Q_0 \epsilon^{-Kt/S}. \quad \ldots \ldots \quad (2)$$

The energy was wholly electric to begin with, but in each of the resulting waves it is half electric and half magnetic, because they are pure waves. As they attenuate, their energy is wasted, and this occurs so that half is wasted by the resistance of the leads, and half by reason of the external conductance. This is necessary in order that a pure wave shall remain pure. The final result is attenuation of the two charges to zero, when they are infinitely widely separated, and with this a complete simultaneous disappearance of the other characteristics, as the magnetic force. In the exceptional case of no resistance and no leakage, there is, of course, an everlasting persistence of the two charges and of the waves of which they form a feature. They go out of range, but not out of existence. In all other cases but the distortionless one, the charge (or the remains of the original charge) is only partly in the plane waves at the two ends of the disturbed region, the rest being diffused between them. But this may happen in two ways. The total charge at time t is expressed by (1) above. But the charge in the terminal waves, half in each, is given by

$$Q_0 \cdot \epsilon^{-(R/2L+K/2S)t}, \quad \ldots \ldots \quad (3)$$

reducing to (2) doubled in the distortionless case. Therefore the excess of (1) over (3) is the amount of the diffused charge. It is

$$Q_0 \cdot \epsilon^{-Kt/S} \left(1 - \epsilon^{-(R/2L-K/2S)t}\right), \quad \ldots \quad (4)$$

and is positive or negative according as R/L is greater or less than K/S.

That is, when the resistance of the leads is in excess, the charge is positive between the waves as well as in them. But when the leakage is in excess the intermediate charge is negative. This happens equivalently when the inductance is in excess, or the permittance in deficit. How this reversal comes about has been already explained in the exact theory. In the present circumstances we may say that an intermediate resistance in the circuit of the leads reflects V positively and C negatively, whilst an intermediate leak reflects C positively and V negatively. The same applies to the elements of the circuit, and from this follows the reversal of V when leakage is in excess. Whether the tailing be of the usual positive or of the negative kind (referred to the charge), the maximum or minimum density is in the middle; that is, at the origin. The density decreases symmetrically both ways to the two ends, where the plane waves are to be found. When they have attenuated to practically nothing, there is left merely the widely diffused charge, unless it has attenuated similarly. With the resistance largely in excess, there may be nearly all the charge left when the terminal waves have become of insensible significance. But if the leakage be in excess, the disappearance of the terminal waves is accompanied by a similar disappearance of the intermediate negative charge, because the total charge is always positive.

So far regarding the spreading of a charge, next consider the spreading of induction. The magnetic analogue of Q is $P = LC$, the magnetic momentum per unit length of circuit. So, as L is constant, the spreading of P is represented by that of C. Suppose initially there be current C_0 in unit length at the origin, with momentum $P_0 = LC_0$, and no charge there or elsewhere. This also splits immediately into two plane waves, which only differ from those arising from a charge in having similar C's and opposite V's instead of similar V's and opposite C's.

The total induction remains constant only if the leads have no resistance. In general it decreases, so that at time t we have

$$P = P_0 \cdot \epsilon^{-Rt/L} \quad . \quad . \quad . \quad . \quad (5)$$

At the same time the amount in the terminal waves, half in each, is

$$P_0 \cdot \epsilon^{-(R/2L + K/2S)t} \quad . \quad . \quad . \quad (6)$$

The amount between the terminal waves is, therefore, the excess of (5) over (6), which is

$$P_0 \epsilon^{-Rt/L} \left(1 - \epsilon^{-(K/2S - R/2L)t}\right) \quad . \quad . \quad (7)$$

These equations (5), (6), (7), may be instructively compared with the set (1), (3), (4), their analogues. Notice that in transforming from one set to the other, we exchange V and C, Q and P, R and K, L and S. In the exact theory the R in these formulæ would also be a conductance, as before described.

By (7) we see that in the distortionless case all the induction left is in the terminal waves, which is to be expected. But we require the leakage to be in excess for the intermediate induction to be positive, and conversely the resistance must be in excess for it to be negative. Of course, the initial induction is assumed to be positive. Also observe that leakage by itself has no effect on the total induction except to redistribute it, just as the resistance of the leads has no attenuating effect on the total charge. As before, the maximum or minimum density of induction is at the origin, though, of course, the terminal inductions may have greater or smaller density than this, according to the extent of attenuation at the moment concerned. Remember that the destruction of induction by the resistance of the leads is represented, in the exact theory, by its destruction by the equivalent magnetic conductance externally.

Putting together the preceding results, we may readily see the effect of having both V and C initially at any spot. One case is specially noteworthy. Take $V_0 = LvC_0$ initially. This means a pure plane wave sheet, travelling in the positive direction. There is, therefore, no initial splitting, and the wave just goes on. To what extent this will continue depends upon the relation the constants of the circuit bear to the distortionless state. When the latter obtains, the wave is transmitted without spreading out behind, so that at time t the initial state, if existent over unit length, will be found over unit length at distance vt to the right, attenuated to

$$V = LvC = V_0 \, \epsilon^{-Rt/L}, \quad . \quad . \quad . \quad . \quad (8)$$

where for t may be substituted its equivalent x/v. In all other cases there is reflection in transit, and the reflected portions travel back, besides getting mixed together, thus making a tail of length $2vt$, half on each side of the origin, but now, of course, without a head at the negative end to balance the one at the positive end.

If the resistance be in excess, the charge in the tail is positive and the current is negative, at least initially; whilst if the leakage be in excess, the current is positive and the charge initially negative. *See* §§ 195 to 199 for more details, which may be readily translated to suit the present case. The total charge subsides according to equation (1) above, and the total induction according to (5). Similarly (3) represents the charge in the one terminal wave (instead of both, as there), and (6) shows the induction in the terminal wave.

If initially $V_0 = -LvC_0$, this means a negative wave (going to the left) to start with. This needs no separate notice in detail.

The Distortionless Circuit. No limiting Distance set by it when the Attenuation is ignored.

§ 209. The distortionless state forms a simple and natural boundary between two diverse kinds of propagation of a complicated nature, in each of which there is continuous distortion which is ultimately unlimited. Given time enough, and a circuit of infinite length to work with, the least departure from the distortionless condition would be sufficient to allow all the natural changes to be gone through, with ultimate unlimited distortion. The attenuation is a separate matter in this connection, though itself of great importance. If, however, we are only concerned with the distortion produced in a finite interval of time, which may be quite small, then the matter may be differently regarded. There is much or little distortion, generally speaking, according as the ratios R/L and K/S are nearly equal or widely different. But this needs to be understood with caution, for obviously there may be next to no distortion of signals even when the distortionless state is widely departed from, provided the changes of charge and current are made slowly enough. But when this is not the case, and there is marked distortion of waves in transit, then we shall increase it by making the time-constants more unequal than they are, and decrease it by a tendency to equalisation. Furthermore, this process is a continuous one, so that, starting from the distortionless state connected with equal time-constants, we may, by continuously increasing one time-constant, bring on continuously increasing distortion of one kind, or else, by increasing the other, whilst the first is kept constant, bring on continuously increasing distortion of the other kind. And this is true by variation of any one of the four quantities concerned, the resistance, leakage, inductance, and permittance, to prove which we need only remark that with any values given to these constants we can equalise the time-constants and abolish the distortion altogether by altering any one only of the four.

When a circuit has been brought to this state, then arbitrary signals of any size and manner of variation, originated at the beginning of the circuit, will run along it at the speed of light, and in doing so suffer no alteration save a weakening according to the exponential law given above. The voltage and gaussage at any spot will be always in the same phase, and the succession of values will faithfully repeat those at the origin. If the circuit be infinitely long, the train of disturbances will run out to infinity, of course, though it may be still a long way from infinity when the attenuation is so great as to make the disturbances insensible. Therefore, disregarding the attenuation, there is no limiting distance of possible signalling. More than that, it is not only possible, but perfect signalling, according to this theory.

But there is no perfection in this world. Even if a circuit were constructed with constant R, K, L, and S, and with R/L = K/S exactly, so that it should be truly distortionless in this practical theory, it would not be so in reality. There are several disturbing causes, which, though they might not be of much importance in general, would serve to prevent the attainment of the clean-cut perfection of the theory which ignores them. Instead of zero distortion, then, it is practically only a state of minimum distortion that is attainable. But it may be a very good imitation of the theoretical ideal. It is unnecessary to enlarge upon the necessary failure of a professedly approximate theory when pushed to extremes. It is sufficient to say that the ignored disturbing influences alone would serve to set a practical limit, even if the condition R/L = K/S were truly attained, and attenuation were of no importance.

But there is an influence in full action in the practical theory itself which must not be overlooked, namely, the attenuation due to resistance and leakage. Of what use would it be to have a distortionless circuit if it took nearly all the life out of the current in the first 100 miles, when you want it to go 1,000? Now, on paper nothing is easier than to increase the battery power to any extent you want, till, in fact, the minute fraction which got to the desired distance became magnified up to a recognisable size. But practically there would usually be strong reasons against this process. At any rate, it is impossible to overlook the attenuation. It is not merely enough that signals should arrive without being distorted too much ; but they must also be big enough to be useful. If an Atlantic cable of the present type were made distortionless by the addition of leakage, there would be no sign of any signals at the distant end with any reasonable battery power. Nor can we say that telephony is possible on a circuit of given type to this or that distance merely because a certain calculable and small amount of distortion occurs at that distance. Nor can we fix any limiting distance by consideration of distortion alone. And even if we could magnify very weak currents, say a thousandfold, at the receiving end, we should simultaneously magnify the foreign interferences. In a normal state of things interferences should be only a small fraction of the principal or working current. But if the latter be too much attenuated, the interferences become relatively important, and a source of very serious distortion. We are, therefore, led to examine the influence of the different circuit constants on the attenuation, as compared with their influence on the distortion.

The two Extreme Kinds of Diffusion in one Theory.

§ 210. Although the equations connecting the voltage V and gaussage C, through the line constants R, S, K, and L, are of a symmetrical character, so that, abstractedly considered, a study of the nature of their propagation on either side of the distortionless state is as desirable as on the other side; yet when we examine the conditions prevailing in practice, we see at once that one side only presents itself actively. The reason is easily to be seen. It is in all cases desired to get plenty of received current, because that is the working agent, and leakage is detrimental to this consummation. There are, besides, practical inconveniences connected with leakage. Thus it comes about that we are mainly concerned with the state of things existing when R/L is greater than, or at the least equal to, K/S, and scarcely at all with the exceptional case of an excess of

leakage, unless it be as a curiosity, and for the sake of its
theoretical connection with other electromagnetic problems.

Now this concerns the propagation of V and C through the
external dielectric. The influence of R prevails over that of
K, and controls matters. It is, however, somewhat remarkable
that the other side of the propagation problem does actually
present itself actively, when we regard the important secondary
effect of the propagation into the guides of the cylindrical
waves, which result from the passage over them of the plane
waves in the external dielectric. Suppose, for example, there
is no external conductance. This gives one extreme form of
propagation of V and C, the quantity K/S (or equivalently k/c)
being zero, and R/L finite. Now, remember that in the exact
plane-wave theory R stands for the magnetic conductance of
the external medium, so that R/L is the same as g/μ, and therefore we are, when considering V and C, also discussing the
influence upon the plane waves of the fictitious property of
magnetic conductance.

On the other hand, in the guides themselves, which are electrical conductors, and perhaps, and most probably, dielectrics
as well, it is k/c that is greater than g/μ, since the latter is
zero. Here, then, we have the other extreme form of the
theory. In the external dielectric k/c is zero and g/μ (equivalent to R/L) finite, whereas in the guides g/μ is zero and k/c
finite. The distortion of the external plane waves by the
resistance of the guides is, therefore, of the opposite kind to
that of the cylindrical waves in the guides, for E and H change
places when the form of the distortion is discussed.

Now, in the guides, on account of their being very good conductors, the permittivity is swamped, and may be ignored
in calculating resultant effects. Then we have the theory of
pure diffusion (as of heat according to Fourier), controlled by
the *two* constants k and μ, since both g and c are now zero.
Similarly, in the external dielectric, it may happen that the influence of μ (or equivalently of L) is insensible, as in the slow
working of long cables ; then the propagation of V and C is
controlled by the *two* constants g and c (or actually and equivalently by R and S), whilst the other two, k and μ, are zero.
Here, then, we have a curious exchange of active properties
between the dielectric and the conductor. In both, the resul-

tant propagation follows the diffusion law, but is controlled by different properties. The propagation of V or E outside is like that of H inside, and the propagation of C or H outside is like that of E inside.

We see by the above illustrations that in spite of the absence of magnetic conductivity in reality, we are, nevertheless, obliged to consider both sides of the wave-theory it involves, even in one and the same actual problem, provided it be treated comprehensively.

The Effect of varying the Four Line-Constants as regards Distortion and Attenuation.

§ 211. As, however, we are not immediately concerned with the internal state of the guides, and, so to speak, eliminate it by treating R and L as constants, we may now dismiss its consideration, and return to V and C only, on our understanding that the quantity K/S is always less than R/L, or does not exceed it in the limit.

Now we have shown that the distortion in transit depends upon the difference

$$\sigma = \frac{R}{2L} - \frac{K}{2S}, \quad \ldots \ldots (9)$$

whilst the attenuation of the front of an advancing wave depends upon the sum

$$\rho = \frac{R}{2L} + \frac{K}{2S}. \quad \ldots \ldots (10)$$

Both σ and ρ should be as small as possible, of course. The four line-constants act upon ρ and σ in different ways, and it is rather important to understand them. So, remembering that the first term of the right members of (9) and (10) is bigger than the second, observe the effect of varying the line-constants, one at a time.

[R]. By increasing R we increase both ρ and σ. Thus the resistance of the guides is wholly prejudicial, since it not only weakens signals in transit, but distorts them. It is, therefore, a fundamental notion that the resistance of the guides should be reduced if possible, but never increased, if the object be to facilitate signalling.

[L]. By increasing L we decrease both ρ and σ. Thus the inductance of the circuit (other things being equal) is wholly beneficial, since it diminishes both the attenuation and the distortion in transit, or makes the received signals both larger and plainer. These are not necessarily insignificant effects, but may be very large when applied to very rapid vibrations on a long circuit.

[K]. Increasing K increases ρ and reduces σ. Thus leakage is both beneficial and prejudicial, according to circumstances, since although it reduces the distortion, it simultaneously increases the attenuation in transit. This action is most pronounced when the signalling is slow.

[S]. Increasing S increases σ and reduces ρ. It may be, therefore, both prejudicial and beneficial. But practically the important effect of S, notably on cables, is the prejudicial one of distorting the signals, which is a large effect.*

Now the above statements, though obviously true enough as deductions from (9) and (10), are merely qualitative. The practical import of varying one quantity when three others are constant, depends upon the actual immediate values of all four. The length of the line also comes in as an important factor, and likewise the frequency of the waves, to settle under what circumstances this or that variation of the line constants is important in a special case. Nevertheless, the above properties are important, and should be most usefully borne in mind in general reasoning, as well as in the detailed examination of formulæ, for instance, the important solutions for simply periodic waves.

Some of these effects have been known and understood from the earliest times, or, more correctly, since the time when William Thomson taught telegraph-cable engineers the principles of their business (in its electrical aspects), and, to a great extent, the practice too. They were previously in quite a benighted state, generally speaking, though there were a few, whose names it is needless to mention, who combined enough

---

* Every rule is said to have an exception. It is, at any rate, possible for the above rules to fail, or appear to fail, when the making and reception of the signals is not of a simple kind. We may then attribute the failure to instrumental peculiarities. An example will occur a little later in which K reduces working speed largely.

electrical science with their practical knowledge to have fairly correct ideas about retardation in submarine cables. I do not mention Faraday in this connection, for that great genius had all sorts of original notions, wrong as well as right, and not being a mathematician, could not effectually discriminate, especially as he had so little practical experience with cables.

A complete history of the subject of the rise of Atlantic telegraphy, including the scientific side, written by a man fully and accurately acquainted with the facts, preferably with personal knowledge, and devoid of personal bias, would be of great value and of permanent utility. It will not be very long before no contemporary will be left to do it. Perhaps, on the other hand, it might be more fairly done at second hand. In the meantime, there is a somewhat bulky volume, the Report of the Submarine Cable Committee of 1859, which serves to show the state of knowledge at that date. But it is the history of the march of knowledge in the several preceding years that is important.

The two effects referred to are that resistance and permittance (R and S) act conjointly in producing "retardation," so that both R and S should be reduced to improve the electrical efficiency of a cable. This is embodied in Lord Kelvin's theory of 1855, and gave rise to a rather neat law of the squares, something that practicians could grasp. This law soon became matter of common knowledge amongst cable engineers and electricians generally. But when once grasped they could not let go, and in later times the law has been most grossly abused and misapplied. The fact still remains, however, that R and S are prejudicial. But allowance has to be made for other influences, and it may be very large allowance.

But, very curiously, there has been an exception to the rule. There have not been wanting electricians who, when telephony was extended to circuits containing underground work, supposed that high, not low resistance, was good for telephony. This was giving up the law of the squares with a vengeance. But there was really no warrant for such a conclusion, either theoretical or practical. As this view turned out manifestly wrong by practical failure, it was followed by a precipitate revulsion, the law of the squares being again seized, and grasped more firmly than ever, and against all reason. The fact that R and S are prejudicial does not make the law of the squares.   To have that, R and S must be the only controlling electrical factors.

All the other effects are of later recognition, and were mostly discovered, and their laws investigated, and consequences worked out by myself.   They completely remodel the subject. The old views remain right, in their proper application, but have a very limited application in general, as in modern telephony particularly.   My extended theory opened out a wide field for possible improvements in signalling over long circuits in advance of what could be possible were R and S the only controlling factors, and so far as it goes, modern telephony already goes far beyond the supposed restrictions of the law of the squares.

We should not only consider the effects of varying the four line-constants one at a time, but also of combinations.   Thus, K may be counteracted by L.   The former is beneficial, inasmuch as it lessens the distortion, but it is prejudicial in the attenuation it causes in transit.   But if we simultaneously increase L, we counteract the attenuation caused by K; moreover, we still further lessen the distortion.   This is corroborative of the preceding remark as to the unreserved benefit due to L.   We see, therefore, that [K] and [L] above point the way to salvation, if that be by means of increasing the distance through which signalling can be carried on with a cable of given type as regards R and S.   But let it not be forgotten in studying K and L, that R and S themselves might perhaps (in some particular scheme, for example) be reduced as cheaply as K and L could be carried out.   Be liberal with the copper, for one thing. Look out for some practical way of insulating the wires which shall not make the permittance be so great as it is at present with gutta-percha or india-rubber.   Something of this sort has been done for telephone underground cables, though I do not know the practical merits.   But, as regards Atlantic cables, it should be something very good and trustworthy, or the reduced permittance may be bought too dearly.

There has been some difficulty in getting people to grasp the ideas embodied in [K] and [L], especially the latter, of course, on account of its novelty; whilst [K] was already partially recognised by knowing ones, and is capable of very ready explanation.   [L] was the trouble.   Now, remembering the state

of fog that prevailed in the fifties as regards the effects of
R and S alone, there is no reason for wonder that more recondite effects, as of L, unaccompanied by prior experience, should
not be readily recognised when pointed out, though that was
no excuse for burking the matter. [K] and [L], especially the
latter, require to be studied to be understood. But if people
will not take the trouble to study the matter, how can they
expect to understand it? Now I find that there has been too
much of an idea that I have merely given some very complicated formula which no fellow can understand, and have made
some dogmatic and paradoxical statements about the consequences. But a reference to my "Electrical Papers" will show
that I have treated the matter of propagation very fully, in
the elementary parts as well as in the advanced, and from
many points of view, and that I have not been above pointing
the moral by numerical examination of results, whilst at the
same time I have been particularly careful in elaborating the
theory of the distortionless circuit, on account of its casting
so much light upon the subject generally. The notion that
has been promulgated, that I cannot make myself understood
by physicists, recoils upon its authors with great force. Letting
alone the fact that some physicists have understood me, and
moreover, the fact that various students have understood me
(as I know by my correspondence), there is the very notable
fact that the promulgators of the false idea (which must be
most discouraging to students) have evidently not taken the
trouble to examine my writings to see what my views really
are, but have only made acquaintance with fragments thereof.
I have advertised the distortionless circuit so often that I have
become ashamed of doing so. Nevertheless, since it is, as I
have often pointed out, the royal road to an inner understanding of the subject in general, I do not hesitate now to
add some more remarks in connection therewith, and leading
up thereto, in attempted elucidation of former work. We
should first consider [K] alone, or in conjunction with [R] and
[S], and then [L], and finally both [K] and [L].

The Beneficial Effect of Leakage in Submarine Cables.

§ 212. What leakage does is to increase the attenuation of
waves in transit, but at the same time to lessen the distortion

they suffer. Now there is, by elementary knowledge, no difficulty in seeing why leakage on the line should diminish the strength of the received current, in spite of the fact that the strength of the current sent is increased by the leakage. Ohm's law applied to a simple circuit with a shunt attached will suffice to give the reason so far as one fault is concerned ; and the principle is the same for any number of faults, or for distributed leakage.

As regards the lessening of the distortion, that is not quite so elementary. It does not present itself for consideration at all in common telegraph circuits. When, however, a cable is in question, and there is manifest distortion of signals, the influence thereupon of leakage becomes an obvious matter of inquiry. We may do it on the basis of the electrostatic theory. In fact, in his telegraph theory, Lord Kelvin did insert a term in the differential equation to express the effect on the potential of uniform leakage, though it was merely in passing, and I do not think there was any development of the matter in his later papers. The obvious attenuating effect of leakage, and the manifest evils of faults, caused high insulation to be aimed at, and possibly caused the full effect of leakage to be overlooked, and its theory to remain uninvestigated.

Now a cable is a condenser, or leyden, and a very big one too, although its permittance is so widespread that the permittance of a yard or so is very small. The discharge of a condenser against resistance takes time, and so does its charge, and the greater the resistance the greater the time needed. In the extreme, a perfectly insulated charged condenser keeps its charge unchanged, and so does a well-insulated cable, if also insulated at the ends. Even when it is earthed (or grounded, as the Americans say) at both its ends, only the portions close to the end can discharge at once. Away from the ends, the discharge has to take place through resistance, to a varying extent though in different parts, when we regard the long cylindrical condenser as a collection of small condensers side by side in parallel, with all the similar poles joined together by resisting wires. This is how it comes about that the resistance acts conjointly with the permittance in causing "retardation" in cables, so that the electrostatic time-constant is proportional to their product, or RSl^2. It is obvious, then, that we can facilitate the discharge by lessening the resistance to discharge, that is, by leakage in the cable, either distributed or in detached lumps. At the same time the charging is similarly facilitated or expedited.

Now, in making signals from end to end, the operation of charging the cable is necessarily precedent to that of receiving current at the distant end, which, in fact, represents a part of the charge coming out. Therefore, to expedite changes in the charge in the cable, not necessarily or usually complete charges or discharges, results in expediting changes in the current at the distant end. That is, it lessens the distortion at a given working speed, and so expedites the signalling. There is, in the electrostatic theory, no limit to this process, and by sufficiently reducing the insulation resistance, or, which is the same, increasing the leakance (which is, perhaps, an admissible short name for leakage conductance), we might signal through the cable as fast as we pleased, were there no other effect produced. This other effect is the extra attenuation. Whilst the possible speed of signalling can be raised apparently *ad lib.*, there is only an infinitesimal current left to do it with. (Remember that the influence of L is not now in question.) It is important to recognise that it is a consequence or natural extension of the old electrostatic theory, without any appeal to self-induction, that you can signal as fast as you like through an Atlantic cable of the present type (with leakage added, of course) simply by throwing away most of the current, and utilising the small residuum that manages to survive the ordeal.

Thus, a proper Atlantic cable, suitable for rapid signalling, should not have high insulation resistance, but the very lowest possible consistent with getting enough current through to work with, and other considerations, mechanical and commercial, &c. The theory, moreover, is a sufficiently clear one, and does not need elaborate mathematics to make it at least probably accurate, though it may, of course, be desirable to thoroughly verify the consequences mathematically.

It is now about 23 years since I came to see, substantially as above, the theoretical absurdity of a system of signalling which deliberately and seemingly of purpose places in the way the greatest possibly hindrance to the efficient transmission of signals, viz., by the use of the highest insulation possible, causing them to be regularly strangled, choked, and mutilated, to say nothing of their sometimes dying of inanition.

But this was mere theory. As a matter of fact, the effect of leakage presented itself to me at that time in practice on real live cables in a surprisingly different manner. The opposition between plausible theory and live practice was so glaring that it will be instructive to briefly describe it next.

Short History of Leakage Effects on a Cable Circuit.

§ 213. First, as regards long land lines, of say 400 miles in length (which is a long distance in this small country), leakage presented itself as an unmitigated nuisance. The worst case of all was during a continuous soaking rain all over the country. When this had gone on for a few hours the insulation resistance fell so low, by leakage over the dirty wet insulators mostly, though assisted by contacts with wet foliage, &c., that many of the important circuits were completely invalided, the received current (even with increased battery power) being too small for practical use with the instruments concerned. This was usually the case in the U.K. on any very wet day. But it was also the same in effect in the E. and I., although not so many lines were invalided, nor did they break down so quickly.

Incidentally, it may be mentioned here that the so-called "earth-currents" (meaning currents *in the line* itself), occurring during magnetic storms, such currents being a manifestation thereof, did not seem to mind the leakage much, perhaps because their E.M.F. was not like that of a battery at one end of a line, but acted continuously along the circuit, by a continuous slow change going on in the earth's magnetic force, or, rather, in the magnetic force in the air surrounding the earth. Probably the "earth-current" goes the same way in the line as in the earth (provided the line is not set at right angles to the real earth-current), whereas when a battery is used, the earth and line-currents are opposed. But it is not easy in wet weather to get a good magnetic storm, so that close observations can be made regarding the influence of leakage on earth-currents. We may dismiss them, and sum up that, as regards land-lines, leakage does not seem to have a single redeeming feature.

The effects of leakage next presented themselves to me on
cable circuits with land-lines attached, and were exceedingly
curious. There were two circuits of the kind, but one in particular had a varied history as regards leakage. It was a cable
of about 400 or 450 miles running east and west, having a land
line of 120 miles at the east end, and one of 20 miles at the
west end. The apparatus used was the Wheatstone automatic,
that is, an automatic transmitter and a polarised receiver,
recording dots and dashes, and the working was done without
condensers, by utilising the earth direct to complete the circuit.

The best results were obtained in fine dry weather with a
hard frost—the harder the better. The insulation of the long
(east) land line was then at its highest, and the signalling could
be done up to 30 words per minute or a little more from west
to east, though considerably less the other way, for a peculiar
reason. But this was quite exceptional, and did not last long.
The next best was in any very dry weather. But in ordinary
average fair weather, not rainy or damp, the speed was much
lower, say 24 or 20 words per minute. The effect of rain was
always to lower the speed, perhaps down to 15 words per
minute. A leakage fault on the land line or a contact with
another wire had a similar effect, only worse, provided it was a
bad fault, and the speed would go down to 10 words per minute
or less. Now it is possible that the extra good results obtained
in a hard dry frost arose partly from the reduced resistance of
the east land line (the usual iron wire), but there was no doubt
at all that the main cause of the wide variations of speed was
leakage on the land lines, especially on the long one, of course.

Now it is to be noted here that in the former case of long
land lines the leakage was prejudicial by the attenuation it
caused, whereby the received currents were made too feeble to
suit the apparatus. The working speed, though relatively fast,
was too slow for any notable distortion of signals in transit by
the condenser action so marked in cables. But in the latter
case, of a cable with terminal land lines upon which similar
leakage occurred, the prejudicial effect was not due to the
attenuation. There was plenty of current—in fact, a super-
abundance—but it did not come out at the receiving end
properly shaped for making dots and dashes, except at a
greatly lowered speed. It should also be understood that the

actual speed of working was always pushed up to the greatest possible (the press of business being such as to make 25 hours' work per day all too short), and that the automatic character of the signalling made the fixation of a limiting speed, without complete failure of any marks, quite definite at any particular time. Practically, however, the working speed was pushed a good deal higher than this limit by making regular use of mutilated signals. Now it was proved by daily experience that there was a remarkable sensitiveness to damp, or leakage of any kind, the speed of working, conditioned by the reception of good marks, going up and down like the barometer, though with a far wider range than that instrument. The east land line was so manifestly the offender that it came in for a great deal of blame. Yet it was not a bad line by any means, and its leakage would have passed unnoticed, had there not been such a peculiar lowering effect on the speed connected there-with.

But how about leakage in the cable itself, seeing that the above results were so out of harmony with what they should have been, by general reasoning regarding retardation? This question was soon answered by the automatic development of a first-class leakage fault in the cable. Its effect on the speed was substantially the same as that of leakage on the land-line. It lowered the speed, and since it was always on, the effect was permanent. The ordinary fluctuations disappeared in a great measure, as of course did the fine weather high speeds. It was like continuous heavy wet. As the fault got very bad the speed went down to 10 words per minute or less, in spite of increased battery power.

On describing this prejudicial effect of leakage to the electrician of the cable ship which came to remove the fault (it was the late Mr. S. E. Phillips, of Henley's), he assured me I was quite wrong. It was impossible for a leak to act like that. So said the captain, too. It was known, said they, that a leak was an excellent thing. It had been found so on the French Atlantic not long before, and in other cases. I fully believed this, because it seemed theoretically sound, and I had heard similar statements before. But they did not assist me to an understanding of the patent fact that leakage could be highly prejudicial instead of beneficial. In corroboration of this, when the fault was removed, the speed went up again, and the normal state of things was restored.

The next stage in this history was the arrival of that wonderful instrument, Thomson's recorder, which had been recently brought out. It was the original Atlantic cable pattern, with 20 big tray cells for the electromagnets, and a batch of smaller ones for the mousemill and driving gear. This instrument simplified matters greatly, for the current could be considered as and seen to be a continuous function of the time, whereas with the polarised recorder of Wheatstone (an instrument quite unsuitable for cable work) the variations of the current could be only roughly guessed.

Now, a peculiarity of the siphon recorder was that no definite limiting speed was fixable. If one person thought 40 words per minute fast enough for good marks ; another, more practised, would as easily read them at 50 or 60, and even manage to make them out at 70 words per minute, though the last was undoubtedly rather troublesome. The automatic transmitter being still used, the dashes became big humps, and the dots little ones, on the recorder slip, and at 70 words per minute, although the big humps were plain enough, the little ones were mostly wiped out, though it could be seen where they ought to have been, so that some reading was possible. The absence of a definite limiting speed under given circumstances introduced a fresh complication. Nevertheless, a careful study of the recorder slips taken at various speeds, but particularly in the neighbourhood of the critical speed for the polarised recorder, proved that the plain result of general reasoning that leakage should be beneficial in general was certainly true, and, moreover, furnished a partial explanation of the anomalous behaviour of the polarised recorder.

The siphon (Thomson) recorder, with its shifting zero, was sublimely indifferent to the vagaries of the current which were so destructive of intelligible marks in a dot and dash instrument, so that ordinary leakage passed quite unobserved as a rule. Nevertheless, it was plainly to be seen that when there was much leakage the result was to improve the distinctness of the marks by reducing the range of variation of the current from the zero line, and improving the dots, or small humps, relatively to the dashes, or big humps, which were reduced in size.

But, to clinch the matter, a second leakage fault appeared in the cable whilst the siphon recorder was on, and rapidly developed. Now, it was only at the west end that there was a siphon recorder; at the east end was the polarised recorder. The result was that the speed from west to east went down greatly as the fault developed, in accordance with prior knowledge. But the other way, from east to west, receiving with the siphon recorder, the working was greatly improved. The ultimate result was that when the fault got very bad no marks at all could be got at the east end, even by key-sending, owing to the great attenuation, whilst we could receive first rate on the siphon recorder at the west end. On the very last day, when communication from west to east had to be done by a roundabout course, we were able to receive a long batch of, I think, 150 messages at high speed (about 55 or 60 words per minute) by using 80 Leclanché cells at the sending end, and putting on the electromagnets all the tray batteries available, including most of the small ones, to magnify the marks. Now this might seem to be merely a testimony to the sensitiveness of the instrument. But the significant point was that although the marks were very small, in spite of magnification by increased battery power at both ends, they were exceedingly clear. They were the best ever got at the speed. We were practically working past a fault of no resistance, for the cable soon gave up the ghost. The tail end of the big batch finally degenerated to a straight line, indicative of a very dead earth.

On the removal of this fault, the received current was largely increased of course. But, simultaneously, the distortion of the marks on the siphon recorder became huge at the speed 60. It needed to be much lowered to give conveniently readable marks.

Explanation of Anomalous Effects. Artificial Leaks.

§ 214. Now the above facts gave a decisive and cumulative proof that leakage was beneficial even when concentrated at a single spot, provided the current-variations were recorded in a simple manner. But a complete explanation of the very large decrease in the speed possible when the polarised recorder was used, turned out to be rather a complicated matter in detail, although sufficiently simple in the fundamental principle.

This effect, so detrimental to an efficient use of the cable, was mixed up with another one, viz., a difference in the working speed according to direction, of say 25 per cent. (with the same instruments), but variable with circumstances. This effect, again, seemed to have two causes, one depending upon the circuit, and the uncentrical position of the cable in it, independent of the kind of instruments used ; whilst the other was special and peculiar to the automatic transmitter and receiver when combined with the uncentrical position of the cable. It may be readily imagined that the resultant effect on the working speed requires some study to be understood. It will be sufficient here to mention the instrumental peculiarity in its relation to the detrimental effect of leakage.

The original automatic transmitter of Wheatstone was constructed to make dots only upon the slip of a special receiver ; there were two rows of dots, one set being for the dots, and the other for the dashes of the Morse code. But when the inventor wanted the E. and I. Co. to take it up, they demanded dots and dashes, as usual. So the instruments had to be modified. Now I have personal reason to know that a few years later the inventor was quite at sea as to the reason of the peculiar effects observed in working his instruments on composite circuits and cables. It is therefore pretty safe to assume that previously, when the conversion was made, the manner of the conversion, from making double dots to dots and dashes, was dictated purely by considerations of mechanical convenience, and this conclusion may be made nearly certain by a comparison of the original with the modified transmitter. But Mr. Stroh knows all about it. At any rate, the peculiar way of making the dash and the long space was the cause of all our trouble with leakage, which cost the Company an immense sum of money, viz., what they might have gained otherwise, for they were blocked with work, and might have had much more.

To make a dot, a short positive current was sent, immediately followed by a similar short negative current. To make a dash, after sending the first short positive current, there was an interval of no current lasting twice as long as the first current, followed by a short negative current which terminated the dash. Similarly, to make a long space, a short negative current came first, then an interval of no current, and finally a short

positive current. Now in these intervals of no current, the
sending end of the line was insulated. When sending current,
on the other hand, the line was to earth, through the battery.
Two distinct electrical arrangements were therefore operative
in turn, viz., a cable with both ends earthed, and a cable with
only one end (the receiving end) to earth. If the former system always prevailed, the manner of transmission of electrical
changes would follow one law. If the latter system prevailed,
it would follow another law, and would give much greater
retardation. In real fact, the two ways were in alternate action
irregularly. This has to be considered carefully in examining
the influence of the uncentrical position of a cable in a composite circuit, and also of leakage.

Now the practical effect worked out thus. There was a real
positive and material advantage in making the dash and long
space of the Morse code by a short initial current instead of a
long one, as in ordinary working with reversing keys. I estimated the gain as amounting to from 100 to 125 per cent.
under favourable circumstances. This implies that the limiting
speed is really worked up to. But it was vitally necessary
to secure this advantage, that the insulation should be very
good. A peculiar balance had to be preserved between the
positive and the negative currents. The currents were of equal
duration and of alternate signs, but necessarily not equidistant
in time. Now, with exceedingly good insulation, the short
current sent at the beginning of a dash was sufficiently prolonged
in reception as to not merely make a good dash firmly, but to
keep the balance for the next currents. But if you put a fault
on the line, and let the charge leak out during the dead
interval of the dash, then the next following negative current,
terminating the dash, would be too strong, so that an immediately succeeding positive current (with a negative current after
it) to make a dot would miss fire altogether. Similar remarks
apply to the space. If the negative current that initiated the
space were allowed to die away by the action of a leak, the next
following positive current would be too strong, and cause an
immediately following negative current to miss fire. Remember
that the reception of the signals was with a polarised recorder,
the zero of which could not be continuously altered to catch the
missing currents except by a very artful person who knew what

was coming, and was quick enough to alter the regulation exactly at the right moments.   It will, by the above, become roughly intelligible how a leak, actually quickening signals in a remarkable degree, can yet be most prejudicial in a special system of signalling requiring the preservation of a sort of balance.   If the siphon recorder be substituted the necessity of the balance disappears; but the marks would be very hard to read at the same speed.   To make them readable the speed should be at least doubled.   The dashes will then become well-formed humps.

The reader who may desire to pursue the subject of leakage effects further may be referred to my "Electrical Papers," Vol. I., especially p. 53 for continuous leakage, p. 61 for more details relating to the above-described case, and p. 71 for the general theory of leaks, which is fully worked out in several cases to show the expediting effect.   I also there propose artificial leaks.   The following extract (p. 77) is easy to read, and so may be worth quoting :—

"When a natural fault, or local defect in the insulation, is developed in a cable, it tends to get worse—a phenomenon, it may be observed, not confined to cable-faults.   Under the action of the current the fault is increased in size and reduced in resistance, and, if not removed in time, ends by stopping the communication entirely.   Hence the directors and officials of submarine cable companies do not look upon faults with favour, and a sharp look out is kept by the fault-finders for their detection and subsequent removal.   But an artificial fault, or connection by means of a coil of fine wire between the conductor and sheathing, would not have the objectionable features of a natural fault.   If properly constructed it would be of constant resistance, or only varying with the temperature, would contain no electromotive force of polarisation, would not deteriorate, and would considerably accelerate the speed of working.   The best position for a single fault would be the centre of the line, and perhaps $\frac{1}{32}$ of the line's resistance would not be too low for the fault."

The effect of one leak is marked, perhaps 40 per cent.   What does this represent in £ sterling?   How many thousands or tens of thousands of pounds' worth of copper and g.p., &c., would be equivalent?   If one leak would not quite do it, then two would.

But to have only one or two leaks is to look upon the leak as a dangerous nuisance.   Although the best position for a given amount of leakance *in a lump* is in the middle of the cable, yet to have the given leakance in a lump is the worst arrangement as regards its accelerative efficiency. The best arrangement is uniformly-spread leakance, as much as possible, with, however, a thoroughly useful residual terminal current to work with, which current could be varied far more rapidly than is possible with a single leak.

Is it impossible to find an insulator of comparatively low resistance which should be suitable in other respects? Telegraph cable manufacturers would probably say, No. That, however, need not be considered conclusive, because, as is well known, when an industry or institution is once established, it always gets into grooves, and has to be moved out of them by external agency, if at all. But should such an insulator be unattainable, then the only alternative is to have artificial leaks, as many as possible. The theory is, I think, quite plain, and I know that the practice is also, up to a certain point. But here be grooves again. For people working in established ways have their own proper work to do, and have no time to waste upon "fads." So much is this the case that it often happens that they have exceedingly little knowledge of how things would work out if they departed from the ruts they are accustomed to run along. That is to say, the spirit of scientific research, which was to some extent present in the industry in its early stages, has nearly all evaporated, leaving behind regular rules and a hatred of fads. I am no wild enthusiast, having been a practician myself, but it has certainly been a matter of somewhat mild surprise to me that cable electricians, who have had such unexampled opportunities in the last 20 or 30 years, say, should have done nothing in the matter of leaks. Perhaps there may turn out to be objections at present unthought of. The same could be said of anything untried. Several cases of resistance to change of established practice, where the changes ultimately turned out to be of great benefit, have come under my personal observation. The proposed changes were (I believe quite honestly) first scouted as fads, and the innovator was snubbed; next, under pressure, they were reluctantly tried; thirdly, adopted as a matter of course, and the innovator snubbed again. But this is the nature of things, and can be seen everywhere around—very notably in the political world.

Self-Induction imparts Momentum to Waves, and that
carries them on. Analogy with a Flexible Cord.

§ 215. We have now to consider in what way self-induction comes in to influence the transmission of waves along
wires. The first step in this direction was made by Kirchhoff
as long ago as 1858, but it is only in recent years that self-
induction has become prominent as an active agent, and its
real resultant effect worked out and understood, more or
less. This effect is, on first acquaintance, a somewhat surprising and paradoxical one. Everyone knows the impeding
effect of self-induction in apparatus. How, then, in the name
of Faraday, can it be beneficial in a telegraph or telephone
circuit, and make signals bigger and clearer? To answer this,
it will be convenient to make use of an analogy in the first
place.

There is, in a great measure, a formal resemblance between
the problem of a telephone circuit along which electromagnetic
disturbances are being propagated, and the mechanical problem
of the transverse motions of a stretched flexible cord. But to
make the formal resemblance be also a practical resemblance,
several little things have to be attended to, and reservations
made. Now there are a great many mechanical problems concerning wave-motion subject to friction which make better
analogies for the use of the mathematical inquirer ; but the
case of a flexible cord is more or less familiar to everyone, and
is, up to a certain point, easily realisable.

The commonest knowledge of the transverse motions of a
stretched cord is concerning its normal vibrations, the fundamental mode and its harmonics. But we do not want them in
our analogy. For although we may have something similar
with short waves, and especially Hertzian waves, it would not
be easy to set up such a state of things on a telegraph circuit,
which is usually so long as to prevent the assumption of normal
modes by reflection at the ends. For the normal vibrations of
a cord are complex affairs, depending not only upon propagation
of motion along it, but also on the constraints to which it is
terminally or intermediately subjected. We desire the propa-
gational effect to show itself distinctly. But if we disturb a
stationary stretched wire at one end, the pulses produced run

to and fro too quickly for practical observation. Take, then, a long india-rubber cord, suspended from a height, as is sometimes done by lecturers. The speed of transmission of pulses may be made small enough to allow the disturbances to be conveniently watched. Then, if the cord hang vertically at rest to begin with, and its lower end be momentarily jerked, bringing it immediately back to its original position, a hump is generated, which runs along to the fixed end. It is not much changed in transit, and is reflected back, and re-reflected many times. But we only want to consider the passage of a hump one way. Properly speaking, then, we should have so long a cord that reflection is (by frictional loss) insignificant. Perhaps a very long cord laid upon a surface of ice would be more suitable for the purpose. But if the "fixed end" be movable transversely, and its motion be resisted by a force proportional to its velocity, their constant ratio may have such a value given to it (depending upon the mass and tension of the cord), that the arriving disturbances do not produce any reflected waves, but are absorbed by the terminal resistance. The cord then behaves as if it were infinitely long. This is theory, of course, but I dare say a sufficiently good practical imitation could be got by means of a terminal block sliding upon a wire with friction. An imperfect cancellation of the reflection is easily got.

Disregarding the slow change of type and the attenuation, so that there is undistorted propagation of a hump, or of a succession of humps making a train of waves going one way only, this case is analogous to the transmission of waves along a uniform circuit of no resistance. The two constants L and S in the electrical case are replaced by another two, viz., the density (linear, or mass divided by length) and tension of the cord. Thus, large permittance, that is, small elastance, means small tension, or a slack cord ; and large inductance means a massive cord. If I said a "heavy" cord, it might be thought that its weight was concerned in the matter in question, but it is not, save as a disturbing factor. The transverse displacement of the cord may be compared with the transverse voltage in the electrical circuit. Other comparisons (theoretically better) may be made, but this is perhaps the most convenient if the motion of humps be actually watched.

Now to imitate the action of the resistance (assumed to be constant) of an electrical circuit, the transverse motion of the cord must be frictionally resisted, and the frictionality (or coefficient of friction) should be constant. That is, the frictional force should be proportional to the transverse velocity. The motion of the cord is frictionally resisted, both internally and externally. It is very unlikely that this should give rise to a constant frictionality, but there is a similarity insofar as the waves sent along the cord do attenuate and are distorted as they progress.

But the friction is not anything like enough in the case of a cord in air to imitate the submarine cable; besides that, the mass is too great, and likewise the tension. Even to copy telephonic waves going along very long copper wires of low resistance, we should increase the friction in the mechanical analogue if we desire to have an equal amount of attenuation in a wave length. The cord should, therefore, be in a viscous medium, and the viscous resistance to the cord's motion should be so great as to produce a marked attenuation during the transit of a hump, though it should not be so great as to prevent its distinct propagation as a hump, without great distortion of shape and spreading out. Under these circumstances we may exemplify low-resistance, long-distance, overland telephony, where the inductance is relatively large and the permittance relatively small.

But to imitate a submarine cable we should have a slack cord of small mass in a highly viscous medium. We can then send a hump only a little way along it as a hump, for it becomes greatly distorted and diffused. Consequently, if we compared the displacements at a great distance along the cord with those impressed upon it at the free end, we should find a state of things quite different from what occurs in the case of an india-rubber cord in air. The main features of the originated disturbances might be recognised, unless the attenuation were too great, but they would be out of all proportion. In particular, there is a tendency to obliterate small humps, or ripples whilst the big ones persist. The propagation is of the sluggish kind characteristic of diffusion, quite unlike that of purely elastic waves. If the slack cord could have no mass at all, we should then imitate a submarine cable wholly without inductance.

Under these circumstances we could convert the diffusive waves to something like elastic waves by giving mass to the cord. Without changing either its tension or the frictionality, we could, by a continuous increase of density, cause the nature of propagation to pass continuously from the sluggish, diffusive kind to the elastic kind with approximate preservation of form of waves. We might also do the same by increasing the tension and reducing the frictionality, keeping the density small (like, by reducing the permittance and the resistance in an electrical circuit, bringing the inductance into importance). But these are not immediately in question. By increasing the mass, or equivalently the inertia of the cord, we impart momentum to the waves, which allows them to be carried forward, in spite of the little tension, against the resistance of the surrounding medium.

So it is in the electrical case. By increasing the inductance we impart momentum to the waves, and that carries them on. This statement is independent of the particular mechanical analogy that may be employed for illustration. It is probably representative of the actual dynamics of the matter, associating L with inertia, and LC with momentum. The soundness of this dynamical explanation (made by me some years since in asserting and proving the fact concerned) was endorsed by Lord Kelvin. And I am proud to add, for the benefit of those who may excuse themselves from understanding me by the plea that I am a Schopenhauer (!), that Lord Kelvin expressed his appreciation of the very practical way in which I had reduced my theory to practice.

Many other dynamical arrangements might be alluded to, exemplifying the power of inertial momentum to overcome frictional resistance, but it is not easy to get a good one exhibiting elastic waves plainly.

To illustrate the nature, and to emphasise the possible magnitude of the effect, suppose we have the power of continuously increasing the inductance of an Atlantic cable from a small quantity up to a very large one, without altering its resistance or permittance, and without introducing any other actions than those to be controlled by resistance, permittance, and inductance. Starting with $L = 0$, and making, say, 1,000 waves per second at one end, there will be 1,000

waves per second of workable magnitude only in a small part of
the cable near the sending end; the amplitude will attenuate
to insignificance, not exactly in the shore-end, but before 100
miles is traversed.   But let waves of the same size (which will
need greater battery power to produce) be kept up at the
sending end whilst L is continuously increased all along the
cable.   The region of practical waves will continuously extend
itself right through the cable, by a partial conversion to elastic
waves, with little attenuation in transit.   With very big
inductance the waves would reach the distant end nearly in
full size.   Even then they could be doubled in size by the
reflection, though there would be no useful purpose served by
that.   The amount of inductance required, however, to produce
these results would be out of all reason.   Nevertheless, the
illustration serves its purpose in showing the action of self-
induction, which is, as before stated, wholly beneficial, as it
increases the amplitude and lessens the distortion.   Some
remarks on increasing inductance will come later.

**Self-Induction combined with Leaks.   The Bridge System
of Mr. A. W. Heaviside, and suggested Distortionless
Circuit.**

§ 216. In the meantime, some few remarks on self-induction
combined with leakage should come in here.   Self-induction
came in with the telephone.   That was, originally, a mere
house-to-house affair.   At any rate, lines were only a few miles
long.   So condenser action was out of practical question then.
On the other hand, the frequency of the currents concerned is
very great, and that makes inertia important.   So the theory
of telephones is a magnetic affair when the lines are short
enough, just as if they were lines in a laboratory.

But the telephone soon got beyond that, and it became
necessary, with the use of longer and longer lines and underground lines, to examine the influence of their permittance in
conjunction with their inductance.   In hardly any case of
telephony can we ignore the self-induction, and take the
permittance alone.   The resistance would need to be so large
that the lines would be wholly unsuitable for telephony, except
for mere local lines.   So permittance and inductance usually

go together in considering telephony, and if one of them is
ignorable it is the permittance rather than the inductance.

Following the telephone came exchanges. But there was an
exchange before telephones. My brother, Mr. A. W. Heaviside,
had an intercommunication exchange at Newcastle, using
alphabetical indicators, whereof the manipulation resembled
that of the hurdy-gurdy. These were beautiful instruments
mechanically; but, of course, the telephone soon stopped their
manufacture. The exchange referred to was then turned into
a telephone exchange. Now, in certain cases it was needed to
have many instruments in one circuit, with local intercommunication; a sort of family arrangement, so to speak; for
example, a number of coal pits and colliery offices belonging
to one firm. The custom was to set all the intermediates in
sequence (otherwise, but perhaps not so well, called in series)
in the circuit. Why was this done? Simply because they
knew no better. It was the custom to do so in the telegraphs
generally, being an obvious arrangement—not so much in
England, though, as elsewhere. But when this was done with
telephones, or with call-instruments replacing them, it was
found that the intermediates had a very deleterious influence,
which very soon set a limit to the admissible number of intermediates on a circuit and to the length of the circuit, especially
when underground wires were included. My brother, who took
up the telephone with ardour from the first (making the first
one made in England, I believe), found by experiment that it
was quite unnecessary to put the instruments in sequence
according to old practice, and that an immense improvement
was made by taking them out of the circuit, and putting them
across it, as shunts or bridges. Like lamps in parallel, in fact.
It seems very easy and obvious now. But it is not so easy to
conceive the state of mind of old stagers. Anyhow, the proposition to put the instruments in bridge was pooh-poohed at
first, not being understood. As, however, it was demonstrably
a wonderful improvement it was adopted, first at Newcastle,
and then elsewhere, and became known as the Bridge System.

When this system was brought to my knowledge, and I was
asked for theoretical explanation, it presented itself in a double
aspect. First of all, there was the complete removal from
the circuit of intermediate impedance in big lumps. This was

manifestly the main reason of the increase in working distance possible, and why a far greater number of intermediates could be put on. For they were not put in the circuit, but on it, so to speak.

The other aspect of the matter was this. Considering that the bridge system consists electrically of a circuit with a number of leaks on it, and bearing in mind my old investigations concerning leaks, I inquired whether the intermediates were not actually beneficial to through communication, independently of the previous effect. Was not the articulation better, not merely than if the intermediates were in series in the circuit, but better than if the shunts or bridges were removed altogether, leaving insulation? My brother's answer was Yes, certainly, in certain cases, but doubtful in others. It appeared, however, to be only a minor effect, of not much importance compared with the major one. Besides, the bridges caused a weakening of the intensity of the speech received when many bridges were passed. To prevent this weakening becoming inconveniently great, the intermediate call-instruments in bridge were purposely made to have considerable resistance and inductance, far more than would be needed or desirable for their natural use. This tended to prevent the currents passing along the line from entering the shunts, and especially so as regards the currents of high frequency, and allowed them to be transmitted in greater magnitude.

It will be observed here that the expediting effect of leakage considered before (§§ 213, 214) is merely incidental, and is, in fact, partly destroyed by the large impedance of the bridges, in order to mitigate the attenuative evil. The shunts are inductive shunts, intentionally very much inductive. Now, my proposed leaks (§ 214) for submarine cables were noninductive, and intentionally so, though it may be remarked that it would make very little difference with the slow signalling of Atlantic cables if they were inductive. On the other hand, a recent proposal of Prof. Silvanus Thompson is to use inductive shunts on cables. The arrangement is electrically as in the bridge system, but the object is different. This point will be returned to.

The investigation of this bridge system of telephony suggested to me the distortionless circuit, as I have before acknowledged

("Electrical Papers," Vol. II., p. 402).   It may be of interest
to state how it came about, since it is not presented to one by the
above-described experiences with the bridge system.   The question was, What was the effect of a bridge, or of a succession of
bridges, when the self-induction of the line was not negligible ?
Now the influence of the bridges can be very readily stated for
steady currents, of course.   There is no difficulty, either, in
finding the full formulæ required to express the voltage and
current and their variations in any part of the system when
the inductance and permittance are operative, as well as a set
of leaks, when one knows how to do it.   But the general
formulæ thus obtained are altogether too complicated to be
readily interpretable.   I was, therefore, led to examine the
effect produced by a leak on a wave passing it of the elastic
type.   This implies that the resistance of the line is sufficiently
low for waves of the frequency concerned to retain that type
approximately.   (Ln should be greater than R.)   Now this was
comparatively an easy problem, and the result was to show that
a leak (or conductance in bridge) had the opposite effect on a
wave passing it along the line to that of a resistance inserted
in the circuit at the same place, in this sense :—A resistance in
the circuit reflects the charge positively and the current
negatively, or increases the charge and reduces the current
behind the resistance.   On the other hand, a leak reflects
the current positively and the charge negatively, or increases the current and reduces the charge behind the
leak.   (See also §§ 197, 209.)   A resistance and a leak
together at the same place therefore tend to counteract one
another in producing a reflected wave, leaving only an attenuating effect on the transmitted wave.   This led to the
distortionless circuit.   For although the compensation of the
resistance in the circuit and the conductance across it is imperfect when they are finite, it becomes perfect when they are
infinitely small.   They may then be a part of the circuit itself,
namely, the resistance a part of the resistance of the line, and
the conductance a part of the leakage.   We now have perfect
compensation with this one, or any number of successive resistances and leaks of the same kind, provided their ratio is properly chosen.   This occurs when R and K, the resistance and
conductance per unit length of circuit, of the line and of the

leakage respectively, are taken in the same ratio as the two other constants, L and S. Or, when L/R = S/K, indicating equality of the electric and magnetic time-constants. There is now an undistorted transmission of any kind of signals. And since every circuit has R, L, and S, it follows that it can be made distortionless by the addition of leakage alone of a certain amount, or distortionless to the extent permitted by the approximate constancy of R, L, and S. (*See* "Electrical Papers," Vol. II., pp. 119 to 168, for detailed theory of the distortionless circuit.)

In § 213 we saw that even in the electrostatic theory leakage would allow us to signal as fast as we liked provided we could do it with an infinitesimal current. This was not because the circuit was made distortionless, but on account of the infinitely rapid charge and discharge. Now the inclusion of the influence of self-induction shows that we can do the same with a finite leakage conductance, and have finite distortionless signals, due to waves of the purely elastic type. Given R and S then, if L should be so small that the introduction of K of the proper amount to reach the distortionless condition should produce unreasonable attenuation, then one way of curing this would be to increase L. For then a smaller amount of leakage will serve. The bigger L, the smaller need K be, and when it is very big, then very little K is needed. This brings us round to the case at the end of § 215, where we had elastic waves with little attenuation on an Atlantic cable got without any leakage, and since we see now that very little leakage is needed to bring about the distortionless condition, we conclude that it is approximately distortionless without the leakage.

Evidence in Favour of Self-Induction. Condition of First-Class Telephony. Importance of the Magnetic Reactance.

§ 217. Let us now ask what is the evidence in favour of the beneficial action of self-induction. There was plenty in 1887; it is overwhelming now. But there are two sorts of evidence, the theoretical and the experiential. The two are not fundamentally different, however; only the former is more indirect than the latter. We have certain electrical laws established, mainly by laboratory work followed by mathematical inves-

tigation, and there is no reason whatever to suppose that they
do not hold good out of doors as well as inside. The important
thing is to correctly recognise the prevailing conditions and
determine their results. Now, if you can be pretty sure that
you have done this, and have confidence in your investigations,
then you may consider that you prove a thing, even though
there may be no direct evidence. Discoveries are frequently
made in this way. "A foundation of experimental fact there
must be; but upon this a great structure of theoretical deduction can be based, all rigidly connected together by pure
reasoning, and all necessarily as true as the premises, provided
no mistake is made. To guard against the possibility of mistake and oversight, especially oversight, all conclusions must
sooner or later be brought to the test of experiment; and if
disagreeing therewith, the theory itself must be re-examined, and
the flaw discovered, or else the theory must be abandoned."
(Oliver Lodge.) But to convince other people is quite another
matter. They may not be competent to understand the
evidence. Or they may be fully competent, but not have sufficient acquaintance with the subject; or not have time to
examine it; or not have the energy; or have no interest in it.
Then, as Elijah said to the priests of Baal, you must "Call him
louder!"

Now, the foundation of fact is the experimental facts of electricity and the laws deduced therefrom, though these laws are
sometimes more comprehensive than the facts upon which they
are based. But there was, in 1887, a certain amount of direct
evidence as well, quite enough to make the theoretical conclusions practically certain, by the confirmation they afforded.
Assuming the electrostatic theory to be true in telephony (for
which, however, there was never any sufficient warrant) we can
calculate its consequences. They indicate that the product
RSl^2 of the total resistance and the total permittance limits
the distance l through which telephony is possible. How big
it may be is questionable, on account of the very complicated
nature of the problem when the human is taken in as a part
of the mechanism. We cannot say certainly beforehand how
much distortion is permissible before the human will fail to
recognise certain sounds as speech. But the electrostatic
theory shows such very considerable distortion even when

THEORY OF PLANE ELECTROMAGNETIC WAVES.   439

RSl^2 is as low as 5,000 or 7,000 ohms × microfarads, that it is scarcely credible that any sort of telephony could be practicable if it were much bigger, as 10,000 for example. Perhaps 5,000 might be taken as the practical limit for rough purposes. But experience showed that good telephony was got up to 10,000 and higher, in this country with circuits including underground cables, as well as in America, where long-distance telephony had already made a good beginning. Manifestly the electrostatic theory was erroneous by direct as well as indirect evidence. Now the inclusion of the influence of self-induction in the theory explained these results (and also the failure of iron wires and the success of copper) and pointed out how to improve upon them, viz., by directing us to lower R and increase L in order to increase the value of RSl^2 possible. So there was sufficient direct evidence then to satisfy anyone competent to judge who considered it fairly without being prejudiced by the law of the squares. Since that time, the amount of this direct evidence has been greatly multiplied, and RSl^2 has got up to 50,000, which may be about ten times as great as the probable value under purely electrostatic conditions. But there is nothing critical about 50,000. It might be 500,000 if the circuit were long enough. We may need to reduce R when we increase l, but not in proportion to the inverse square of l, or anything like it.

Any long circuit may be made approximately distortionless if the magnetic side of the phenomena can be made important, and without the inconvenience of the leakage needed to remove the distortion, which leakage, however, is essential when the magnetic side cannot be made important. The useful guide in considering telephony is the size of the quantity R/Ln, the ratio of the resistance to the reactance, the two terms appearing in $(R^2 + L^2n^2)^{\frac{1}{2}}$, the impedance of unit length of circuit, apart from the permittance.

The term "reactance" was lately proposed in France, and seems to me to be a practical word. It may be generalised to signify the value of Ln, positive or negative, of any combination of coils and condensers, L being then the effective inductance. When L and the reactance are positive, the magnetic side prevails and the current lags; and when it is negative, the electric side prevails and the current leads.

At present, however, R and L are as before. Now the ratio
R/Ln may be very large. If so, the influence of L is small,
and we shall have something like the electrostatic theory,
provided the range of n be such as to make the ratio always
large. But this state of things is quite unsuitable for long-
distance telephony. The most suitable state is when R/Ln is
a proper fraction throughout the whole range of n; that is,
when R is less than the smallest value of Ln. We shall then
have approximately distortionless propagation of the type

$$f(x) = \epsilon^{-Rt/2L} f(x - vt),$$

as may be seen by examining the solution for simply periodic
waves, and observing that when R/Ln is small we have nearly
the same attenuation in a given distance at all frequencies (not
too low), so that there is little distortion, and the element of
frequency tends to disappear altogether, and we reduce the
formula to the above type.

Practice shows, however, that this state of things is unnecessarily good for commercial telephony, which admits of a
considerable amount of distortion. Then R/Ln is not less
than 1 throughout the whole range of n, but is greater at the
low frequencies and less at the higher.* For example, if the
lowest value of n be 625, and L = 20 (centim. per centim., usual
units), the least value of Ln is 12,500. If also the resistance
is 5 ohms per kilometre, then R is 50,000. The value of
R/Ln is then 4 at the lowest frequency, and 2, 1, $\frac{1}{2}$, $\frac{1}{4}$, $\frac{1}{8}$ at
the five successive octaves, the last being at n = 20,000. Here
we have an excellent state of things from n = 2,500 up to
20,000, and only at the low frequencies does R prevail over
Ln. If R be halved, or L be doubled (for it is indifferent
which is done), then R/Ln is less than 1 from n = 1,250
upwards. Divide n by 2π to obtain the frequency. This 2π
has nothing to do with the ridiculous 4π which the B.A.
Committee want to consecrate. The significant point in calculations of this kind is the position of n in the scale of frequency
which produces equality of R and Ln. When it is low we
have good quality; when high, then relatively bad. Perhaps
the American electricians may be able from their extended

* *See* my "Electrical Papers" for details; especially App. C, p. 339,
Vol. II.

experience to give practical information on this and other
points concerned in telephony, and construct an empirical
formula of a sound character for determining the limiting distance roughly in terms of the electrical data, with given kinds
of instruments for transmission and reception.   It should be
understood that the above remarks do not apply to short lines,
at least in general.   We may, indeed, so arrange matters that
the above type is still preserved (with R/Ln less than 1) by
practical annihilation of reflected waves, but otherwise serious
modification may be needed.   And with very short lines we
may apply a purely magnetic theory, of course.

Various Ways, good and bad, of increasing the Inductance of Circuits.

§ 218. Supposing, now, that practicians have come to discard
the supposed limitations so persistently asserted by official
electricians in England, and to open their eyes to the part
played by self-induction, they will naturally recognise the wide
possibilities that are suggested in future developments, and
want to know whether such possibilities can be converted into
actualities, and whether the conversion is practicable.   For
myself, I am not much concerned in this part of the question.
It is for practicians to find out practical ways of doing things
that theory proves to be possible, or not to find them if they
should be impracticable.   Nevertheless, since I am responsible
for these views concerning the parts played by self-induction
and leakage, I add a few remarks on the applicational side.

The ideal is the distortionless circuit, requiring both self-
induction and leakage. If we utilise the present existent
inductance only, then the needed leakage may be too great by
far in many cases.   On the other hand, if the existent inductance is big enough to make R/Ln small, in the sense described
in § 217, we do not want the leakage particularly, or not at all
practically, it may be. It is clear, then, that we should aim
at making L/R, the magnetic time-constant, as big as possible.

Now, here is one way of doing it. If, a few years ago, an old
practitioner was asked to put up a long distance telephone circuit,
he, being under the belief that R should be high, or at least
that it need not be small, and that L should be as small as

possible, in order to prevent the deleterious action of self-induction, would probably carry out his principles by putting up a circuit, several hundred miles long, consisting of a pair of fine wires, put quite close together.   Or, if he was free from the error as regards R, he would use larger wires.   But still close together, for that makes L the smallest.   Also, he would not consider this to be bad as regards S, the permittance, being under the belief that the permittance of the circuit of two wires was just one half that of either with respect to the earth, so that their proximity did no harm in increasing condenser action, whilst it destroyed the deleterious self-induction.   Now, such a circuit might do very well for local telephony, but for the purpose required there could not be a worse arrangement. The supposed case is not an extravagant one, since it involves the carrying out of official views thoroughly.   The cure would be quite easy.   Increase the inductance by separating the wires as widely as possible or convenient.   If necessary, reduce R as well.   The permittance will also be reduced in about the same ratio as the inductance is increased.   By this simple process L may be largely multiplied, say from 2 to 20 (cm. per cm.), a tenfold magnification, without separating the wires so much as to require a double line of poles.   This way of doing it is the secret of the success of the long-distance telephony which arose in America when iron was discarded.   L is made important, and R is kept low.   But although L increases very fast on initial separation of wires, it does not continue to increase fast on further separation, so that practical limitations are soon set to the process.   The above example shows how a very bad line may be turned into a very good one by increasing the inductance.

Another way is suggested, but only to be immediately rejected.   It is, nevertheless, instructive to study failures. We can increase L to any extent by using finer and finer wires. This would be admirable if it were all.   But, unfortunately, this process alters R, and faster still ; so that R/L increases instead of diminishing.   Consider, for simplicity, merely a wire with a concentric tube for return.   Then R/L is nearly proportional to RS.   Now Lord Kelvin gave a formula long ago, showing that this quantity was made a minimum when the radii of the tube and wire bore a certain ratio.   Then L/R is a maximum, approximately, and on decreasing the radius of the wire it falls off, ultimately very rapidly, which is the reverse of what is wanted.

A third way must also be immediately rejected. By using iron wires we can increase L largely with steady currents. But then the skin effect, or imperfect penetration, becomes important with telephonic currents in iron wires. It does two things, both harmful. It lowers L and increases R. So we do not get the large increase of L suggested at first, but only a comparatively small one with currents of great frequency, and we do get a lot of extra resistance that should particularly be avoided. Iron wires are failures for long-distance telephony, as was found long ago.

The case of copper-coated iron or steel wires is doubtful. Such wires were used in America, and it was said that they were successful for long-distance telephony. But here, evidently, the copper is very important in giving conductance. As regards the interior iron core, it must be considered very questionable whether it can be so good as an amount of copper of the same resistance (steady). It is not well placed for increasing the inductance, and the imperfect penetration will operate injuriously.

If we put the iron outside the copper the iron will be better situated for increasing L for steady currents (which is no good), but will act injuriously with rapidly oscillating currents, and partly prevent the utilisation of the conductance of the copper, if it be thick enough to make the increased L important.

The above considerations show that if iron be introduced to increase the inductance it should not be in the main circuit, but external to it. There are, then, two principal ways suggested. First, to load the dielectric itself with finely-divided iron, and plenty of it. This is very attractive from the theoretical point of view, as it results in the production of a strongly magnetic insulator, which is practically homogeneous in bulk, being, therefore, a sort of non-conducting iron of low permeability. In this way L, as compared with the same without the iron, may be multiplied greatly. There is a partially counteractive influence, however, due to increased permittance. This plan is interesting, in view of the wave theory. It is like multiplying the μ in ether, and lowering the speed of propagation, but at the same time allowing waves to keep up against the resistance of wires they may be running along. But as regards the practicability of this plan I say nothing, except that I have often smiled at this ironic insulator (in more than one sense ironic), and the idea of telephoning through a core resembling a poker with a copper wire run through the middle.

The other way is to put the divided iron outside the working dielectric. Then it had better not be divided into particles, but in the way well known to dynamo and transformer people, cut up so as to facilitate the flux of induction, whilst being electrically non-conducting transversely thereto.

Now it may be said that submarine cables of the present type, having cores surrounded by a large quantity of iron, have large inductance already. But it does not follow. If the iron was in a solid tube, then undoubtedly L would be much magnified with steady currents. But it is not a solid tube, as it consists of spirally laid separate stout iron wires. The magnetic continuity is interrupted, and the permeance is much reduced, as Prof. Hughes pointed out (equivalently) some years ago. But besides that, when the current is not steady, we do not utilise the iron even in the above imperfect manner. This is particularly the case with telephonic currents, when imperfect penetration will still further reduce the effective inductance. So the increased L due to the sheath must be far less than large quantity of iron might suggest. Again, there is the the increased resistance due to the sheath not being properly divided to prevent it. The case is therefore a considerably mixed one. Whilst we can say with confidence that with the very slow signalling which obtains through an Atlantic cable the sheath cannot make much difference, it is not easy to decide whether its action is beneficial with telephonic currents to any great extent, though it would be if there were no increased resistance. It is not even easy to say what is the L of a submarine cable under the circumstances. A least value is readily obtained, but the magnitude of the increase due to the iron and the outside return current is rather speculative. It may be well deserving of attention to change the type of the iron sheathing, so as to increase L and keep down R. Remember here, that although there is plenty of evidence of the beneficial (and very important) influence of this principle when it is carried out in a direct and natural manner, there is, so far as I am aware, no experimental evidence yet published of the equivalent success of indirect ways involving the use of iron.

Another indirect way is this. Instead of trying to get large uniformly spread inductance, try to get a large average inductance. Or, combine the two, and have large distributed inductance together with inductance in isolated lumps. This means the insertion of inductance coils at intervals in the main circuit. That is to say, just as the effect of uniform leakage may be imitated by leakage concentrated at distinct points, so we should try to imitate the inertial effect of uniform inductance by concentrating the inductance at distinct points. The more points the better, of course. Say m coils in the length l, or $2m$ coils of the same total inductance, and therefore each of half the inductance; or mn coils each of one n^{th} the inductance of the first. The electrical difficulty here is that inductance coils have resistance as well, and if this be too great the remedy is worse than the disease. But it would seem to be sufficient if the effect of the extra resistance be of minor importance compared with the effect of the increased inductance. This means using coils of low resistance and the largest possible time-constants. For suppose 4 ohms per kilom. is the natural resistance, and there be one coil per kilom. having a resistance of 1 ohm. This will raise the average resistance to 5 ohms per kilom.; and if the time-constant be big enough, the extra inductance may far more than nullify the resistance evil. The same reasoning applies to coils at greater intervals, only of course in a more imperfect manner. To get large inductance with small resistance, or, more generally, to make coils having large time-constants, requires the use of plenty of copper to get the conductance, and plenty of iron to get the inductance, employing a properly closed magnetic circuit properly divided to prevent extra resistance and cancellation of the increased inductance. This plan does not belong to the category of those mentioned before, which a moment's consideration showed to be worse than useless. It is a straightforward way of increasing L largely without too much increase of resistance, and may be worth working out and development. But I should add that there is, so far, no direct evidence of the beneficial action of inductance brought in in this way.

The combination with leaks does not need any particular mention in detail after what has been already said relating to the distortionless circuit, which is the theoretical ideal, any approach to which is desirable if it be done without too much loss ; bearing in mind, too, that we should increase L first in preference to putting on leaks first.* I have confined myself so far entirely to my theory of 1886-87 and results thereof. Inductive shunts involve some other considerations, and a modified theory.

Effective Resistance and Inductance of a Combination when regarded as a Coil, and Effective Conductance and Permittance when regarded as a Condenser.

§ 219. In any electromagnetic combination the effects of electric and magnetic energy are antagonistic in some respects. Or we may say that magnetic inductances tend to neutralise electric permittances. Or, more generally, that the effects of elastic compliance and of inertia tend to neutralise one another. Thus, when a coil is under the action of an impressed simply periodic voltage, the current lags, owing to the self-induction. But if a condenser be introduced in sequence with the coil, the lag is diminished, and may be reversed or converted into a lead. The result on the current is the same as if we substituted for the coil another of the same resistance and of reduced inductance, or even of negative inductance. Thus, the resistance operator of the coil being $R + Lp$, where R and L are its resistance and inductance, and p is the differentiator d/dt; and that of the condenser being $(Sp)^{-1}$, where S is its permittance, when the two are in sequence the resistance operator of the combination is

$$Z = R + Lp + (Sp)^{-1}, \quad . \quad . \quad . \quad . \quad (1)$$

which, in a simply periodic state of frequency $n/2\pi$, making $p^2 = -n^2$, or $p = ni$, becomes

$$Z = R + Lp - \frac{p}{Sn^2}$$

$$= R + \left(L - \frac{1}{Sn^2} \right)p. \quad . \quad . \quad . \quad (2)$$

---

* But remember the influence of the frequency in conjunction with the inductance. Leakage can be very beneficial when self-induction does next to nothing, as before described.

This shows that the effective inductance of the condenser, which is $-(Sn^2)^{-1}$, is additive with real inductance when in sequence therewith, and being negative, reduces the effective inductance of the combination from L to $L - (Sn^2)^{-1}$.

But when the condenser and coil are in parallel, it is more convenient to use the conductance operators. Thus

$$Y = (R + Lp)^{-1} + Sp \quad . \quad . \quad . \quad . \quad (3)$$

is the conductance operator of the coil and condenser in parallel, being the sum of the conductance operators of the coil and condenser taken separately. And when $p = ni$, we convert Y to

$$Y = \frac{R - Lp}{R^2 + L^2n^2} + Sp$$

$$= \frac{R}{R^2 + L^2n^2} + \left(S - \frac{L}{R^2 + L^2n^2}\right)p, \quad . \quad . \quad (4)$$

from which we see that the combination is equivalent to a condenser of conductance $R(R^2 + L^2n^2)^{-1}$ and of permittance $S - L(R^2 + L^2n^2)^{-1}$. The steady conductance being R^{-1}, we see that the effective conductance is reduced by the self-induction. At the same time the effective permittance is reduced, and may become negative. Remember that in (3) and (4) the standard of comparison is a condenser, not a coil; whereas in the former case (1), (2) the standard of comparison is a coil, not a condenser. For further information regarding resistance and conductance operators *see* " Electrical Papers," Vol. II., p. 355, and elsewhere. The present remarks are merely definitive and introductory to the theory of waves sent along a circuit, either naturally, or partly controlled by subsidiary arrangements. The comparison with a coil is, I believe, more generally useful. Then we imagine any combination to be replaced by a simple coil whose resistance and inductance are those (effectively) of the combination. But there are cases when the other way is preferable. Then we imagine the combination replaced by a condenser whose conductance and permittance are those (effectively) of the combination. And in the following it will be convenient to use both ways in the same investigation.

Inductive Leaks applied to Submarine Cables.

§ 220. Now imagine the permittance of a submarine cable to be concentrated in lumps at a number of points. Let S be the

permittance of one of the equivalent condensers. If it be
shunted by a coil of such R and L that $S = L (R^2 + L^2 n^2)^{-1}$, we
know by the above that the state of the cable, when simply
periodic, will be the same as if the condenser and coil were
removed and replaced by a simple leak of conductance
$R (R^2 + L^2 n^2)^{-1}$. This reasoning applies to every condenser, if
it have its appropriate coil, or to any selected group we please.
But if the frequency be changed the permittance will come into
sight again, of the real or positive kind, or of the fictitious
negative kind, as the case may be. We have, therefore, the
power of reducing the effective permittance of a cable under
simply periodic forces, by means of numerous auxiliary inductive shunts or leaks, and of practically cancelling it at a particular frequency. How will this work out in the transmission
of telephonic currents through the cable?

From the purely theoretical point of view one may be somewhat prejudiced against the system at first. For, obviously,
it is not a distortionless arrangement. That is got by balancing
the self-induction of the main circuit against the lateral permittance, which can, subject to practical limitations, be done perfectly. That is, there is a balance at any frequency, or the
idea of periodic frequency does not enter at all. I doubt
whether there can be any other distortionless circuit than that
which (in the absence of another) I always refer to as *the* distortionless circuit. With inductive shunts we produce a partial neutralisation of the permittance, the amount of which may
vary very widely during the transmission of telephonic currents.
Remember, too, that during the change from one frequency
to another there are other phenomena than those concerned
during the maintenance of a simply periodic state of variation.

Nevertheless, we should be careful not to be prejudiced against
an imperfect plan merely because it is so manifestly imperfect.
It may perhaps be more easily applied than a theoretically
perfect system, and may possess practical advantages of importance. Now this is a matter for practical experiment and experience. But it is just here that there is an almost complete
dearth of information. For although Prof. Silvanus Thompson*
has mentioned that he has made many experiments, yet he has

---

* "Ocean Telephony," Chicago Congress, 1893.

only described one case in which an inductive shunt was found good.   It was an artificial cable of 7,000 ohms and 10 micro-farads.   Presuming that there was a fairly good spreading of the permittance, we see that the permittance is that of about 40 kilometres of a submarine cable, whilst the resistance is more like that of 1,000 or 2,000 kilometres.   The inductance would probably be very small.   We see that the example does not well represent a submarine cable.   We should rather compare it with a very long overland circuit of very much smaller permittance per kilometre than the cable.   Only, such a circuit, if properly put up, would have a large inductance, so that there is a failure here.   However, it is mentioned that one shunt of 312 ohms and a time-constant of 0·005 sec. rendered tele-phonic transmission possible except for shrill sounds.   I cannot help thinking that Prof. Thompson has been much too reticent on the experimental side.   He will perhaps furnish us with fuller information later on.   At present there is little to judge by.

General Theory of Transmission of Waves along a Circuit with or without Auxiliary Devices.

§ 221.   In the meantime the following theory and formulæ may perhaps be useful to those who may desire to go into the matter.   I put the theory into such a form that it may be applied to various other cases of auxiliary arrangements.   Let V and C be the transverse voltage and the current, at distance x, at time t.   Let their connections be

$$-\frac{dV}{dx} = ZC, \qquad -\frac{dC}{dx} = YV, \quad . \quad . \quad . \quad (5)$$

where Z is a resistance operator, and Y a conductance operator, as described in § 219, only now belonging to unit length of circuit.   In the simplest case Z reduces to a resistance, and Y to a conductance, per unit length.   But in general Z and Y may have various forms, in unlimited number, being then functions of electrical constants and of the time differentiator p.   When there are no auxiliary devices, and the natural resist-ance R, inductance L, leakance K, and permittance S (all per unit length) are constants, then we have

$$Z = R + Lp, \qquad Y = K + Sp, \quad . \quad . \quad . \quad (6)$$

as before, § 201—4.   But whatever the forms of Y and Z may be, in finite terms or transcendental, we may reduce them to the simple standard forms (6) when the simply periodic state of vibration is maintained, by employing the transformation $p = ni$, or $p^2 = -n^2$.   We see from this that we can practically examine cases of simple periodicity which might at first seem beyond all bounds of practicability.   Note, however, that "forces" and "fluxes" should be in constant ratios, so that such phenomena as hysteresis, magnetic or electric, are excluded.

From equations (5) we see that the characteristic equation of V is

$$\frac{d^2 V}{dx^2} = YZV = q^2 V, \text{ say,} \quad . \quad . \quad . \quad (7)$$

provided Y and Z are independent of x.   This restriction is not a necessary one, but, of course, it makes an important practical simplification to have uniformity along the circuit.   In (7) q^2 represents the operator YZ, and in passing we may notice an interesting matter connected with partial differential equations of the type (7), or of the more general type

$$\nabla^2 V = q^2 V, \quad . \quad . \quad . \quad . \quad (8)$$

appropriate to three dimensions in space.   This type of equation occurs in all sorts of physical problems, V being some variable which is propagated through a medium in a manner depending upon the nature of the operator q^2, which involves the time-differentiator.   It is usually a very simple rational function of p, such, for example, as to bring in only the first and second derivatives of V with respect to the time, and, so far as I know, no physical problems have presented themselves which involve an irrational partial differential equation for characteristic, as, for example,

$$\nabla^2 V = (a + bp + cp^2)^{\frac{1}{2}} V. \quad . \quad . \quad . \quad (9)$$

It might, indeed, seem at first sight that a characteristic of this form was physically impossible, being meaningless. Nevertheless, it is quite easy to see by the way equation (7) was constructed out of the components (5), that real physical problems may involve irrational characteristic equations, such as are exemplified by (9).   For if either Y or Z be irrational, so is YZ. And we can easily (in imagination, not so easily in execution) choose auxiliary devices making Y or Z irrational. Similar remarks apply to (8). It results from the union of two distinct equations involving two variables, one of which is then eliminated to make the characteristic. If either of the component equations involves an irrational operator, the characteristic resultant will be irrational. It must not be imagined that the solutions of such equations obtained by physical reasoning are impossible or beyond the range of mathematics, or that the results are physically meaningless. Some partial characteristics of the form (8), with irrational right members, I have examined and solved. I should imagine that there will probably be a field for equations of the kind in the future study of the complicated influence of matter upon phenomena which occur in their simplest manner in the ether away from matter, as in the theory of dispersion of light, for example, of electric absorption, and in similar subjects.

Returning to equation (7), the general solution involves two arbitrary functions of the time, as in the form

$$V = \cosh qx . V_0 + \frac{\text{shin } qx}{q} ZC_0. \quad . \quad . \quad . \quad (10)$$

But it is preferable to take the case of disturbances propagated from a source into an infinitely long cable, in order to eliminate complications due to reflections and terminal apparatus. Then

$$V = \epsilon^{-qx} V_0 \quad . \quad . \quad . \quad . \quad . \quad (11)$$

is the solution representing V at x, t due to V_0 given as a function of the time at $x = 0$, the origin. And the current is given by

$$C = \frac{q}{Z} V = \left(\frac{Y}{Z}\right)^{\frac{1}{2}} \epsilon^{-qx} V_0, \quad . \quad . \quad . \quad (12)$$

by using the first of (5).

To see the final state due to the continued action of constant V_0, give q, Y, and Z in (11), (12), the values they assume when p is put $= 0$ in their general expressions, viz., for Y the steady leakance and for Z the steady resistance per unit length, and for q their geometric mean. To find how this state is arrived at, and, more generally, to interpret (11), (12) when V_0 is any given function of the time, demands the conversion of these equations to proper algebraical form by execution of the analytical operations implied in q, Y, and Z. This we are not concerned with here, except as regards the case arising when V_0 is a simply periodic function. Then put $p = ni$, and reduce Z to $R + Lp$, and Y to $K + Sp$, as before stated, equations (6), making (11), (12) become

$$V = \epsilon^{-qx} V_0, \quad \dots \dots \quad (13)$$

$$C = \left(\frac{K + Sp}{R + Lp}\right)^{\frac{1}{2}} \epsilon^{-qx} V_0, \quad \dots \quad (14)$$

where $\qquad q = (R + Lp)^{\frac{1}{2}} (K + Sp)^{\frac{1}{2}}. \quad \dots \quad (15)$

These expressions have next to be reduced to the form $(A + Bp)V_0$, which is a matter of common algebra. When done we shall find that if

$$\text{P or Q} = (\tfrac{1}{2})^{\frac{1}{2}} \left[(R^2 + L^2 n^2)^{\frac{1}{2}} (K^2 + S^2 n^2)^{\frac{1}{2}} \pm (RK - LSn^2) \right]^{\frac{1}{2}}, \quad (16)$$

then the V solution is

$$V = e\, \epsilon^{-Px} \sin(nt - Qx), \quad \dots \quad (17)$$

due to $V_0 = e \sin nt$ at the origin. From this, by the use of the first of (5), giving

$$C = \frac{-\dfrac{dV}{dx}}{R + Lp} = -\frac{R - Lp}{R^2 + L^2 n^2} \frac{dV}{dx}, \quad \dots \quad (18)$$

we may derive the C solution. Or we may derive it by developing (14). The result is

$$C = \frac{e\, \epsilon^{-Px}}{P^2 + Q^2} \left[(KP + SnQ) \sin + (SnP - KQ) \cos \right] (nt - Qx), \quad (19)$$

where the value of $P^2 + Q^2$ is given by

$$P^2 + Q^2 = (R^2 + L^2 n^2)^{\frac{1}{2}} (K^2 + S^2 n^2)^{\frac{1}{2}}. \quad \dots \quad (20)$$

It may be only necessary to consider the amplitudes of V and C, and, of course, C is most important. That of V is obviously $e\, \epsilon^{-Px}$, by (17). That of C may be got from (19), and is expressed by

$$(C) = e \left(\frac{K^2 + S^2 n^2}{R^2 + L^2 n^2}\right)^{\frac{1}{2}} \epsilon^{-Px}. \quad \dots \quad (21)$$

The wave speed is n/Q, the wave length $2\pi/Q$, and the periodicity $n/2\pi$. Another convenient form of (21) is

$$(C) = \frac{e}{Lv} \epsilon^{-Px} \left(\frac{1 + K^2/S^2 n^2}{1 + R^2/L^2 n^2}\right)^{\frac{1}{2}}, \quad \dots \quad (22)$$

where $v = (LS)^{-\frac{1}{2}}$.

Application of above Theory to Inductive Leakance.

§ 222. The above simply periodic solution is discussed in my " Electrical Papers," Vol. II. (especially p. 396 and p. 339), when R, L are the natural effective resistance and inductance, and K, S the leakance and permittance per unit length, the latter pair being supposed to be the same at all frequencies. (A paper by Prof. Perry, *Phil. Mag.*, August, 1893, may also be consulted, but I do not think he is right in some of his conclusions.) But the same formulæ are applicable when auxiliary devices are employed, if they are numerous enough, by a process analogous to the obvious one of representing large numbers of separate leaks by uniform leakance.

Thus, in the case of inductive shunts, if we treat them as a set of separately located shunts, the full theory is very complicated, because it requires a separate formula for every section of the line. The best course to take is to distribute the inductive leakance uniformly, in imagination, of course. The result will represent the fullest possible carrying out of the principle concerned, and more. Thus, let ρ be the resistance, and λ the inductance of the leak belonging to the unit length of cable. The reciprocal of ρ is of course the steady leakance per unit length. But it does not operate fully when the state is changing, owing to λ. We shall now have

$$Y = \frac{1}{\rho + \lambda p} + \sigma p, \quad \cdots \quad (23)$$

where $(\rho + \lambda p)^{-1}$ is the leakance operator, taking the place of ρ^{-1}, whilst σ is the steady permittance per unit length. We have to use this special Y in the above formulæ to represent the effect of inductive leakage. In the simply periodic case Y develops to

$$Y = \frac{\rho}{\rho^2 + \lambda^2 n^2} + \left(\sigma - \frac{\lambda}{\rho^2 + \lambda^2 n^2} \right) p. \quad \cdots \quad (24)$$

The first term is the effective leakance, and the coefficient of p is the effective permittance per unit length.

In the developed formulæ (16), (17), (19) to (22), therefore, we must give K and S the values

$$K = \frac{\rho}{\rho^2 + \lambda^2 n^2}, \qquad S = \sigma - \frac{\lambda}{\rho^2 + \lambda^2 n^2}. \quad \cdot \quad (25)$$

to suit the case of inductive leakage, whilst R and L remain
as before, the effective resistance and inductance per unit
length of line.   The solutions should be examined through a
wide range of frequency in order to arrive at a good knowledge of how the circuit would behave to telephonic currents,
with different values of λ and ρ.   It is, of course, necessary
that ρ should not be so small as to produce too great a loss in
transit.

Notice that λ and ρ are immensely big when the unit length
is a centimetre.   Remember that m equal coils in parallel
behave like one coil of one m^{th} the resistance and inductance
of the individual coils.   Thus, when ten shunts at the rate of
one per kilometre take the place of one shunt per ten kilometres, each of the ten should have ten times the resistance and
inductance of the one.   This leads to physically monstrous
results when we do as was done above, and make the action
uniform, as it implies infinitely large inductance of an infinitely
small section of the leakance.   But that need not cause any
alarm when we are employing the ideal case for calculating
purposes.   What is more important is that a stop would soon
be set to the multiplication of the shunts.

At the particular frequency making the effective permittance
vanish, the solution may be obtained simply from the fundamental formulæ, since Y reduces to a constant, viz., the value
of K in (25), S being zero ; and only the rationalisation of Z
has to be attended to.   If, further, we assume that $L = 0$, then
Z is also constant, viz , R, and the solutions are simply

$$V = \epsilon^{-x(RK)^{\frac{1}{2}}} V_0, \quad \cdots \quad (26)$$

$$C = \left(\frac{K}{R}\right)^{\frac{1}{2}} V. \quad \cdots \quad (27)$$

These represent a very remarkable state of affairs throughout
the cable.   But it is thoroughly deceptive, because even if the
inductance of the line were quite unimportant under normal
conditions, the effective cancellation of the permittance would
make it important under the present circumstances ; that is, L
cannot be neglected.   Including it, when $S = 0$, the value of P
reduces to

$$P = (\tfrac{1}{2})^{\frac{1}{2}} \left(\frac{\rho}{\rho^2 + \lambda^2 n^2}\right)^{\frac{1}{2}} \{R + (R^2 + L^2 n^2)^{\frac{1}{2}}\}^{\frac{1}{2}}, \quad . \quad (28)$$

and the current amplitude, according to (21), becomes

$$(C) = \frac{\rho^{\frac{1}{2}}\epsilon^{-Px}e}{(\rho^2 + \lambda^2 n^2)^{\frac{1}{4}}(R^2 + L^2 n^2)^{\frac{1}{4}}} \quad \cdot \quad \cdot \quad \cdot \quad (29)$$

in which, of course, ρ and λn are not independent, having a relation fixed between them by the vanishing of S in (25). By giving a suitable value to σ, for instance, $\frac{1}{4}$ microfarad per kilometre (which is a suitable unit of length to employ), and also fixing the frequency, we may readily apply the formulæ to estimate the attenuation with different values of the insulation resistance or its equivalent.

APPENDIX B.

A GRAVITATIONAL AND ELECTROMAGNETIC ANALOGY.

Part I.

To form any notion at all of the flux of gravitational energy, we must first localise the energy. In this respect it resembles the legendary hare in the cookery book. Whether the notion will turn out to be a useful one is a matter for subsequent discovery. For this, also, there is a well-known gastronomical analogy.

Now, bearing in mind the successful manner in which Maxwell's localisation of electric and magnetic energy in his ether lends itself to theoretical reasoning, the suggestion is very natural that we should attempt to localise gravitational energy in a similar manner, its density to depend upon the square of the intensity of the force, especially because the law of the inverse squares is involved throughout.

Certain portions of space are supposed to be occupied by matter, and its amount is supposed to be invariable. Furthermore, it is assumed to have personal identity, so that the position and m tion of a definite particle of matter are definite, at any rate relative to an assumed fixed space. Matter is recognised by the property of inertia, whereby it tends to persist in the state of motion it possesses ; and any change in the motion is ascribed to the action of force, of which the proper measure is, therefore, the rate of change of quantity of motion, or momentum.

Let ρ be the density of matter, and e the intensity of force, or the force per unit matter, then

$$F = e\rho \quad . \quad . \quad . \quad . \quad . \quad (1)$$

expresses the moving force on ρ, which has its equivalent in increase of the momentum.　There are so many forces nowadays of a generalised nature, that perhaps the expression "moving force" may be permitted for distinctness, although it may have been formerly abused and afterwards tabooed.

Now the force F, or the intensity e, may have many origins, but the only one we are concerned with here is the gravitational force.　This appears to depend solely upon the distribution of the matter, independently of other circumstances, and its operation is concisely expressed by Newton's law, that there is a mutual attraction between any two particles of matter, which varies as the product of their masses and inversely as the square of their distance.　Let e now be the intensity of gravitational force, and F the resultant moving force, due to all the matter.　Then e s the space-variation of a potential, say,

$$e = \nabla P, \quad . \quad . \quad . \quad . \quad . \quad (2)$$

and the potential is found from the distribution of matter by[*]

$$P = \text{pot} \frac{\rho}{c} = \Sigma \frac{\rho}{4\pi c r},$$

where c is a constant.　This implies that the speed of propagation of the gravitative influence is infinitely great.[†]

Now when matter is allowed to fall together from any configuration to a closer one, the work done by the gravitational forcive is expressed by the increase made in the quantity $\Sigma\frac{1}{2}P\rho$.　This is identically the same as the quantity $\Sigma\frac{1}{2}ce^2$ summed through all space.　If, for example, the matter be given

---

[*] *See* § 133.

[†] The density is expressed in terms of the intensity of force by $\rho = \text{conv}\, c e$. This is also the case in the extended analogy later, when the lines of e are slightly shifted.

initially in a state of infinitely fine division, infinitely widely separated, then the work done by the gravitational forcive in passing to any other configuration is $\Sigma \frac{1}{2} P \rho$ or $\Sigma \frac{1}{2} c e^2$, which therefore expresses the "exhaustion of potential energy."[*] We may therefore assume that $\frac{1}{2} c e^2$ expresses the exhaustion of potential energy per unit volume of the medium. The equivalent of the exhaustion of potential energy is, of course, the gain of kinetic energy, if no other forces have been in action.

We can now express the flux of energy. We may compare the present problem with that of the motion of electrification. If moved about slowly in a dielectric, the electric force is appreciably the static distribution. Nevertheless, the flux of energy depends upon the magnetic force as well. It may, indeed, be represented in another way, without introducing the magnetic force, but then the formula would not be sufficiently comprehensive to suit other cases. Now what is there analogous to magnetic force in the gravitational case? And if it have its analogue, what is there to correspond with electric current? At first glance it might seem that the whole of the magnetic side of electromagnetism was absent in the gravitational analogy. But this is not true.

Thus, if u is the velocity of ρ, then ρu is the density of a current (or flux) of matter. It is analogous to a convective current of electrification. Also, when the matter ρ enters any region through its boundary, there is a simultaneous convergence of gravitational force into that region proportional to ρ. This is expressed by saying that if[†]

$$C = \rho u - c \dot{e}, \quad . \quad . \quad . \quad . \quad . \quad . \quad (4)$$

then C is a circuital flux. It is the analogue of Maxwell's true current; for although Maxwell did not include the convective term ρu, yet it would be against his principles to ignore it. Being a circuital flux, it is the curl of a vector, say

$$\operatorname{curl} h = \rho u - c \dot{e}. \quad . \quad . \quad . \quad . \quad (5)$$

This defines h except as regards its divergence, which is arbitrary, and may be made zero. Then h is the analogue of mag-

---

[*] "Thomson and Tait," Vol. I., Part II., § 549.

[†] Observe that it is $-e$ that is the analogue of electric force, and $-c \dot{e}$ of displacement current.

netic force, for it bears the same relation to flux of matter as magnetic force does to convective current. We have*

$$\mathbf{h} = \text{curl pot } \mathbf{C}, \quad \cdots \cdots \quad (6)$$

$$= \text{curl } \mathbf{A},$$

if $\mathbf{A} = \text{pot } \mathbf{C}$. But, since instantaneous action is here involved, we may equally well take

$$\mathbf{A} = \text{pot } \rho\mathbf{u}, \quad \cdots \cdots \quad (7)$$

and its curl will be $\mathbf{h}$. Thus, whilst the ordinary potential P is the potential of the matter, the new potential $\mathbf{A}$ is that of its flux.

Now if we multiply (5) by e, we obtain

$$e \text{ curl } \mathbf{h} = e\rho\mathbf{u} - ec\dot{\mathbf{e}}, \quad \cdots \cdots \quad (8)$$

or, which is the same,†

$$\text{conv } V e\mathbf{h} = \mathbf{F}\mathbf{u} - \dot{\mathbf{U}}, \quad \cdots \cdots \quad (9)$$

if $\mathbf{U} = \frac{1}{2}ce^2$. But $\dot{\mathbf{U}}$ represents the rate of exhaustion of potential energy, so $-\dot{\mathbf{U}}$ represents its rate of increase, whilst $\mathbf{F}\mathbf{u}$ represents the activity of the force on ρ, increasing its kinetic energy. Consequently, the vector $V e\mathbf{h}$ expresses the flux of gravitational energy. More strictly, any circuital flux whatever may be added. This $V e\mathbf{h}$ is analogous to the electromagnetic $V\mathbf{EH}$ found by Poynting and myself.‡ But there is a reversal of direction. Thus, comparing a single moving particle of matter with a similarly-moving electric charge, describe a sphere round each. Let the direction of motion be the axis, the positive pole being at the forward end. Then in the electrical case the magnetic force follows the lines of latitude with positive rotation about the axis, and the flux of energy coincides with the lines of longitude from the negative pole to the positive. But in the gravitational case, although $\mathbf{h}$ still follows the lines of latitude positively, yet since the radial e is directed to instead of from the centre, the flux of energy is along the lines of longitude from the positive pole to the

---

* *See* § 134, page 208.

† By the transformation (178), § 132.

‡ *See* § 70, equation (13). Take $e_0 = 0 = h_0$.

negative. This reversal arises from all matter being alike and attractive, whereas like electrifications repel one another.

The electromagnetic analogy may be pushed further. It is as incredible now as it was in Newton's time* that gravitative influence can be exerted without a medium ; and, granting a medium, we may as well consider that it propagates in time, although immensely fast. Suppose, then, instead of instantaneous action, which involves†

$$\text{curl } e = 0, \quad \ldots \quad \ldots \quad (10)$$

we assert that the gravitational force e in ether is propagated at a single finite speed v. This requires that

$$v^2\nabla^2 e = \ddot{e}, \quad \ldots \quad \ldots \quad (11)$$

for this‡ is the general characteristic of undissipated propagation at finite speed. Now,‖

$$\nabla^2 = \nabla \operatorname{div} - \operatorname{curl}^2,$$

so in space free from matter we have

$$- v^2 \operatorname{curl}^2 e = \ddot{e}. \quad \ldots \quad \ldots \quad (12)$$

But we also have, by (5),

$$- \operatorname{curl} h = c\dot{e}, \quad \ldots \quad \ldots \quad (13)$$

away from matter. This gives a second value to $\ddot{e}$, when we differentiate (13) to the time, say

$$\ddot{e} = - \frac{1}{c}\operatorname{curl} \dot{h}. \quad \ldots \quad \ldots \quad (14)$$

So, by (12) and (14), and remembering that we have already chosen h circuital, we derive

$$cv^2 \operatorname{curl} e = \dot{h}. \quad \ldots \quad \ldots \quad (15)$$

---

* To Newton himself, as shown in his often-quoted letter to Bentley.

† This equation is equivalent to (2) above, and it implies that the gravitational force is *exactly* dependent on the configuration of the matter.

‡ As shown by Poisson, the value of a quantity at a given place and time, when controlled by this equation, depends solely on the state of things at distance vt and time t earlier. That is, disturbances travel at speed v. But it is much simpler to understand this property through plane waves.

See (193), § 132.

Or, if μ is a new constant, such that

$$\mu c v^2 = 1, \quad \ldots \ldots \quad (16)$$

then (15) may be written in the form

$$\text{curl } e = \mu h. \quad \ldots \ldots \quad (17)$$

To sum up, the first circuital law (5), or

$$\text{curl } h = \rho u - c\dot{e}, \quad \ldots \quad (5) \ bis.$$

leads to a second one, namely (17), if we introduce the hypothesis of propagation at finite speed.* This, of course, might be inferred from the electromagnetic case.

In order that the speed v should be not less than any value that may be settled upon as the least possible, we have merely to make μ be of the necessary smallness. The equation of activity becomes, instead of (9),†

$$\text{conv } Veh = Fu - \dot{U} - \dot{T}, \quad \ldots \quad (18)$$

if $T = \frac{1}{2}\mu h^2$. The negative sign before the time-increase of this quantity points to exhaustion of energy, as before. If so, we should still represent the flux of energy by Veh. But, of course, T is an almost vanishing quantity when μ is small enough, or v big enough. Note that h is not a negligible quantity, though the product μh is. Thus results will be sensibly as in the common theory of instantaneous action, although expressed in terms of wave-propagation. Results showing signs of wave-propagation would require an inordinately large velocity of matter through the ether. It may be worth while to point out that the lines of gravitational force connected with a particle of matter will no longer converge to it uniformly from all directions when the velocity v is finite, but will show a tendency to lateral concentration, though only to a sensible extent when the velocity of the matter is not an insensible fraction of v.

The gravitational-electromagnetic analogy may be further extended if we allow that the ether which supports and propagates the gravitational influence can have a translational‡

---

* These equations are analogous to (4) and (5), §35.

† This is analogous to (12), §70, with the impressed forces made zero.

‡ This does not exclude rotational motion, which is, in fact, a differential effect in a special kind of translational motion.

motion of its own, thus carrying about and distorting the lines of force. Making allowance for this convection of **e** by the medium, with the concomitant convection of **h**, requires us to turn the circuital laws (17), (18) to

$$\mathrm{curl}\,(\mathbf{e}+\mu V q \mathbf{h})=\mu \mathbf{h}, \quad . \quad . \quad . \quad (19)$$

$$\mathrm{curl}\,(\mathbf{h}+c V \mathbf{e}q)=\rho \mathbf{u}-c\dot{\mathbf{e}}, \quad . \quad . \quad . \quad (20)$$

where q is the velocity of the medium itself.*

It is needless to go into detail, because the matter may be regarded as a special and simplified case of my investigation of the forces in the electromagnetic field, with changed meanings of the symbols. It is sufficient to point out that the stress in the field now becomes prominent as a working agent. It is of two sorts, one depending upon **e** and the other upon **h**, analogous to the electric and magnetic stresses. The one depending upon **h** is, of course, insignificant. The other consists of a pressure parallel to **e** combined with a lateral tension all round it, both of magnitude $\frac{1}{2}ce^2$. This was equivalently suggested by Maxwell. Thus two bodies which appear to attract are pushed together. The case of two large parallel material planes exhibits this in a marked manner, for **e** is very small between them, and relatively large on their further sides.

But the above analogy, though interesting in its way, and serving to emphasise the non-necessity of the assumption of instantaneous or direct action of matter upon matter, does not enlighten us in the least about the ultimate nature of gravitational energy. It serves, in fact, to further illustrate the mystery. For it must be confessed that the exhaustion of potential energy from a universal medium is a very unintelligible and mysterious matter. When matter is infinitely widely separated, and the forces are least, the potential energy is at its greatest, and when the potential energy is most exhausted, the forces are most energetic !

Now there is a magnetic problem in which we have a kind of similarity of behaviour, viz., when currents in material circuits are allowed to attract one another. Let, for completeness, the initial state be one of infinitely wide separation of infinitely small filamentary currents in closed circuits. Then, on concentration to any other state, the work done by the attractive forces is represented by $\Sigma\frac{1}{2}\mu H^2$, where μ is the inductivity and H the magnetic force. This has its equivalent in the energy of motion of the circuits, or may be imagined to be so converted, or else wasted by friction, if we like. But, over and above this energy, the same amount, $\Sigma\frac{1}{2}\mu H^2$, represents *the* energy of the magnetic field, which can be got out of it in work. It was zero at the beginning. Now, as Lord Kelvin showed, this double work is accounted for by extra work in the batteries or other sources required to maintain the currents constant. (I have omitted reference to the waste of energy due to electrical resistance, to avoid complications.) In the gravitational case there is a partial analogy, but the matter is all along assumed to be incapable of variation, and not to require any supply of energy to keep it constant. If we asserted that $\frac{1}{2}ce^2$ was stored energy, then its double would be the work done per unit volume by letting bodies attract from infinity, without any apparent source. But it is merely the exhaustion of potential energy of unknown amount and distribution.*

Potential energy, when regarded merely as expressive of the work that can be done by forces depending upon configuration, does not admit of much argument. It is little more than a mathematical idea, for there is scarcely any physics in it. It explains nothing. But in the consideration of physics in general, it is scarcely possible to avoid the idea that potential energy should be capable of localisation equally as well as kinetic. That the potential energy may be itself ultimately kinetic is a separate question. Perhaps the best definition of the former is contained in these words:—Potential energy is energy that is not known to be kinetic. But, however this be, there is a practical distinction between them which it is found useful to carry out. Now, when energy can be distinctly localised, its flux can also be traced (subject to circuital indeterminateness, however). Also, this flux of energy forms a useful working idea when action at a distance is denied (even though the speed of transmission be infinitely great, or be

---

* It would appear that we must go to the ether to find the source of all energy.

assumed to be so). Any distinct and practical localisation of energy is therefore a useful step, wholly apart from the debatable question of the identity of energy advocated by Prof. Lodge.

From this point of view, then, we ought to localise gravitational energy as a preliminary to a better understanding of that mysterious agency. It cannot be said that the theory of the potential energy of gravitation exhausts the subject. The flux of gravitational energy in the form above given is, perhaps, somewhat more distinct, since it considers the flux only and the changes in the amount localised, without any statement of the gross amount. Perhaps the above analogy may be useful, and suggest something better.

Part II.

In the foregoing I partly assumed a knowledge on the part of the reader of my theory of convective currents of electrification* ("Electrical Papers," Vol. II., p. 495 and after), and only very briefly mentioned the modified law of the inverse squares which is involved, viz., with a lateral concentration of the lines of force. The remarks of the Editor† and of Prof. Lodge‡ on gravitational aberration, lead me to point out now some of the consequences of the modified law which arises when we assume that the ether is the working agent in gravitational effects, and that it propagates disturbances at speed v in the manner supposed in Part I. There is, so far as I can see at present, no aberrational effect, but only a slight alteration in the intensity of force in different directions round a moving body considered as an attractor.

Thus, take the case of a big Sun and small Earth, of masses S and E, at distance r apart. Let f be the unmodified force of S on E, thus

$$f = \frac{SE}{4\pi r^2 c}, \quad \cdot \quad \cdot \quad \cdot \quad \cdot \quad \cdot \quad (1)$$

using rational units in order to harmonise with the electromagnetic laws when rationally expressed. Also, let F be the

---

* See also §§ 52 to 62, and §§ 163, 4.

† The Electrician, July 14, p. 277, and July 28, p. 340.

‡ The Electrician, July 28, p. 347.

modified force when the Sun is in motion at speed u through the ether. Then*

$$F = f \times \frac{1-s}{(1-s\sin^2\theta)^{\frac{3}{2}}}, \quad \cdots \quad (2)$$

where s is the small quantity u^2/v^2, and θ is the angle between r and the line of motion. ("Electrical Papers," Vol. II., pp. 495, 499).

Therefore, if the Sun is at rest, there is no disturbance of the Newtonian law, because its "field of force" is stationary. But if it has a motion through space, there is a slight weakening of the force in the line of motion, and a slight strengthening equatorially. The direction is still radial.

To show the size of the effect, let

$$u = 3 \times 10^7 \text{ centim. per sec.} \atop v = 3 \times 10^{10} \text{ ,, ,, ,,} \Big\} \quad \cdots \quad (3).$$

This value of u is not very different from the speed attributed to fast stars, and the value of v is the speed of light itself.

So we have

$$s = \frac{u^2}{v^2} = \frac{1}{10^6}, \quad \cdots \quad (4)$$

i.e., one-millionth. All perturbing forces of the first order are, therefore, of the order of magnitude of only one-millionth of the full force, even when the speed of propagation is as small as that of light.

The simplest case is when the common motion of the Sun and Earth is perpendicular to the plane of the orbit. Then $\theta = \frac{1}{2}\pi$ all round the orbit, and

$$F = f(1 + \tfrac{1}{2}s), \quad \cdots \quad (5)$$

showing an increase in the force of attraction of S on E of one two-millionth part, without alteration of direction or variation in the orbit.†

---

* This is the case of steady motion. There is no simple formula when the motion is unsteady. Equation (2), above, is immediately derivable from equation (40), § 163.

† But Prof. Lodge tells me that our own particular Sun is considered to move only 10·9 miles per second. This is stupendously slow. The size of *s* is reduced to about 1/360 part of that in the text, and the same applies to the corrections depending upon it.

But when the common motion of the Sun and Earth is in their plane, θ varies from 0 to 2π in a revolution, so that the attraction on E, whilst towards the Sun's centre always, undergoes a periodic variation from

$$F = f(1 - s) \quad . \quad . \quad . \quad . \quad . \quad (6)$$

when $\theta = 0$, to $\qquad F = f(1 + \tfrac{1}{2}s), \quad . \quad . \quad . \quad . \quad . \quad (7)$

when $\theta = \tfrac{1}{2}\pi$. The extreme variation is, therefore, $\tfrac{3}{2}sf$, according to the data used. The result is a slight change in the shape of the orbit.

But, to be consistent, having made v finite by certain suppositions, we should carry out the consequences more fully, and allow not merely for the change in the Newtonian law, as above, but for the force brought in by the finiteness of v which is analogous to the "electromagnetic force." This is very small truly, but so is the above change in the Newtonian law, and since they are of the same order of magnitude, we should also count the auxiliary force. Call it G. Then*

$$\mathbf{G} = \mathbf{F} \times \frac{xqu}{v^2} \times V\mathbf{q}_1 V\mathbf{r}_1 \mathbf{u}_1, \quad . \quad . \quad . \quad (8)$$

where F is as before, in (2) above, q is the actual speed of the Earth (not the same as u), and in the third vectorial factor, $\mathbf{q}_1$, $\mathbf{u}_1$, and $\mathbf{r}_1$ are unit vectors drawn parallel to the direction of the Earth's motion, of the Sun's motion, and from the Sun to the Earth. We see at once that the order of magnitude cannot be greater than that of the departure of F from f, before considered, because u and q will be of the same order, at least when u is big. As for x, it is simply a numerical factor, which cannot exceed 1, and is probably $\tfrac{2}{3}$.

The simplest case is when the motion of the Sun is perpendicular to the orbit of the Earth. Then

$$G = F \times xs \quad . \quad . \quad . \quad . \quad . \quad (9)$$

gives the tensor or size of the auxiliary force. It is radial, but outwards, so that the result is merely to reduce the size of

---

* See § 46, equation (9), and §§ 82 and 85. To apply to the present case, we have to note that G is the "electromagnetic force" of the Sun's "magnetic induction" on the Earth's "electric current," or rather, the analogue thereof ; and also that the direction must be reversed, just as the ordinary "electric force" is taken reversed.

the previous correction, viz., the difference of F from f in the same motional circumstances.

But when the line of motion of the Sun is in the plane of the orbit, the case is much more complicated. The force **G** is neither constant (for the same distance) nor radial, except in four positions, viz., two in the line of motion of the Sun, when the auxiliary force vanishes, and two when $\theta = \pm \frac{1}{2}\pi$, when it is greatest. But this force is still in the plane of the orbit, which is an important thing, and is, moreover, periodic, so that the tangential component is as much one way as the other in a period.

All we need expect, then, so far as I can see from the above considerations, are small perturbations due to the variation of the force of gravity in different directions, and to the auxiliary force. Of course, there will be numerous minor perturbations.*

If variations of the force of the size considered above are too small to lead to observable perturbations of motion, then the striking conclusion is that the speed of gravity may even be the same as that of light. If they are observable, then, if existent, they should turn up, but if non-existent then the speed of gravity should be greater. Furthermore, it is to be observed that there may be other ways of expressing the propagation of gravity.

But I am mindful of the good old adage about the shoemaker and his last, and am, therefore, reluctant to make any more remarks about perturbations. The question of the ether in its gravitational aspect must be faced, however, and solved sooner or later, if it be possible. Perhaps, therefore, my suggestion may not be wholly useless.

---

* The solution for steady rectilinear motion has been employed. The justification thereof is the smallness of u/v and the large periodic time. If we allow for the small curvature of path of an attracting body, we shall introduce corrections of the second order of small quantities.

Printed in the United States
By Bookmasters